Auction Theory for Computer Networks

Do you have the tools to address recent challenges and problems in modern computer networks? Discover a unified view of auction theoretic applications and develop auction models, solution concepts, and algorithms with this multidisciplinary review. Devise distributed, dynamic, and adaptive algorithms for ensuring robust network operation over time-varying and heterogeneous environments, and for optimizing decisions about services, resource allocation, and usage of all network entities. Topics including cloud networking models, MIMO, mmWave communications, 5G networks, data aggregation, task allocation, user association, interference management, wireless caching, mobile data offloading, and security. Introducing fundamental concepts from an engineering perspective and describing a wide range of state-of-the-art techniques, this text is an excellent resource for graduate and senior undergraduate students, network and software engineers, economists, and researchers.

Dusit Niyato is a professor in the School of Computer Science and Engineering at Nanyang Technological University, Singapore, and a Fellow of the IEEE.

Nguyen Cong Luong is a senior lecturer in Faculty of Information Technology at PHENIKAA University, Hanoi, Vietnam. He is also a researcher in PHENIKAA Research and Technology Institute (PRATI), A&A Green Phoenix Group JSC, Hanoi, Vietnam.

Ping Wang is an associate professor in the Department of Electrical Engineering and Computer Science, York University.

Zhu Han is a John and Rebecca Moores Professor in both the Department of Electrical and Computer Engineering and the Computer Science Department at the University of Houston, and a Fellow of the IEEE and the AAAS.

Auction Theory for Computer Networks

DUSIT NIYATO
Nanyang Technological University

NGUYEN CONG LUONG
PHENIKAA University

PING WANG
York University

ZHU HAN
University of Houston

CAMBRIDGE
UNIVERSITY PRESS

University Printing House, Cambridge CB2 8BS, United Kingdom

One Liberty Plaza, 20th Floor, New York, NY 10006, USA

477 Williamstown Road, Port Melbourne, VIC 3207, Australia

314–321, 3rd Floor, Plot 3, Splendor Forum, Jasola District Centre, New Delhi – 110025, India

79 Anson Road, #06–04/06, Singapore 079906

Cambridge University Press is part of the University of Cambridge.

It furthers the University's mission by disseminating knowledge in the pursuit of education, learning, and research at the highest international levels of excellence.

www.cambridge.org
Information on this title: www.cambridge.org/9781108480765
DOI: 10.1017/9781108691079

© Cambridge University Press 2020

This publication is in copyright. Subject to statutory exception and to the provisions of relevant collective licensing agreements, no reproduction of any part may take place without the written permission of Cambridge University Press.

First published 2020

Printed in the United Kingdom by TJ International, Padstow Cornwall

A catalogue record for this publication is available from the British Library.

ISBN 978-1-108-48076-5 Hardback

Cambridge University Press has no responsibility for the persistence or accuracy of URLs for external or third-party internet websites referred to in this publication and does not guarantee that any content on such websites is, or will remain, accurate or appropriate.

**To
our families**

Contents

1	**Introduction**		*page* 1
	1.1 A Brief Overview of the History of Auctions		1
	1.2 Auction Theory in Computer Networks		2
	1.3 Organization and Timeliness of This Book		3
		1.3.1 Organization	3
		1.3.2 Timeliness of the Book	7
	1.4 Acknowledgments		10
2	**Overview of Modern Computer Networks**		11
	2.1 Internet of Things		11
		2.1.1 Definitions	11
		2.1.2 IoT Architecture	12
		2.1.3 Resources and Services of IoT	14
		2.1.4 Wireless Sensor Network	15
		2.1.5 Mobile Crowdsensing Network	16
	2.2 Cloud Networking		18
		2.2.1 General Architecture	18
		2.2.2 Cloud Data Center Networking	20
		2.2.3 Mobile Cloud Networking	21
		2.2.4 Edge Computing	22
		2.2.5 Cloud-Based Video-on-Demand System	24
	2.3 5G Wireless Networks		25
		2.3.1 Massive Multiple-Input and Multiple-Output	26
		2.3.2 Heterogeneous Networks	27
		2.3.3 Millimeter Wave Communications	29
		2.3.4 Cognitive Radio	30
		2.3.5 Device-to-Device Communications	31
		2.3.6 Machine-to-Machine Communications	33
	2.4 Data Collection and Resource Management		34
		2.4.1 Data Aggregation	34
		2.4.2 Task Allocation	35
		2.4.3 User Association	37
		2.4.4 Interference Management	38

		2.4.5	Wireless Caching	40
		2.4.6	Mobile Data Offloading	42
	2.5	Wireless Network Security		43
		2.5.1	Users and Attackers in Wireless Networks	44
		2.5.2	Eavesdropping Attack	44
		2.5.3	Denial-of-Service Attack	47
		2.5.4	Information Security Issues	49
		2.5.5	Illegitimate Behaviors in Wireless Networks	50
	2.6	Summary		51
3	**Mechanism Design and Auction Theory in Computer Networks**			**52**
	3.1	Mechanism Design		52
		3.1.1	Mechanism	52
		3.1.2	Mechanism Design	53
		3.1.3	Revelation Principle	55
		3.1.4	Incentive Compatibility	57
		3.1.5	Individual Rationality	58
		3.1.6	Economic Efficiency and Budget Balance	59
	3.2	Optimal Mechanisms		60
		3.2.1	Social Surplus and Profit	60
		3.2.2	Social Surplus Maximization Problem	61
		3.2.3	Profit Maximization Problem	63
	3.3	Auction Theory in Computer Networks		64
		3.3.1	Auction Basics	65
		3.3.2	Auction Theory for Computer Networks	68
		3.3.3	Basic Terminology in Auction Theory	69
	3.4	Summary		71
4	**Open-Cry Auction**			**72**
	4.1	English Auction		72
		4.1.1	English Auction Process	72
		4.1.2	Equilibrium Strategies	74
	4.2	Development of English Auction for Computer Networks		78
		4.2.1	System Model and Problem Formulation	79
		4.2.2	Walrasian Equilibrium	81
		4.2.3	English Auction for Walrasian Equilibrium	82
	4.3	Dutch Auction		83
		4.3.1	Dutch Auction Process	83
		4.3.2	Revenue Equivalence Theorem	85
		4.3.3	Equilibrium in Dutch Auction	86
	4.4	Development of Dutch Auction for Computer Networks		87
		4.4.1	Prevention of Black Hole Attacks in Mobile Ad Hoc Networks	87
		4.4.2	Relay Selection in the Internet of Things	91
		4.4.3	Channel Allocation in 5G Heterogeneous Networks	93

		4.5	English–Dutch Auction	97
		4.6	Summary	99

5	**First-Price Sealed-Bid Auction**			100
	5.1	Definition		100
	5.2	Equilibrium		101
		5.2.1	Strategic Analysis	101
		5.2.2	Bayesian–Nash Equilibrium	102
	5.3	First-Price Sealed-Bid Reverse Auction		104
	5.4	Development of First-Price Sealed-Bid Auction for Computer Networks		105
		5.4.1	Incentive Mechanism for Data Aggregation	105
		5.4.2	Market-Based Adaptive Task Allocation	109
		5.4.3	Market-Based Relay Selection	112
		5.4.4	Denial-of-Service Attack Prevention	114
	5.5	Summary		117

6	**Second-Price Sealed-Bid Auction**			119
	6.1	Second-Price Sealed-Bid Auction		119
		6.1.1	Definition	119
		6.1.2	Dominant Strategy and Nash Equilibrium	121
		6.1.3	Second-Price Sealed-Bid Reverse Auction	123
		6.1.4	Development of Second-Price Sealed-Bid Auction for Computer Networks	124
	6.2	Vickrey–Clarke–Groves Auction		135
		6.2.1	Definition	135
		6.2.2	Description	136
		6.2.3	Dominant Strategy	139
		6.2.4	Examples	140
		6.2.5	Virtues	141
		6.2.6	Development of VCG Auction for Computer Networks	142
	6.3	Summary		157

7	**Combinatorial Auction**			158
	7.1	Introduction		158
	7.2	Substitutable and Complementary Items		159
	7.3	Single-Round Combinatorial Auction		161
		7.3.1	Bidding Language	161
		7.3.2	Winner Determination Problem	163
	7.4	Iterative Combinatorial Auctions		165
		7.4.1	Ascending Proxy Auction	165
		7.4.2	Clock-Proxy Auction	167
	7.5	Development of the Combinatorial Auction for Computer Networks		170
		7.5.1	Spectrum Allocation in Cognitive Radio	170
		7.5.2	Virtualization of 5G Massive MIMO	174

Contents

7.5.3	Mobile Data Offloading in 5G HetNets	180
7.5.4	Resource Allocation in D2D Communication Underlying Cellular Networks	184
7.6	Summary	188

8 Double-Sided Auction 189

- 8.1 Introduction 189
- 8.2 Single-Round Double Auction 189
  - 8.2.1 Uniform Pricing Policy 192
  - 8.2.2 Discriminatory Pricing Policy 193
- 8.3 Continuous Double Auction 194
- 8.4 Development of Double Auction for Computer Networks 196
  - 8.4.1 Sensing Task Allocation in Participatory Sensing 197
  - 8.4.2 Location Privacy in Participatory Sensing 201
  - 8.4.3 Spectrum Allocation in Heterogeneous Networks 204
  - 8.4.4 Cloud Resource Allocation in Edge Computing 209
- 8.5 Summary 214

9 Other Auctions 215

- 9.1 Ascending Clock Auction 216
  - 9.1.1 Auction Process 216
  - 9.1.2 Application of Ascending Clock Auction for Physical Layer Security 217
- 9.2 Share Auction 221
- 9.3 Online Auction 224
  - 9.3.1 Basic Terminologies 224
  - 9.3.2 Development of Online Auction for Cloud Resource Pooling 226
- 9.4 Waiting-Line Auction 231
- 9.5 Summary 235

10 Optimal Auction Using Machine Learning 236

- 10.1 Optimal Auction 236
- 10.2 Machine Learning 238
- 10.3 Machine Learning for Optimal Auction 239
  - 10.3.1 Design 239
  - 10.3.2 Example 244
- 10.4 Machine Learning for Myerson Auction 249
  - 10.4.1 Design 250
  - 10.4.2 Example 254
- 10.5 Summary 259

References 260
Index 278

1 Introduction

In this introductory chapter, we present an overview of the history of auctions. We then present critical issues and challenges of the current and future computer networks. Auctions can be used as effective tools to address these issues and challenges. We also discuss motivations for and significance of the use of auctions in the computer networks. Finally, we present the objectives, organization, timeliness, and potential audience of this book.

1.1 A Brief Overview of the History of Auctions

Auction theory is an applied branch of economics that deals with how people act in auction markets and researches the properties of auction markets. The word "auction" is derived from the Latin word *augeō*, which means "an increase of price" [1]. Auctions have been used for thousands of years for the sale of a variety of objects. In particular, around 500 BC, auctions of women for marriage were conducted in ancient Babylon [1]. During the Roman Empire, auctions were often used by Roman soldiers for trading spoils of war, such as slaves, after military victory. The Romans also used auctions to liquidate the assets of debtors whose property had been confiscated [2]. One infamous auction during the Roman Empire was described by Edward Gibbon in his *Decline and Fall of the Roman Empire* (1776). According to his account, when the Praetorian Guard killed the Roman emperor Pertinax in AD 193, an auction was used to sell the title of Roman emperor to the highest bidder [3]. The winner was a wealthy senator, Didius Julianus.

During the seventeenth and eighteenth centuries, the candle auction, a variation on the typical English auction, was commonly used in England for trading goods and leaseholds. In the candle auction, the end of the auction was signaled by the expiration of a candle flame. This process was intended to ensure that no buyer could know exactly when the auction would end and make a last-second bid.

Auctions of fruits and vegetables became established in the Netherlands around 1880. The farmers' efforts grew to include a vast system of markets for horticultural goods, associated especially with tulips. At the same time, the selling of fish by auction became important in Germany. One main reason for the use of auctions in this case is the fact that fish can be sold fast.

In the modern era, auctions have been used by governments to allocate items for which it is difficult to determine market prices. The items are typically public goods and assets such as electricity, wood, airport time slots, industrial equipment, real estate, bus lines, emission rights, and the use of electromagnetic spectrum. In particular, auctions of rights (i.e., licenses) to use the electromagnetic spectrum for wireless communications are a worldwide phenomenon. In the spectrum auction, the government uses an auction system to assign scarce spectrum resources to parties. With a well-designed auction, the spectrum resources are allocated efficiently to the parties that value them the most, and the government gains revenue. For example, since July 1994, the Federal Communications Commission (FCC), an independent agency of the US government, has raised over $60 billion for the US Treasury by conducting 87 spectrum auctions [4]. Also, in 2000, the Radiocommunications Agency of the UK government (now Ofcom) raised £22.5 billion from an auction of five licenses for radio spectrum to support the 3G mobile telephony standard [5]. In 2018, the Exchequer (i.e., the British government department) received the total of £1.3 billion from the spectrum auction for the future 5G mobile services [6].

In recent years, the rapid growth in the Internet has facilitated conducting auctions. In particular, the Internet facilitates online activities, such as bidding and payment, between buyers and sellers in different locations or geographical areas. This further reduces transaction cost. *eBay* is one example of a company that uses the Internet for auctions. This auction house and business entity directly facilitates the sale–purchase of items between sellers and buyers. In particular, online buyers search for the desired item and then make bids on the item through auctions, and the bidder with the highest bid wins the item. In January 2019, eBay ranked as the 38th most popular website in the world and the 10th most popular website in the United States [7].

With the development of auction theory and experimental research, more sophisticated auction designs have appeared that enable auctions to be applied to various problems in dynamic, uncertain, and complicated environments.

1.2 Auction Theory in Computer Networks

The advancement of computing and telecommunication technologies will enable modern computer networks to become ubiquitous due to the huge demand of pervasive applications. The convergence of cloud computing, mobile networks, and media will allow users to communicate with each other and access any content anytime and anywhere. Modern computer networks will serve a massive number of users and applications with various services with diverse requirements such as high-speed access, real-time Internet games, interactive media, video-on-demand, edge computing, smart homes, intelligent transportation, and autonomous systems. Also, modern computer networks are expected to bring huge new business opportunities for stakeholders including content/service providers, network operators, and infrastructure providers. However, there are many challenges to be addressed to realize the promise of modern computer networks. These challenges come from (i) the heterogeneity and dense deployment of network devices; (ii) the heterogeneous computing and networking resources; (iii) the large degree of uncertainty of the consumer demand for network resources and services; and especially (iv) the rationality and self-interested behaviors of network entities, including both, users and stakeholders. Therefore, one critical issue is devising distributed, dynamic, and adaptive algorithms for (i) ensuring a robust network operation over time-varying and heterogeneous environments and (ii) optimizing decisions on services and resources allocation/usage of all network entities.

Auction theory, also known as a subfield of economics and business management, has been introduced as a tool for computer network design, protocol optimization, and resource allocation. The main reasons are as follows.

- Auctions model interactions among network entities, especially from the economic perspective. This economic aspect of the system becomes more and more important as the network entities are self-interested in their benefits obtained by supplying or consuming network resources and services.
- Auctions provide an incentive mechanism that is important to determine the values and specifically the prices of network resources and services. Auction emerges as a bridge between system design and pricing from engineering and economic perspectives, respectively. The convergence of two fields in computer networks is a nascent approach that can show many advantages over using a classical system or economic solution alone. Therefore, it is inherently suitable for modern computer networks, which consist of multiple autonomous entities.
- Auctions can support different objectives, ranging from revenue maximization for the auctioneer to social welfare maximization for the entities, and provide various desired properties, such as truthfulness, economic efficiency, and individual rationality. This makes designed mechanisms attractive to network entities.
- Auctions allow the allocation of bundles of diverse resources, which satisfies dynamic demands and improves the resource utilization in heterogeneous environments.

As a result, auctions have been explored as means to solve various problems in modern computer networks including the Internet of Things (IoT), cloud networking, and 5G wireless networks. Emerging problems consist of data aggregation, task allocation, user association, interference management, cloud and network resource allocation, wireless caching, and mobile data offloading. More importantly, auctions have been introduced as an effective solution to address network security issues. Practitioners have shown that designing the modern computer networks with self-interested entities can be easily implemented by using auction theory.

1.3 Organization and Timeliness of This Book

1.3.1 Organization

Providing a comprehensive introduction to the basics of auction theory and giving example applications of auction theory for the design of modern and emerging computer networks are the main objectives of this book. Specifically, the first objective is to provide a general introduction to modern computer networks and the most recent developments related to these networks. The second objective is to introduce different auction theoretic models and techniques as well as to present applications of the state-of-the-art auction mechanisms for handling a variety of problems in computer networks. The choices of appropriate techniques for designing computer networks are important. Therefore, we present fundamentals of auction theoretic techniques and then present the developments of these techniques for design, analysis, and optimization of computer networks. To achieve these objectives, this book is organized as follows:

- **Overview of Modern Computer Networks:** Chapter 2 introduces the background, the fundamentals, and the emerging issues of modern computer networks. We first provide definitions of IoT and describe the general architecture, resources, and services of IoT. Important components of IoT, including wireless sensor networks and mobile crowdsensing networks, are further discussed. Second, we introduce cloud networking models that aim to provide on-demand data storage, computing, and network resources to cloud clients/users. The cloud networking models include cloud data center networking, mobile cloud networking, and edge computing. Furthermore, we present the cloud-based video-on-demand system that is known to be a new video content delivery model in the development of cloud networking. Third, we provide an overview of key technologies, such as massive multiple-input and multiple-output (MIMO) and millimeter wave (mmWave) communications, that may potentially be deployed in the 5G wireless networks. Finally, we discuss several emerging issues in the computer networks, including data aggregation, task allocation, user association, interference management, wireless caching, mobile data offloading, and security. In particular, we describe security issues in wireless networks involving eavesdropping attacks, denial-of-service (DoS) attacks, information security issues, and malicious behaviors of users. The motivations of using pricing models as well as auctions for each issue are highlighted.
- **Mechanism Design and Auction Theory in Computer Networks:** Chapter 3 introduces mechanism design and auction theory. Mechanism design aims to determine allocation and payment rules toward objectives or desired objectives. In this chapter, we first define the mechanism as well as the allocation and payment rules of the mechanism, design. The mechanism design task is generally a complicated search problem. Thus, we introduce the revelation principle that can be used for facilitating the mechanism design task. The required properties of the mechanism, such as incentive compatibility, individual rationality, economic efficiency, and budget balance, are also presented. We further discuss optimal mechanisms in terms of social surplus maximization and profit maximization. After that, we introduce the basics of the auction theory and present the motivations as well as the significance of applying auctions to computer networks.
- **Open-Cry Auction:** Open-cry auctions are the most conventional auctions. Chapter 4 discusses two types of open-cry auctions, the English and Dutch auctions, and presents their applications in computer networks. Specifically,

1.3 Organization and Timeliness of This Book

we first introduce the theory of the English auction and demonstrate how to obtain the equilibrium strategies in this type of auction. Second, we discuss the application of the English auction to the spectrum leasing of cognitive radio in the 5G wireless networks. Third, we provide the definition and process of the Dutch auction. In particular, we introduce the revenue equivalence theorem that is used to determine the Nash equilibrium in the Dutch auction. Different from the English auction, the Nash equilibrium in the Dutch auction is the Bayesian Nash equilibrium; that is, a particular bidder determines its equilibrium strategy by knowing the distribution of values of other bidders rather than knowing their actual values. This is a common situation in computer networks where there is no centralized controller to maintain information about users. Fourth, we discuss the applications of the Dutch auction to emerging issues such as black hole attacks, relay selection, and channel allocation. We finally introduce the combination of English and Dutch auctions as a solution for some situations in which the auctioneer has no information about the market price of its item.

- **First-Price Sealed-Bid Auction:** Chapter 5 introduces a common type of kth-price sealed-bid auction, the first-price sealed-bid auction. In the first-price sealed-bid auction, buyers simultaneously submit their bids in sealed envelopes to the seller. The buyer with the highest bid is the winner and pays the seller the price that the buyer submits. In this chapter, we present the strategic analysis of bidders in the first-price sealed-bid auction. We further discuss how to find the equilibrium strategies of bidders in the auction. After that, we introduce the first-price sealed-bid reverse auction, which is a variation of the first-price sealed-bid auction. Finally, we discuss the applications of the first-price sealed-bid auction to address emerging issues in IoT.

- **Second-Price Sealed-Bid Auction:** Chapter 6 presents another type of kth-price sealed-bid auction, the second-price sealed-bid auction or Vickrey auction. Different from the first-price sealed-bid auction, the winner in the Vickrey auction pays the seller the second-highest bid. This chapter contains two main parts. The first part introduces the second-price sealed-bid auction and its application in computer networks. In particular, we provide a definition of the second-price sealed-bid auction and discuss the dominant strategy as well as the Nash equilibrium in this type of auction. Then, we compare the dominant strategy in the second-price sealed-bid auction and the equilibrium strategy in the English auction to show that the two auctions have the same truthful bidding strategy and can achieve the same revenue. To show the efficiency of the second-price sealed-bid auction, we present the applications of this auction for addressing important issues such as the task allocation in IoT, task scheduling in edge computing, and physical layer security in a mobile ad hoc network (MANET). The second part introduces a generalization of the second-price sealed-bid auction with multiple items, the Vickrey–Clarke–Groves (VCG) auction. We formally describe the VCG auction through an example from the computer networks' perspective. Considering some specific cases, we prove the dominant strategy in the VCG auction. Then, we provide examples to show how the VCG auction works.

- **Combinatorial Auction:** A combinatorial auction allows bidders to bid on combinations or packages of multiple items; it both satisfies dynamic demands and improves the resource utilization. Chapter 7 pursues our discussion by introducing the combinatorial auction and its applications in modern computer networks. We first introduce two types of items commonly used in the combinatorial auction, known as substitutable and complementary items. Then, we present types of bidding language, such as atomic bids, OR bids, and XOR bids, that allow bidders to succinctly encode or express common bids in the combinatorial auction. After that, we present the basic definition of the winner determination problem in the combinatorial auction. One of the most challenging aspects of the combinatorial auction is the high computational complexity required to solve the winner determination problem. To address this challenge, we present two iterative combinatorial auctions, the ascending proxy auction and the clock-proxy auction. Finally, we discuss how the combinatorial auction is used to address resource management issues in computer networks.
- **Double-Sided Auction:** Chapter 8 discusses the double-sided auction or double auction in modern computer networks. Different from the aforementioned auctions, the double sided-auction provides a mechanism for multiple buyers and multiple sellers. More specifically, the buyers and sellers first submit their bids and asks, respectively, to an auctioneer. Here, the auctioneer is an entity that conducts the auction. The auctioneer then matches asks from sellers and bids from buyers by assigning items from the sellers to the buyers and payments from the buyers to the sellers accordingly. In this chapter, we first describe the single-round double auction with two pricing policies, uniform pricing and discriminatory pricing. In addition, we show a well-known economic model, the supply and demand model, that works similarly to the single-round double auction. Some specific examples from computer networks are then given and analyzed to show how to determine the winners and the prices in the single-round double auction. The single-round double auction is considered to be the sealed-bid double auction in which the participants – that is, the buyers and the sellers – cannot learn the bidding strategy of their rivals and discover the real values of the resources. To address this issue, we introduce the continuous double auction, which allows the participants to trade the resources in multiple rounds. The applications of the double auction for emerging issues in computer networks are finally reviewed.
- **Other Auctions:** Chapter 9 introduces and discusses some special auctions. We first present the ascending clock auction, which has some similarities to the English auction. To understand how this auction works, we discuss the use of the ascending clock auction for improving the physical layer security in a cognitive wireless network. Then, we introduce the share auction, which has the low computational complexity and is typically used in markets with divisible goods. We next introduce the online auction, which allows the seller to make decisions about allocation and payments in real time. Finally, we introduce the waiting-line auction, which can be formulated as a non-cooperative game and its Nash equilibrium. The waiting-line auction is known as a non-money auction

in which the winners are determined based on waiting times submitted from the bidders rather than their bidding prices.
- **Optimal Auction Using Machine Learning:** The best way to design the optimal option is still an open issue. Chapter 10 introduces the design of optimal auctions using the deep learning technique. In particular, we first introduce the optimal auction design problems, the deep learning technique, and the motivations of the use of the deep learning technique for designing the optimal auctions. After that, we describe the neural network architectures that are used to derive the optimal multi-item auction and the Myerson auction. To demonstrate the efficiency of the deep learning–based auctions, we apply the auctions to resource management in a blockchain network.

In summary, this book constitutes a complete and comprehensive reference for auction theory and its applications in modern computer networks. Furthermore, owing to the aforementioned structure, which integrates the theory and the applications, this book is easy to follow and understand for general readers.

1.3.2 Timeliness of the Book

Recently, there has been tremendous interest in auction theory and its applications in the computer network research community. Auction and mechanism design has become a means of computer network design, protocol optimization, and resource allocation that takes not only system parameters but also economic implications into account. As such, there is an immediate need for references that provide comprehensive basics, background information, insightful analysis, and example applications of auction theory in computer networks. The references should be organized in such a way as to provide descriptions of theory, analytical derivation of models, and demonstration of real-world problems. The major reasons that this book is timely and impactful can be explained based on the following observations:

- *Promising economic-driven approach for next-generation Internet:* Recently, a number of new network services and applications have been introduced and become hugely popular, such as IoT, cloud computing, software-defined networking (SDN), and next-generation cellular networks. They are driving the consumer demand for network connections and services to an unprecedented level. Furthermore, with stringent quality of service (QoS) requirements, it becomes challenging for stakeholders, including content/service providers, network operators, and infrastructure providers, to deliver their services and manage their network resources by using traditional centralized system optimization approaches.

  The limitations of the system optimization approaches are as follows. First, the next generation networks and Internet involve multiple stakeholders that deploy their own services and resources in an uncoordinated way. They aim to achieve their various kinds of benefits, which may not be translated directly from system performance. Second, the different network resources and services are integrated to support customer demand and meet application requirements. Such an integration and composition have to achieve not only system optimality, but also market efficiency. The latter cannot be directly incorporated in the traditional system optimization approaches. Third, all stakeholders in the next-generation networks and Internet make their decision rationally, motivated by interests and incentives. Moreover, the stakeholders do not know the value of their resources and services. System optimization approaches largely fail to capture rationality and self-interest nature of the network entities.

Alternatively, auction and (economic-driven) mechanism design has emerged as an alternative that offers many benefits:

- Including an economic incentive in network service and resource allocation: Traditional centralized system optimization approaches rely only on network performance factors such as throughput, delay, and loss. However, stakeholders in next-generation networks and the Internet are self-interested in maximizing their own benefits such as revenue and profit. Therefore, incentive mechanisms are required for them to reach an optimal service and resources allocation/utilization. Auctions provide a well-defined incentive structure and allocation method, and hence become very suitable tools.
- Determining the value and market prices of network resources and services: Service providers and users may not know the exact values of network services and resources. Auctions offer a method to derive the values of services and resources by allowing different entities to propose their values. The market prices, meaning the values accepted by all entities, are determined by a set of certain rules.
- Supporting open and objective resource and service allocation methods: Auctions provide standard procedures in network resource and service allocation. Stakeholders and users can optimize their decisions about resource and service usage based on the rules and criteria that are given and open in advance. Moreover, the resource and service allocation is publicly observable, promoting transparency and predictability of the system operations.
- Promoting efficient allocation and investment: Since the allocation and pricing rules in auctions are open and objective, they can be defined to achieve the maximum market efficiency. Moreover, auctions can take the cost and investment into account in the allocation to achieve such a goal.
- Incorporating public policy goals and other constraints: As the next-generation networks and Internet will be used as a platform for many commercial and social applications, public policies for network resource and service allocation – for example, for public safety – will be imposed. Moreover, to ensure fair share and competition of the stakeholders, constraints are imposed, such as to avoid collusion. Auctions can incorporate the policies and constraints in the mechanisms.
- More resilient outcomes: Different auctions can achieve desirable properties such as non-negative utility, efficient allocation, and profit/utility

maximization. These properties enable effective network resource and service allocation.
- Rigorous mathematical tools: Auction theory has a long history. It contains rich analytical tools to investigate important characteristics of the systems and allocation. Moreover, auctions include many variants, such as single-item and multi-item auctions, single-sided and double-sided auctions, and combinatorial auctions, that can be applied with a variety of applications and scenarios.

- *Most existing auction theory books Focus on economics applications:* As auction theory has many benefits, a textbook that provides fundamentals and example applications will be useful to researchers and network engineers. However, the current books are mostly written to fit the economics and mathematics context. In consequence, researchers and engineers working in the computer networks area can find it difficult to comprehend and apply the theory to their work. These books use different terminology, contain many technical terms, and make a large number of domain-specific assumptions that are difficult to connect to physical network operations and resource/service management. Furthermore, developing auction models to address different issues in emerging computer networks by relying on standard textbooks from economics and mathematics fields is not straightforward. It requires multidisciplinary domain knowledge to achieve the system goals. This becomes more challenging as the network technology changes rapidly and quick solutions are needed. As a result, there is an urgent need for an auction textbook that provides content in the computer network context.

- *Emergence of network and Internet services:* As software-based networks become a major trend, a number of network services are emerging rapidly in which models that capture the system performance and economic incentive optimization are required. Some examples of such networks are:
  - IoT and wireless sensor networks, participatory sensing networks, and mobile crowdsensing networks, which provide data and information services, location-based services, and IoT security.
  - Network security services, which provide solutions for the security issues in wireless networks such as physical layer security, DoS attack prevention, and information security.
  - Cloud computing and networking, including cloud data center networking, mobile cloud networking, and edge computing, which provide cloud computing infrastructure and networking capabilities such as Network-as-a-Service (NaaS) and video-on-demand (VoD) services.
  - 5G networks with emerging technologies such as MIMO, mmWave communications, cloud-radio access networks, and heterogeneous networks that support a wide range of services including broadband access services, ultra-reliable communications, high user mobility, and massive IoTs.

This book is intended, primarily, for the following audiences:

- Graduate and senior undergraduate students interested in applying auction theory and developing the models for their dissertations, theses, and projects.
- Network engineers and software developers interested in studying and using new economic tools for network service and resource allocation.
- Economists interested in revenue optimization using auction theory in emerging network services and systems.
- Researchers interested in the state-of-the-art literature on auction theory for solving various issues in computer networks.

1.4 Acknowledgments

This work was supported in part by A*STAR-NTU-SUTD Joint Research Grant Call on Artificial Intelligence for the Future of Manufacturing RGANS1906, WASP/NTU M4082187 (4080), Singapore MOE Tier 1 under Grant 2017-T1-002-007 RG122/17, MOE Tier 2 under Grant MOE2014-T2-2-015 ARC4/15, Singapore NRF2015-NRF-ISF001-2277, and Singapore EMA Energy Resilience under Grant NRF2017EWT-EP003-041. This work was partially supported by US MURI AFOSR MURI 18RT0073, NSF CNS-1717454, CNS- 1731424, CNS-1702850, and CNS-1646607.

2 Overview of Modern Computer Networks

In this chapter, we introduce the background and the fundamentals of modern computer networks that are related to the issues and approaches under the context of auctions. The first part of the chapter introduces modern computer networks including the Internet of Things (IoT), cloud networking, and 5G wireless networks. The second part then introduces and discusses emerging issues in computer networks. The issues considered include data aggregation, task allocation, user association, interference management, wireless caching, mobile data offloading, and wireless security issues.

2.1 Internet of Things

IoT is one of the emerging technologies in IT that allows billions of smart devices to be connected to the Internet. The development of IoT has had a great influence on many areas, and many IoT applications have been implemented to improve the system performance as well as the quality of life, such as apps focused on healthcare, transportation, and manufacturing [8]. In this section, we present (i) definitions of IoT, (ii) the general architecture of IoT, (iii) the resources and services that IoT provides, and (iv) the main components of IoT.

2.1.1 Definitions

The concept of IoT is broad, and numerous definitions of IoT are available in literature. Some of the most popular definitions follow:

- IoT is an intelligent infrastructure that links objects, information, and people through computer networks, and in which radio-frequency identification (RFID) technology found the basis for its realization [9].
- IoT envisions a self-configuring and complex network in which "things" are interconnected to the Internet through standard communication protocols [10]. The interconnected things are uniquely identifiable, and have sensing and actuation capability as well as a programmability feature.
- IoT is defined as a pervasive presence of a variety of things and objects with unique addresses that are able to interact and cooperate with each other to reach common goals [11], [12].

- IoT is a dynamic global network infrastructure with self-capabilities based on standard and interoperable communication protocols in which "things" have identities, physical attributes, and virtual personalities; use intelligent interfaces; and are seamlessly integrated into the information network [13].

In summary, IoT refers to a self-configuring, adaptive, complex, and large network that allows a variety of things or objects through unique addressing schemes to interact and to cooperate with each other to reach common goals. Here, things or objects can be RFID tags, sensors, actuators, mobile phones, cameras, smart meters, and GPS terminals. They are able to collect sensing data and transmit this data to other systems without or with minimal human intervention [10]. The unique addressing scheme is critical since it allows us to uniquely identify and control billions of objects through the Internet.

Based on these definitions, we can list some key features of IoT as follows [14], [15]:

- *Automation:* Automation is a key feature of IoT. Accordingly, IoT is able to autonomously (i) support data collection and processing, (ii) make contextual inferences, and (iii) collaborate with other IoT devices. The automation, sometimes called *self-capability* of IoT, includes (i) self-configuration, (ii) self-organization and self-adaptation to dynamic scenarios, and (iii) self-processing of the huge amount of exchanged data.
- *Intelligence:* To enable the automation feature, objects or devices in IoT are required to be "smarter" or "more intelligent" [16]. In particular, apart from performing normal functions, such as sensing information from the surrounding environment, IoT devices make optimal decisions without or with minimal human intervention given their constrained resources and the dynamic of the environment for the requested IoT services.
- *Heterogeneity:* IoT may support different underlying networks, such as wired, wireless, and cellular, and diverse communication devices, such as access point–based and peer-to-peer (P2P) devices.
- *Dynamicity and adaptability:* The devices in IoT, such as mobile phones, can move from one place to another. Thus, the IoT system should be able to dynamically identify the mobility and act accordingly.
- *Zero-configuration:* To support easy integration of devices in the IoT system, plug-and-play feature should be available. This also enables the scalability of the IoT system. Thus, IoT a large-scale system.

2.1.2 IoT Architecture

To provide these features, several IoT architectures have been proposed [8], [11], [17], [18]. Figure 2.1 shows a general IoT architecture that consists of four tiers/layers [16]: device layer, networking and communications layer, platform and data storage layer, and data management and processing layer.

- *Device:* This layer is composed of low-level devices such as RFID tags, sensors, and mobile phones. The devices in the device layer are typically constrained to

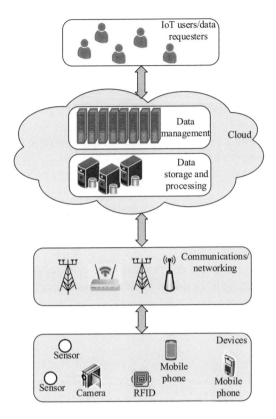

Figure 2.1 A general architecture of IoT [16].

computing, storage, and energy resources. Thus, the functions of the device layer include primitive tasks such as sensing the physical environment (e.g., temperature and humidity, gathering the real-time physical data). For this reason, the device layer is also called the physical layer. Other functions of the device layer are to connect with the Internet gateways for transmitting and forwarding the data to higher layers.

- *Networking and communications:* After data is gathered from devices in the device layer, the data needs to be transmitted to the higher layer – that is, the cloud platform – for further processing. The networking and communications layer is intended to provide device addressing and to perform data delivering. For this, this layer uses heterogeneous network infrastructures, such as access points and cellular base stations, and communication technologies, such as wired and wireless technologies. In particular, compared with the wired technologies, wireless technologies are more flexible, and thus commonly used for IoT. The wireless technologies can be wireless LAN (WLAN), cellular networks, and even satellite.
- *Data storage and processing:* This layer performs data storage from the device layer, data processing, and data forwarding to the data management layer. For the

full realization of IoT, the cloud is typically used in this layer. The cloud refers to virtualized computation and storage resources, and it can be considered to be the natural home for IoT applications for the following reasons [15]:

- Data in IoT is collected from billions of devices.
- The cloud uses data centers that can store and process the high volume of data.
- The cloud provides efficient mechanisms for dynamically and automatically provisioning data.
- By storing data in the cloud, IoT users or data requesters may access and use the data easily.
- The cloud may resolve efficiently several issues such as data processing problems and the extraction of useful information.

- *Data management:* This layer uses the application software to provide access services to IoT users, meaning the data requesters or customers. The data management and processing layer is thus considered to be an application layer.

2.1.3 Resources and Services of IoT

As a heterogeneous large-scale system, IoT provides many different resources and services. Common resources and services in IoT include the following items:

- *Sensing data:* Sensing data, such as the temperature and humidity data, is the output of a device at the device layer when the device detects and responds to some types of input from physical environments. The output is often "raw data" or "source data," so it needs to be further processed at the cloud layer. The output of the cloud layer is *sensing information* that is extracted from the raw data. The sensing data and sensing information are the most important resources in IoT since they can maximize the utility and profit of data owners or providers [19], [20], [21], [22].
- *Power:* IoT is a large-scale system including multiple entities such as sensors, base stations, and servers in the cloud. These entities need energy resources for their operations, and this can significantly increase the operating cost. To reduce the cost, energy harvesting or power harvesting from renewable resources – for example, solar power, thermal energy, and wind energy – has recently been adopted as a viable solution for providing "self-energy recycling."
- *Network resources:* IoT uses wired networks or wireless networks for data transmissions that need to use network resources. When wireless networks are used, the network resources are typically spectrum and network bandwidth. To save spectrum and network bandwidth resources, the sensing data and information can be locally stored and processed at radio base stations. These solutions are known as *edge computing* and *wireless caching*.
- *Cloud services:* The cloud is used in IoT for the data storage and processing. Thus, IoT can provide cloud services to the users. For example, IoT can provide

cloud services such as data storage, information searching, data security, data analysis, and data mining [23].

- *Location-based services:* Sensing data contains real-time geographical information from user devices such as mobile phones. Thus, IoT can provide location services to interested individuals, organizations, and the government. These services involve indoor and outdoor localization, such as identifying a location of a person or an event, or discovering the nearest places (e.g., restaurants, coffee shops, and stores) [24]. Location-based services are primary services of the IoT with an expected revenue of €34.8 billion in 2020 [25].

Since sensing data is an important resource in IoT, in the next sections, we present two common networks that perform the task of collecting sensing data: wireless sensor networks and mobile crowdsensing networks.

2.1.4 Wireless Sensor Network

A wireless sensor network in IoT consists of a large number of intelligent sensors. The intelligent sensors are able to collect, process, analyze, and disseminate valuable data gathered in an area of interest. As shown in Figure 2.2(a), most wireless sensor networks share some common features:

- The sensors transmit their sensor data to a sink node, which then forwards the sensor data to a base station or an access point.
- The sensors can send their data directly to the sink node or through other sensors.

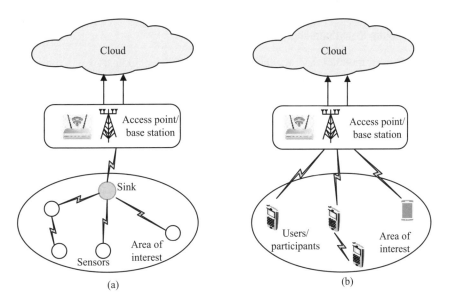

Figure 2.2 Data aggregation using (a) wireless sensor networks and (b) mobile crowdsensing networks [16].

- The coordination among the sensors thus forms a communication network such as a homogeneous multi-hop network, or a hierarchical network.
- The topology and transmission routes of the wireless sensor networks may considerably fluctuate over time, such as when a new sensor joins or leaves the network, or congestion appears at nodes. Thus, the topology should have the capability to heal itself.

The wireless sensor network consists of the following components [8]:

- *Hardware:* Each sensor in the wireless sensor network is composed of sensor interfaces, processing units, transceiver units, and a power supply.
- *Topology stack:* To enable the communication among the sensors as well as the communication between the sensors and the base station (i.e, the sink), networking protocols are integrated into the sensors. Typically, the topology stack consists of a physical layer, data-link layer, network layer, transport layer, and application layer.
- *Efficient and secure data aggregation:* Sensors are constrained by resources, so efficient data aggregation methods are required for improving the lifetime of the wireless sensor network. The data aggregation methods are also secure to protect the network from attackers and from node failure.

In general, wireless sensor networks can be constructed with low cost due to recent technological advances in low-power integrated circuits and wireless communications. However, it is still challenging to deploy the wireless sensor networks in a large-scale system such as IoT. In the next section, we introduce mobile crowdsensing networks, which are expected to be an alternative to wireless sensor networks.

2.1.5 Mobile Crowdsensing Network

Sensor modules are attached to mobile devices, such as smart phones, to perform application-specific sensing tasks; examples include air quality sensors [26] and biometric sensors [27]. Thus, each mobile device can be considered to be one or multiple sensors, and it can be used as a critical component of IoT. When a large number of mobile devices are used to gather data from an area of interest, they constitute a *participatory sensing network* [28] or *mobile crowdsensing network* [29], [30]. Figure 2.2(b) illustrates the IoT model for using a mobile crowdsensing network. As shown in the figure, the data aggregation process of the mobile crowdsensing network is generally similar to that of the wireless sensor network. In particular, each mobile device first performs sensing tasks, such as measuring air pollution, in an area of interest. Then, the raw sensor data is transmitted to the cloud through radio base stations for futher data processing. However, because it leverages mobile devices, the mobile crowdsensing network has a number of characteristics that differ from those of the wireless sensor network [31]:

- The mobile devices have more storage, computing, and communication resources compared with the sensors in wireless sensor networks.

- The mobile devices can be equipped with multiple sensor modules. Thus, each mobile device can be considered to provide multiple sensors in the wireless sensor network.
- The mobile devices have different resources. Thus, the quality – in terms of accuracy, latency, and confidence – of the data collected by the mobile crowdsensing network is more diverse than that of the data collected by a wireless sensor network.
- Due to the high mobility of mobile devices, the mobile crowdsensing network is more dynamic than the wireless sensor network.
- A number of people carry and use mobile devices, meaning that a large number of mobile devices are already deployed. Thus, leveraging these mobile devices significantly reduces the cost and time of deployment of the sensing infrastructure.
- The mobile devices are owned by individual users, and the intelligence of the users can be leveraged to help collect complex data, as when identifying available street parking spots.

Apart from the aforementioned advantages, deploying the mobile crowdsensing network in IoT faces the following challenges [32]:

- *Resource limitations:* Mobile devices have constrained resources, which reduces the sensing capacity and the data quality (e.g., the accuracy and latency).
- *Data inconsistency:* Mobile devices have different resources, which results in a difference in the sensing and computing capability. This leads to the data quality inconsistency issue. Determining an optimal set of mobile devices to provide the desired data consistency is generally a complex problem.
- *Data redundancy:* A large number of mobile devices are required to sense the same area of interest. This can cause data redundancy, which requires increased storage resource and bandwidth resource and incurs a high traffic load and cost. However, detecting the redundant data or the similarity among the collected data is challenging.
- *Motivation and incentives:* Mobile devices are owned by individual users (i.e., mobile users), and the sensing data contribution requires users to consume resources (e.g., storage, energy, and bandwidth). Users are rational, and one key challenge is to design incentive mechanisms that can entice users to contribute their data. Pricing and payment strategies can be adopted for the incentive mechanisms that guarantee the stable scale of participants and improve the accuracy, coverage, and timeliness of the sensing results.
- *Privacy and security:* Mobile users are selected for performing sensing tasks in an area of interest, and these users are usually required to specify their locations. This leads to location information leakage for the users; even worse, there may be attackers in the network who harm the users [33]. This may discourage users from participating in the mobile crowdsensing applications. Incentive mechanisms, along with privacy preservation mechanisms, need to be well designed to address this issue.

2.2 Cloud Networking

Cloud computing is becoming the platform of choice for a number of applications due to several advantages, including high computing and storage resources, high availability and accessibility, and low service cost. Popular areas supported by cloud computing include education [34], commerce [35], healthcare services [36], transportation [37], and social networks [38]. Typically, cloud computing provides on-demand resources and services including data storage and computing hosted in data centers. Common cloud services are platform-as-a-service (PaaS) [39], infrastructure-as-a-service (IaaS) [40], and software-as-a-service (SaaS) [41].

With the growing demands of data storage, computing, and network resources, the EU-funded Scalable and Adaptive Internet Solution project (https://sail-project.eu/) is investigating and developing a cloud model called *cloud networking*. Cloud networking can be considered to be a combination of cloud computing infrastructure and networking capabilities with the aim of providing on-demand data storage, computing, and network resources to cloud clients/users. In particular, the network resources include virtual routers, virtual firewalls, and network bandwidth. To provide on-demand cloud and network resources and services, network virtualization technologies (e.g., network function virtualization) are adopted beyond the data centers. Note that the use of the network virtualization is also a key difference between the cloud networking and traditional computer networks [42].

2.2.1 General Architecture

The cloud networking architecture needs to be constructed to guarantee an efficient composition of cloud and network resources. For this purpose, different architectures have been proposed in different scenarios for cloud networking:

- *Cloud data center networking [43], [44], [45], [46], [47]:* This cloud networking model is based on the fact that (i) the current data centers are connected with each other through high-speed networks and (ii) the cloud and network resources can be provisioned from the data center interconnection.
- *Mobile cloud networking [48], [49]:* This cloud networking model is proposed when the cloud can be integrated with mobile networks to provide combinations of cloud and radio network resources to mobile users.
- *Edge computing [50], [51], [52], [53]:* This cloud network model is designed to meet the growing demands on both cloud and network resources with low latency, low network traffic load, and low overall cost.

Based on these cloud networking models, the authors in [42] provide a general architecture for cloud networking. The architecture illustrated in Figure 2.3 is composed of three major components: (i) cloud data center networking, (ii) mobile cloud networking, and (iii) edge computing. The descriptions of the components are given in the next sections. Note that these components can work independently from each other. For example, the mobile cloud networking may operate without the cloud

2.2 Cloud Networking

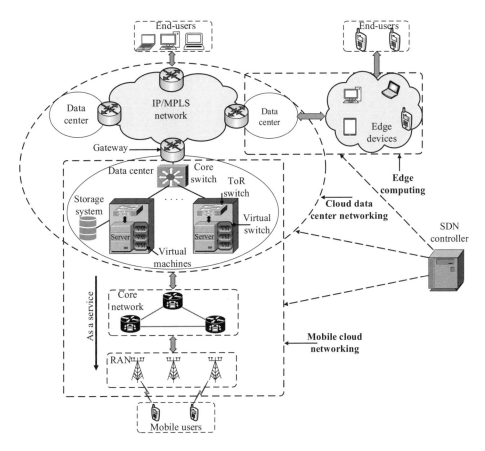

Figure 2.3 A general architecture of cloud networking [42]. IP/MPLS stands for Internet protocol/multi-protocol label switching, ToR stands for top-of-rack, and SDN stands for software-defined networking.

data center networking. However, to adapt diverse requests of users (e.g., some requests in the mobile cloud networking should be allocated among data centers to obtain an optimization), these models should be integrated with each other. The complex architecture of cloud networking introduces various actors or stakeholders. For convenience, we define the stakeholders that commonly participate in cloud networking as follows [42]:

- *Cloud provider:* A cloud provider or cloud service provider is a company or an individual who owns and manages cloud infrastructures, such as data centers. The cloud provider delivers cloud resources and services to other businesses or individuals. For example, Amazon Web Services (https://aws.amazon.com/) and Google Cloud Platform (https://cloud.google.com/) are popular cloud providers that offer the cloud IaaS.
- *Network provider:* A network provider provides the network connectivity among data centers of cloud providers or between the data centers and users. Network providers may be communication service providers, Internet service providers,

network service providers, or wireless service providers. In cloud networking, the network providers tend to cooperate with cloud providers to offer combinations of cloud and network resources to users.

- *User:* There are two types of users in cloud networking:

  - *End-users:* These users generate resource and service requests or workloads that need to be processed using cloud resources. For example, mobile users are typically end-users.
  - *Cloud tenant:* A cloud tenant or cloud user may be an organization or an enterprise that uses cloud and network resources to host applications offered to its end-users. For example, Netflix (www.netflix.com/) is a cloud tenant that provides video-on-demand services. Cloud tenant can be thus considered to be *cloud service brokers*.

2.2.2 Cloud Data Center Networking

A data center is a physical facility that houses a group of computer systems and associated components such as network and storage systems. Cloud data center networking may refer to *intra-data center networking* or *inter-data center networking* [42].

For intra-data center networking, the computer systems within a data center are connected with each other through a network system. An intra-data center networking system has the following major components [42]:

- *Virtual machine:* By using virtualization techniques, a physical server of a real computer can be divided into *virtual machines*. Each virtual machine is able to perform tasks, such as storing data and running applications and programs, as the physical server [54]. In cloud networking, the virtual machines can be migrated among servers within a data center or between data centers.
- *Network slicing:* The virtual machines can be grouped in a virtual network – for example, to support a common application [55]. To do this and to guarantee virtual resource isolation and virtual network performance, network slicing is used.
- *Virtual switch:* A virtual switch is a software program that allows a virtual machine to communicate with other machines.

For inter-data center networking, data centers can be interconnected across the wide area network. The inter-data center networking consists of the following entities [42]:

- *Gateway:* Gateways of data centers provide connectivity (i) among data centers and (ii) between data centers and the Internet and clients/users. They are able to provide virtual routing and switching capabilities.
- *Internet protocol/multi-protocol label switching network:* This packet-switched network employs the Internet protocol enhanced with the multi-protocol label switching standard.
- *Resource pool:* To meet the high demands of users, multiple data centers can constitute a *resource pool*. In other words, a resource pool is a collective set of resources in data centers.

- *Federated cloud networking:* The data centers may be owned by different cloud providers. To create resource pools, the cloud providers are required to cooperate with each other. Such a cooperation is called *federated cloud networking* or *federation cloud networking*. With federated cloud networking, the cloud provider can perform *outsourcing* by borrowing cloud resources from other providers. Also, the cloud provider can perform *insourcing* by renting out its cloud resources to other cloud providers.

2.2.3 Mobile Cloud Networking

Significant growth in mobile data traffic increases the complexity of tasks, such as signal processing and resource allocation, in cellular networks. The mobile cloud networking now being investigated by the EU FP7 Large-Scale Integrating Project (IP) (cordis.europa.eu/fp7/ict/future-networks) has been proposed as a promising solution to address this issue. Mobile cloud networking is actually a virtualization of the cellular networks that aims to integrate cloud and network function virtualization concepts into the cellular networks [56]. By leveraging the cloud and its high computing capacity, mobile cloud networking significantly reduces the complexity of tasks (e.g., signal processing and resource allocation) in the traditional cellular networks, and it is also able to provide on-demand services involving mobile network, storage, and computing. Mobile cloud networking has the following major features [48]:

- By leveraging the high-performance cloud computing infrastructures, mobile cloud networking enhances the real-time performance of cellular network functions such as the baseband unit processing, mobility management, and QoS control.
- Mobile cloud networking thus enables adaptation to the elasticity and the growth of the mobile data traffic demands.
- Mobile cloud networking provides an entirely new mobile cloud application platform by combining infrastructures and services across different domains, including wireless, mobile core networks, and data centers.
- Mobile cloud networking introduces new stakeholders such as mobile virtual network operators and mobile cloud networking providers.

Given these features, mobile cloud networking is able to deliver a wide range of services. One of the most important services is the virtualized network infrastructure service. The virtualized network infrastructure service allows functionalities of radio access networks (e.g., digital processing functions) to be partially or fully moved into a data center in the cloud depending on the actual needs and network characteristics [57]. In the case that the functionalities of radio access networks are fully moved in the cloud and only radio-frequency functions, such as signal receiving and transmitting, are performed at base stations, the mobile cloud networking system is called a *cloud radio access network* or *centralized radio access network* [58]. The cloud radio access network consists of two major components:

- *Baseband processing unit pool:* This component is placed in the cloud and performs baseband processing functions of the traditional base station such as channel coding, modulation, and fast Fourier transform.
- *Remote radio head:* This component is placed at the remote site and performs radio-frequency functions of the traditional base station such as radio-frequency amplification, up/down conversion, filtering, and digital-to-analog/analog-to-digital conversion. Multiple remote radio heads can be served by one baseband processing unit pool. The remote radio heads are typically connected to the baseband processing unit pool by using fronthaul links.

The cloud radio access network has several advantages [42], [59]:

- *Adaptability to nonuniform traffic:* Remote radio heads perform only the simple tasks of the traditional base stations, so the information exchange among the remote radio heads becomes easier with low latency. This enables the cloud radio access network to dynamically manage mobile users and adapt their demand elasticity.
- *On-demand resource provision:* Leveraging the high-performance cloud, the cloud radio access network is able to provide on-demand resources and services.
- *Low handover time:* Handover tasks are performed in the baseband processing unit pool instead of in the remote radio heads. This significantly reduces signaling information exchange among the remote radio heads and, therefore, the handover time.
- *Low capital expenditure and operational expenditure:* The signal processing functions are centralized in the baseband processing unit pool. This reduces the capital expenditures, such as for equipment rooms and power supplies, at the remote site. The result is lower operational expenditures, such as maintenance costs.

2.2.4 Edge Computing

Cloud data center networking and mobile cloud networks can provide vast storage and computing resources to users. However, users often suffer from high network resource costs (e.g., the bandwidth cost) and long latency due to the long-distance communication between users and the remote data centers. In the worst case, users may may experience service interruptions – for example, due to natural disasters such as fires, earthquakes, and power outages [60]. To address these issues, *edge computing* allows data storage and computing resources to be moved toward users or the data sources. As shown in Figure 2.4, edge computing leverages the resources of edge devices distributed at the edge network instead of at the remote data centers in the cloud. Edge computing is thus a type of distributed cloud computing. It pushes the frontier of storage and computing resources and services away from the data centers to the periphery or edges of the

2.2 Cloud Networking

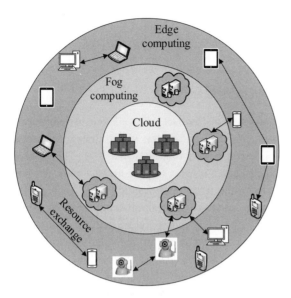

Figure 2.4 Edge, fog, and cloud computing [42].

network [61]. In edge computing, the edge devices can be smartphones, IoT devices, laptops, and vehicles.

Note that compared with fog computing, edge computing pushes the storage and computing resources closer to the network edge. As shown in Figure 2.4, fog computing is located "between" the edge computing and the cloud computing.

Edge computing has the following benefits [62]:

- Edge computing significantly reduces the data traffic, cost, and latency and improves QoS since cloud resources and services are located close to users. As a proof, the authors in [63] ran a face recognition application that decreased the response time from 900 ms to 169 ms by moving computation from the cloud to the edge nodes.
- Edge computing addresses the bottleneck issue and avoid the risk of a potential single point of failure since it is regarded as a distributed cloud paradigm. Edge computing thus enhances the reliability.
- Edge computing enhances data security since the data does not need to be sent to the cloud over a long distance. Also, edge computing preserves the data, and the ability to process data without ever putting it into a public cloud adds a useful layer of security for sensitive data.
- Edge computing provides high levels of scalability and automation.

Further details of edge computing architecture, case studies, advantages, and challenges are well presented in [63], [64], [65]. Here, we highlight just one key challenge of deploying edge computing: the edge devices are typically owned by individual users, so incentive mechanisms need to be well designed to entice users to contribute their resources and to guarantee the stable scale of participants and QoS.

2.2.5 Cloud-Based Video-on-Demand System

Traditional video-on-demand systems such as Internet Protocol Television (IPTV) [66], enable users/clients/viewers to select and watch video content whenever they want, instead of watching only at a specific broadcast time [67]. The traditional systems are typically based on client–server or peer-to-peer architectures. With the large and imbalanced demands of users, the traditional systems face two big challenges: load bottleneck and limited storage capacity. This significantly reduces the QoS of users. To address the challenges, a cloud data center network that is composed of data centers with high storage capacities can be used to support the video-on-demand systems. Such a system, called a *cloud-based video-on-demand system* [68], is considered to be a new video content delivery model in the development of cloud networking. Figure 2.5 depicts a general model of the cloud-based video-on-demand system that consists of the following components [69]:

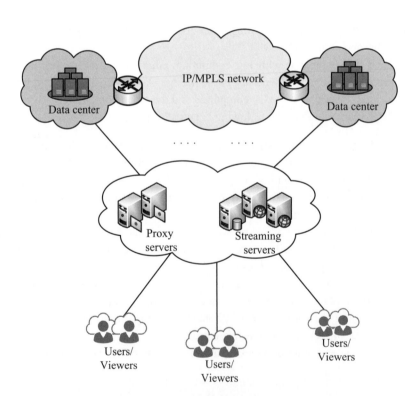

Figure 2.5 An illustration of cloud-based video-on-demand system [42]. IP/MPLS stands for Internet protocol/multi-protocol label switching, and VoD stands for video-on-demand.

- *Cloud data center networking:* This component stores a massive amount of media content.
- *Users:* Users or viewers generate content requests to servers owned by video-on-demand providers. The servers consist of proxy servers and streaming servers.
- *Proxy and streaming servers:* Proxy servers receive the requests from users and deliver the requests to those streaming servers with a lighter traffic load. Then, the streaming servers forward the requests to a nearby data center to obtain the content from the data center.

By leveraging the cloud data center networking, the cloud-based video-on-demand system is able to serve a large number of users with various requirements [70]. However, high-quality video content typically consumes, a large amount of bandwidth, so the bandwidth cost is still significant in the cloud-based video-on-demand system and needs to be addressed.

2.3 5G Wireless Networks

An increasing proliferation of smart devices has introduced various emerging multimedia applications that are leading to an exponential growth in mobile data traffic [71]. Along with the development of IoT, the wireless data demand and usage have already placed a significant burden on existing 4G (Long-Term Evolution) LTE cellular networks. This has triggered the investigation of fifth-generation (i.e., 5G) cellular networks. The 5G networks are expected to provide high data rates, low latency, high capacity, and high QoS improvement for users, compared with the current 4G LTE networks. In particular, some key parameters – indeed, key requirements – of the 5G systems are as follows [72]:

- *Peak data rate:* The achievable data rate is up to 20 Gbit/s.
- *User data rate:* The achievable data rate per user is from 100 Mbit/s to 1 Gbit/s.
- *Latency:* The packet travel time is 1 ms.
- *Mobility:* The speed for handoff and QoS requirements is up to 500 km/h.
- *Capacity:* The total traffic across coverage area is 1,000 Mbit/s/m^2, which can support a high density of mobile broadband users, ultra-reliable, and massive machine-type communications (MTC).

To meet these requirements, the 5G networks will various emerging technologies. The primary technologies proposed for 5G networks include [73] massive multiple-input and multiple-output (MIMO), dense heterogeneous networks (HetNets), mmWave communication, cloud-based radio access networks (C-RANs), full-duplex communication, device-to-device (D2D) communication, massive machine-to-type (M2T) communication, energy-aware communication and energy harvesting, and the virtualization of network resources. Figure 2.6 shows a general model covering technologies that are expected to have the most significant impact on the progress made toward 5G. These technologies are discussed in the next sections.

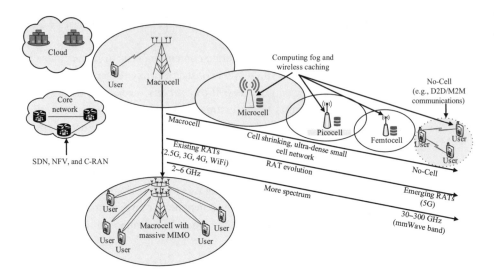

Figure 2.6 Emerging technologies proposed for 5G wireless networks [74]. Arrows indicate the potential technologies deployed in the 5G wireless networks. SDN and NFV stand for software-defined networking and network functions virtualization, respectively. RAT stands for radio access technology.

2.3.1 Massive Multiple-Input and Multiple-Output

The development of new applications such as ultra-high-definition video conferencing and IoT has significantly increased the demands on data rates in cellular systems. The conventional Massive Multiple-Input and Multiple-Output (MIMO) technologies, in combination with the limited number of service antennas, are not able to effectively respond to these demands [75]. Thus, massive MIMO has been proposed to provide high data rates to meet the demand. Massive MIMO, also known as large-scale antenna systems, very large MIMO, or full-dimension MIMO, is seen as a key enabler on the path to 5G. In massive MIMO cellular systems, a very large number of antennas (e.g., a few hundred) are used at each base station to serve multiple single-antenna users. The antenna system of the base station is designed in a two-dimensional grid, and each single antenna in the antenna system is positioned such that directivity in transmission can be achieved. Also, coherent superposition of wave-fronts is employed to improve the capacity by several times [71]. Compared with conventional MIMO, massive MIMO has the following advantages [75]:

- *High throughput:* Massive MIMO can enhance the capacity 10 times due to the use of the aggressive spatial multiplexing. Simultaneously, based on the principle of coherent superposition of wavefronts, energy can be concentrated with extreme sharpness into small regions in the space. This can improve the energy efficiency on the order of 100 times.
- *Low latency:* In wireless communication systems, fading is one of the factors that limits system performance. In particular, the signal from a transmitter needs to

travel through multiple paths before reaching the receiver. This can increase the communication latency when a fading dip happens. Based on the law of large numbers and beamforming, massive MIMO can avoid the fading dips and reduce the latency.
- *Robustness against interference and jamming:* The massive number of antennas enables the massive MIMO system to have a large surplus of degrees of freedom. These degrees of freedom can be used to cancel interference or jamming.
- *High spectral efficiency:* By employing antenna arrays with hundreds or thousands of active elements and performing coherent transceiver processing, massive MIMO systems can simultaneously serve a number of users using the same time-frequency resource [76]. Thus, the spectral efficiency is substantially improved.

Apart from the aforementioned benefits, massive MIMO systems can satisfy the dynamic demands of users. In particular, users in these systems can specify the number of serving antennas. Moreover, by adopting advanced resource management approaches, such as combinatorial auction [77], users can explicitly request bundles of physical resources, including spectrum, power, and antennas, from the base stations.

2.3.2 Heterogeneous Networks

Heterogeneous Networks (HetNets) are a key technology in 5G cellular networks. HetNets are composed of different types of cells, such as macrocells and small cells (e.g., picocells and femtocells), and different radio access technologies (RATs). In particular, each macrocell is covered by one macro base station, and each small cell is covered by a small cell base station. The small cells typically have a coverage radius less than 200 m. Commonly used small cell base stations include femto access points, pico cell base stations, and relay stations. Table 2.1 shows key features of the small cells. The coexistence of different cells forms a *multi-tier HetNet*. Apart from multiple tiers, HetNets exploit the spectrum of different RATs, such as Wi-Fi, evolved high-speed packet access, LTE, and new 5G RATs. The *Coordinated Multi-Point (CoMP)* technology [78] can also be adopted; it enables users in 5G to simultaneously access different base stations with different RATs. By using multiple tiers and multiple RATs, HetNets provide the following advantages.

- *High capacity:* Apart from the macrocells, a large number of small cells are deployed in HetNets. The deployment of cells leads to *ultra-dense networks* (i.e., $\geq 10^3$ cells/km^2) [79]. This significantly improves the network capacity and enables it to serve users with high density (i.e., up to 600 active users/km^2). Moreover, HetNets further improve the network capacity by exploiting the spectrum of multiple RATs.
- *High spectrum efficiency:* HetNets improve the spectrum efficiency in two ways. First, HetNets can leverage unused spectrum from multiple tiers and RATs through the use of the coordinated multi-point technology. Second, small cells in HetNets have small coverage, and network operators can increase the spectrum reuse.

Table 2.1 Cells deployed in 5G HetNets [74].

Cells	Transmit power (W)	Coverage (km)	Deployment scenarios	Users
Femtocell	0.001–0.25	0.01–0.1	Indoor	Up to 30
Picocell	0.25–1	0.1–0.2	Indoor/outdoor	30–100
Microcell	1–10	0.2–2	Indoor/outdoor	100 to 2,000
Macrocell	10–50	8–30	Outdoor	>2,000

- *Coverage improvement:* Small cells, such as relay stations, are deployed at the edge of the network. This expands the coverage of macrocells. Moreover, local area networks such as D2D communications are adopted to further enhance coverage and spectrum efficiency. The adoption of these networks leads to another tier with the elimination of cells.
- *High reliability:* HetNets allow each user to access different tiers and different RATs. This increases the network access rate as well as the reliability of network connections.
- *Energy efficiency:* By deploying a dense array of small cells, HetNets enable the users to access those small cells rather than macrocells. The short communications reduce the transmitting power of users and improve the energy efficiency. The short communications also improve the user QoS since the service request and response time are low.
- *Low network operational and capital expenditures:* Small base stations are cheaper than macro base stations. Also, the small base stations are easier to deploy, and they consume less energy. Thus HetNets reduce both operational and capital expenditures.

However, deploying the heterogeneity and density of wireless devices in HetNets imposes radio resource management issues. These issues include user association, resource allocation, mobility, and interference management. Advanced radio resource management schemes need to be designed that consider the following performance metrics:

- *Outage probability:* Outage probability, also known as coverage probability, refers to the probability that the data rate will drop below a certain threshold, known as the required data rate. This is regarded as a fundamental metric for network performance analysis and network planning.
- *Spectrum efficiency:* Spectrum efficiency, also known as spectral efficiency or bandwidth efficiency, is defined as the data rate that can be transmitted over a given bandwidth in the wireless network.
- *Energy efficiency:* In general, energy efficiency refers to the use of less energy to perform the same task. In wireless networks, it is defined as the ratio of the total network throughput to the total energy consumption.
- *Load balancing:* Load balancing refers to efficiently distributing network traffic across, for example, base stations, in the network.

- *Fairness:* A fairness metric is used to determine whether the users in the network are receiving a fair share of network resources. To measure the fairness, Jain's fairness index [80] is often used. Assume that there are N users in the networks. Then, the transmission rate fairness among the users is defined as $\frac{\left(\sum_{i=1}^{N} r_i\right)^2}{N \sum_{i=1}^{N} r_i^2}$, where r_i is the rate of user i.
- *Interference:* The density of wireless devices in HetNets raises a variety of interferences. Thus, the interferences in the network need to be mitigated. The effectiveness of mitigating the interferences can be measured by the signal-to-noise ratio. Two types of interferences occur in HetNets: *intra-tier interferences* (i.e., the interferences caused by cells within the same tier) and *inter-tier interferences* (i.e., the interferences caused by cells in different tiers).

To improve these performance metrics, it may not be efficient to adopt the traditional resource management approaches. For example, for the user association, the received signal strength indication (RSSI) [81] is not efficient since it is based only on the received signal strength from base stations rather than on the traffic load of the base stations. This leads to an unbalanced traffic load among base stations and low network throughput. Also, for the interference management, optimization methods may not be used since they often require central controllers that significantly increase the signaling and computational overhead. Alternatively, economic and pricing models can provide decentralized solutions to efficiently address the resource management issues.

2.3.3 Millimeter Wave Communications

The 5G network system is expected to support various emerging services such as the high-definition television (HDTV) and ultra-high-definition video (UHDV) [82]. Such a service requires multi-gigabit communications with huge bandwidth, but the frequency bands used in the existing cellular networks, 3G and 4G, provide only a bandwidth to 40 MHz. There is a vast amount of unused spectrum in the Millimeter Wave (mmWave) band ranging from 30 to 300 GHz with wavelengths of 1 to 10 mm. Therefore, mmWave communications have been proposed to be a potential candidate for 5G. The mmWave band is expected to provide multi-gigabit communication services [83].

In general, compared with existing communication systems, mmWave communications suffer from higher propagation loss and have shorter range. The reason is that the mmWave communications have higher rain attenuation and atmospheric absorption. However, as shown in Table 2.2 [83], for short distances, the rain attenuation and atmospheric absorption do not create significant additional path loss for mmWaves. In particular, at 28 GHz and for a distance of 200 m, the rain attenuation is only 0.9 dB, and the attenuation caused by atmospheric absorption is only 0.012 dB.

Since 5G is expected to deploy a high density of small cells with a coverage radius smaller than 200 m, mmWave communication appears to be a promising technology for future 5G cellular networks. In particular, the use of the mmWave band in 5G has the following benefits:

Table 2.2 Common mmWave bands in 5G and propagation characteristics at a precipitation rate of 25 mm/h and a distance of 200 m [74], [83]. LOS stands for line-of-sight.

Freq. band	Propagation loss (LOS)	Rain attenuation	Oxygen absorption	Available bandwidth
28 GHz	1.8–1.9	0.9 dB	0.04 dB	500 MHz
38 GHz	1.9–2	1.4 dB	0.03 dB	1 GHz
73 GHz	2	2.4 dB	0.09 dB	2 GHz

- *High data rate:* mmWave communication leverages the large bandwidth potentially available in the mmWave frequency band, so it is able to provide a high data rate. In particular, mmWave can support high-quality video and multimedia services. Also, mmWave is able to provide high-speed wireless communications such as backhaul solutions for small cells.
- *Immunity to interference:* mmWave communication relies on short-distance transmission, so it mitigates the interference among cells. Also, mmWave communication significantly reduces the interference among active mmWave links due to the high directivity.
- *Effective integration with massive MIMO:* The mmWave band has an extremely short wavelength that is easily integrated with antenna elements of the massive MIMO system.
- *Effective integration with ultra-dense small cell networks:* The small cells deployed in 5G have a small coverage radius. Since the mmWave band has a short range, it is very suitable for the massive deployment of small cells. In return, the use of mmWave in small cell networks solves the path loss problem.

However, exploiting high-frequency bands with narrow beams imposes several challenges. In particular, mmWave communications require a dense deployment of base stations, and this creates problems for mobility management. Specifically, frequent handovers and dynamic user association schemes are required in mmWave communication systems. While existing user association schemes are only suboptimal for mmWave communication systems [84], economic and pricing models have been recently adopted to provide fast and dynamic association schemes [85].

2.3.4 Cognitive Radio

Along with the introduction of new spectrum bands, such as the mmWave spectrum, cognitive radio can be used to solve the spectrum shortage problem by improving the spectrum utilization. Cognitive radio is an adaptive, intelligent radio and network technology that can automatically and autonomously detect available or unused spectrum and change transmission parameters, such as the data rate and transmit power. With this technology, users can opportunistically access unused radio resources. The typical cognitive radio system shown in Figure 2.7 is composed of three components [86]:

2.3 5G Wireless Networks

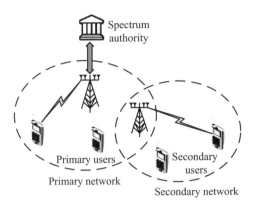

Figure 2.7 A cognitive radio system [86].

- *Spectrum authority:* Spectrum can be a commodity, and the spectrum authority is a government organization, such as the ministry of science and technology, that owns and manages the spectrum rights.
- *Primary networks:* Primary networks are often telecommunications companies that obtain licenses to use the spectrum from the spectrum authority. The primary network's users are called primary users.
- *Secondary networks:* Secondary networks lease available radio resources (i.e., the unused spectrum) from the primary network to serve the secondary network's users. These users are called secondary users. Each secondary user has a *cognitive radio capability*, or sensing spectrum, to check whether the spectrum is being used by a primary user and to change the radio parameters. Thus, compared with the primary users, the secondary users have a lower priority on the usage of the spectrum. Also, when using the spectrum, the secondary users should not cause interference to the primary users.

Given the layered structure of cognitive radio system, two resource trading markets exist for this system. The first trading market involves the spectrum authorities and the primary networks, and the second market involves the primary network and the secondary networks. Pricing and economic models such as auctions have been widely used to efficiently trade the spectrum in cognitive radio systems. A survey of these approaches is given in [86].

2.3.5 Device-to-Device Communications

Device-to-device (D2D) communication is considered to be one of the key technology components of the evolving 5G network. D2D communication refers to the radio technique that enables direct communication between two mobile devices, or mobile users, without the involvement of a base station or a core network (see Figure 2.8). D2D communication has the following advantages:

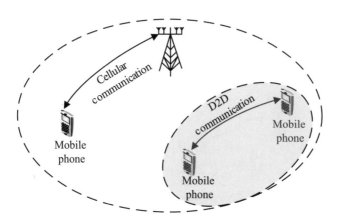

Figure 2.8 Cellular communication and D2D communication.

- *System capacity gain:* D2D communication improves the system capacity due to the possibility of sharing spectrum resources between the cellular links and D2D links.
- *User data rate gain:* D2D communication can achieve high peak rates due to the short propagation.
- *Low energy consumption:* The short propagation allows D2D users to transmit data with low power while achieving the target QoS.
- *Base station traffic offloading:* Since there is no involvement of the base station, D2D communication reduces the traffic load at the base station.
- *Low latency:* As the users communicate over a direct link, the end-to-end latency is reduced.

D2D devices can communicate with each other by using the unlicensed spectrum or by using the cellular spectrum (i.e., the licensed spectrum). The former case is called *outband D2D*, and the latter is called *inband D2D* [87].

- *Outband D2D:* In this case, the unlicensed spectrum is used for the D2D links. Since the unlicensed spectrum is not used for the cellular links, the outband D2D avoids the interference between the D2D links and the cellular links. To use outband D2D communication, D2D devices need to be equipped with extra interfaces and wireless technologies, such as Wi-Fi Direct and Bluetooth.
- *Inband D2D:* With outband D2D communications, the interference in the unlicensed spectrum is uncontrollable. For example, nearby D2D communications can use the same unlicensed spectrum and cause interference with each other. This imposes constraints on the QoS provided. To address this issue, inband D2D communication leverages the licensed spectrum (i.e., the spectrum of the base station). There are two types of inband D2D communication. In the first type, called *underlay inband D2D*, the D2D links use the same licensed spectrum as the cellular links; that is, the D2D links reuse the spectrum of the cellular links. In the second type, called *overlay outband D2D*, a part of the licensed

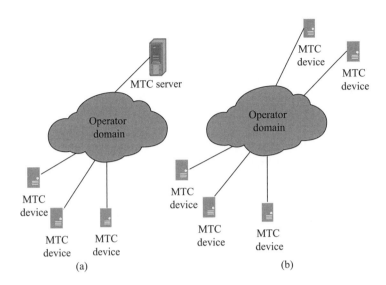

Figure 2.9 M2M communication model (a) with an intermediate server and (b) without an intermediate server [16].

spectrum is dedicated to the D2D devices and the remainder of the licensed spectrum is assigned to the cellular links. The main disadvantage of inband D2D is the interference issue caused by the D2D links to the cellular links, and vice versa. Advanced interference management mechanisms need to be designed to address this issue. However, this results in the high computational overhead of both base station and D2D devices.

2.3.6 Machine-to-Machine Communications

Similar to the D2D communication, Machine-to-Machine (M2M) communication is expected to be a key technology component in the evolving 5G network. M2M communication, also called machine-type communication (MTC), refers to the technology that enables devices such as computers, embedded processors, smart sensors, smart grid devices, actuators, and smartphones to communicate with each other via wired or wireless communication networks with little or no human intervention [88]. Figure 2.9 shows two models of M2M communication. In the first model, M2M devices communicate with each other via one or some intermediate servers (see Figure 2.9(a)). In the second model, M2M devices communicate directly with each other without an intermediate server (see Figure 2.9(b)). As such, we can say that M2M is more general form of D2D communication. In particular, D2D focuses only on mobile phones, while M2M covers not only mobile phones but also other devices, such as smart sensors and actuators.

The major features of the M2M communication include automated data generation, processing, transfer, and exchange between intelligent machines with minimal human intervention. Due to these features, M2M communication can support IoT. The M2M communication model envisions a number of devices with small data, sporadic transmissions, high reliability, low latency, and real-time operation. Serving a massive number of M2M devices, however, imposes several challenges for M2M communication. One main challenge is how to address issues such as congestion management, access control, security, and QoS guarantees. A survey discussing these issues and ways to address them by using economic and pricing models can be found in [16]. Also, a survey of reviews of architectural enhancements, network functionalities, and implementation challenges is given in [89]. Another survey of major reviews of existing M2M research work includes various commercial, hardware and research platforms [90]. The major advances and developments in architecture, protocols, standards, and security for M2M evolution from 4G to 5G are discussed in [91].

2.4 Data Collection and Resource Management

Modern computer networks are expected to fully support various emerging multimedia applications. However, the adoption of emerging technologies introduces issues for resource management. The reasons are (i) the dynamics of the network environment, (ii) the heterogeneity and dense deployment of network devices, (iii) the heterogeneous resources, (iv) the coverage and traffic load imbalance of base stations, (v) the high frequency of handovers, (vi) the constraints of the fronthaul and backhaul capacities, and (vii) the participation of a large number of users and stakeholders with different objectives. In the next sections, we present common issues in the modern computer networks, including (i) data aggregation, (ii) task allocation, (iii) user association, (iv) interference management, (v) wireless caching, and (vi) mobile data offloading.

2.4.1 Data Aggregation

Data aggregation can be considered one of the most important emerging issues in IoT. It is defined as the process of aggregating and collecting sensing data from sensor nodes or mobile devices. Figures 2.2(a) and (b) show data aggregation through the use of sensor nodes and mobile devices, respectively. The goals of data aggregation are to (i) improve the efficiency and accuracy of information, (ii) reduce unnecessary information, (iii) save energy for the nodes/devices, and (iv) decrease the traffic load. To achieve these goals, data aggregation techniques are proposed. The discussions of data aggregation techniques can be found in [92]. In general, to evaluate the efficiency of the data aggregation techniques, the following performance metrics are often used [92]:

- *Data accuracy:* Data accuracy is how close the measured data is to the true data. Data inaccuracy can occur because nodes and devices in IoT may have data sensing and transmission errors that result in the collection of the wrong data.
- *Energy consumption:* Energy consumption refers to the total energy consumed by the nodes in IoT. Due to the energy constraints of the IoT nodes, energy consumption is a critical factor for the data aggregation methods and impacts the network lifetime.

2.4 Data Collection and Resource Management

- *Fault tolerant:* Fault tolerance refers to the ability of the network to continue to perform its task even if a network fault, such as a hardware error related to a sensor node, occurs.
- *Latency:* Latency is the time delay required for transmitting data from the source (e.g., the sensor node) to the destination (e.g., the sink or the base station).
- *Network lifetime:* The network lifetime is defined as the number of data collecting rounds that are performed until the energy of $\alpha\%$ of nodes in the network is exhausted, where α is a parameter defined by the network manager. For example, for a network with 100 sensors and $\alpha = 1$, the network lifetime is the number of data collecting rounds until the energy of the first sensor is exhausted.
- *Security:* IoT devices may be compromised by attackers and provide wrong sensing data. Thus, the data aggregation techniques need to be secure to detect and isolate such a compromised device.
- *Privacy:* Mobile users that are selected for performing sensing tasks are required to specify their locations. This leads to the leakage of sensitive information, such as location information, and opens up users to the possibility of attack actions. The data aggregation methods need to preserve users' privacy.
- *Incentive cost:* Incentive cost refers to the total cost that is paid to the users for their data contribution. From data requesters' perspective, the cost should be at its minimum.

In general, it is challenging to design a data aggregation technique that satisfies all of these metrics. For example, the aggregation approaches based on compressed sensing [93] achieve only energy efficiency; they do not guarantee high data accuracy and low incentive cost. Recently, economic and pricing models based on negotiation mechanisms have been effectively used to design data aggregation techniques that achieve multiple objectives. For example, aggregation approaches based on sealed-bid reverse auction are proposed in [94], [95], and [96]; an approach based on sealed-bid multi-attribute reverse auction is presented in [97]; one based on Vickrey–Clarke–Groves reverse auction is given in [98]; approaches based on bargaining game are presented in [99] and [100]; and a method based on the multi-objective knapsack problem is proposed in [101].

2.4.2 Task Allocation

In computer networks such as IoT, task allocation assigns sensing tasks, such as traffic monitoring, to sensors or mobile devices in the network for execution. In fact, task allocation is related to data aggregation in that it enables the data aggregation operation to improve the performance metrics. For example, by assigning the sensing tasks to sensors with high energy, task allocation achieves a fair energy balance, improves the network lifetime, and minimizes the latency. We consider the following example to understand how allocation of common tasks is implemented in IoT. In the example, a pricing model is adopted to design the task allocation. The pricing model is used because it provides negotiation mechanisms among entities in the network that can optimize

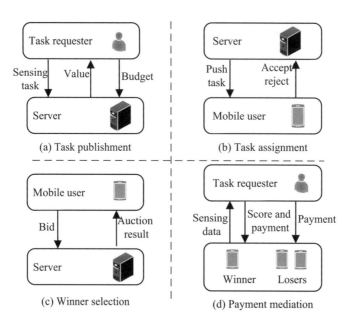

Figure 2.10 An example of task allocation in IoT [16].

both task allocation and resource utilization. This task allocation approach is proposed in [102] and is briefly described here.

The model is an IoT system with a mobile crowdsensing network (see Figure 2.2(b)). The model consists of three entities: (i) the task requester that posts a sensing task, (ii) mobile users that are willing to execute the sensing task, and (iii) the server that is responsible for mobile user selection, task price evaluation, and payment mediation. Here, the pricing model uses a sealed-bid reverse auction. In this type of auction, sellers compete to sell their commodities to the buyer. The sellers submit their asks (i.e., the prices of the commodities) to the auctioneer. The bid of the seller with the lowest price is accepted. In the our example, the task requester is the buyer, the mobile users are the sellers, and the server is the auctioneer. The task allocation process based on the sealed-bid reverse auction is shown in Figure. 2.10 and involves four stages [102]:

- *Task publishment:* The task requester publishes the sensing task description. Upon receiving the task description, the server evaluates the task price according to its sensing region and sensing period. For example, if the task is within a region that makes its execution easy, its value (i.e., price) should be low; conversely, if the task is within an area that makes it difficult to perform, its value should be high. The task value is sent back to the task requester for suggesting a budget.
- *Task assignment:* Based on the budget suggested by the task requester, the server selects those mobile users that satisfy the budgetary constraint and best match the requested sensing context for the task.
- *Winner selection:* The selected mobile users submit their asks or asking prices (i.e., the prices that the mobile users are willing to receive for executing the

requested sensing task) to the server. The server selects the winner based on the mobile users' asking prices and reputation. In particular, the reputation of the mobile user refers to the quality of the sensing data that the mobile user submitted to the server in the past.
- *Payment mediation:* The winner receives a reward and a reputation score for executing the sensing task. The losers also receive a reward for participating in the auction.

The proposed task allocation approach is able to achieve high data quality and a low incentive cost, while preserving privacy. In fact, different objectives can be attained by adopting different pricing models. For example, to guarantee a fair distribution of energy among sensor nodes that prolongs the network lifetime, a first-price sealed-bid auction can be used as proposed in [103]. Also, to reduce message exchange overhead in the network and energy consumption of the sensor nodes, a combinatorial reverse auction may be adopted as presented in [104].

2.4.3 User Association

Before data transmission commences, each mobile user needs to be assigned to a specific base station. This process is called *user association*. In the existing networks, such as 4G LTE systems, user association schemes are typically based on the received signal strength indication (RSSI) [81]. In particular, the schemes are based only on the received signal strength from base stations rather than on the traffic load of the base stations. Such a scheme may not be efficient in modern computer networks such as 5G networks, because the 5G networks involve a massive mobile users and a high density of base stations. The user association schemes that are based only on the received power might result in unbalanced traffic load among base stations and low network throughput. The design of sophisticated radio resource management schemes for the user association is thus needed. To evaluate the user association schemes, the following performance metrics are commonly used [105]:

- *Network throughput:* Network throughput is defined as the average number of successfully transmitted bits per second per hertz. Maximizing the network throughput enhances the spectrum efficiency, so it is an important requirement for user association schemes.
- *Load balancing:* Load balancing refers to efficiently distributing incoming network traffic across multiple base stations in the network. Load balancing aims to optimize resource utilization, maximize network throughput, and avoid traffic overload as well as congestion. Load balancing is thus a mandatory requirement of the user association schemes.
- *Latency:* The latency of the user association schemes needs to be minimized to guarantee the QoS provision.

To improve these performance metrics, the traditional approaches based on centralized solutions [106] can be used. However, such an approach typically requires a large amount of signaling and has high computational complexity. Thus, it may not be a viable solution for large-scale networks, especially for IoT and 5G networks. Recently, economic and pricing models have been used to provide distributed solutions that optimize the aforementioned metrics with low computational complexity. For example, a distributed auction is adopted for the user association to maximize the network throughput in the 5G mmWave network [107] and 5G HetNets [108]. To achieve load balancing and fairness among base stations, the forward auction and congestion-based pricing approaches are used as proposed in [109] and [110], respectively.

2.4.4 Interference Management

Future networks such as 5G wireless networks adopt multiple tiers including macro cells, small cells, relays, D2D communications, and M2M communications (see Figure 2.6). The adoption of multiple tiers significantly improves performance in terms of QoS, capacity, coverage, spectral efficiency, and total power consumption. However, such a multi-tier architecture also raises interference management issues for the following reasons [73]:

- The multi-tier architecture includes the heterogeneity and density of wireless devices that share the licensed spectrum.
- Base stations in different tiers have different transmit power, which results in imbalances in coverage and traffic load.
- The tiers may have different access policies, such as public and private access policies, that lead to diverse interference levels.
- The introduction of carrier aggregation and coordinated multi-point (CoMP) transmission techniques among the base stations further complicates the dynamics of the interference.

There are two common types of interference, co-tier interference and cross-tier interference [111], in the 5G networks, as shown in Figure 2.11.

- *Co-tier interference:* Co-tier interference, also known as intra-tier interference, refers to the interference occurring between network entities that belong to the same tier. The network entities can be base stations or mobile users. Figure 2.11 shows the interference between a femtocell and a femto user belonging to a different femtocell. Co-tier interference occurs when the distance between the two femtocells is small, and they share the same spectrum band.
- *Cross-tier interference:* Cross-tier interference, also known as inter-tier interference, refers to the interference occurring between network entities that belong to different tiers. Figure 2.11 illustrates the interference between a femtocell and a macro user belonging to the macrocell.

Both co-tier interference and cross-tier interference reduce the network performance, and interference management methods need to be designed. Optimization algorithms for power control (e.g., [112] and [113]) are typically adopted for interference management. In general, these algorithms observe and control the transmit power of the base

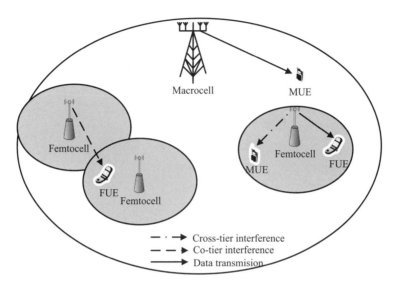

Figure 2.11 Intra-tier interference and inter-tier interference in 5G HetNets. FUE and MUE stand for femto user equipment and macro user equipment, respectively.

stations to keep the interference below a predefined threshold while guaranteeing the QoS requirement. However, the optimization algorithms usually require use of central controllers, which leads to a huge signaling and computational overhead. Inter-cell interference coordination techniques [114] can be also used by applying restrictions to the radio resource management. However, these techniques require a large number of measurement message reports from neighboring cells, which significantly increases the signaling overhead. As such, the traditional approaches for interference management may not be efficient in 5G networks, and a new look into the interference management problem is required.

Recently, economic and pricing models have been efficiently used to address interference management. For example, the distributed auction is proposed for channel allocation and power control [115], the non-cooperative game is adopted for power control as presented in [78], whereas the Stackelberg game is proposed for power control in [116], [117], [118], [119], [120], and [121]. The common idea underlying these interference management approaches is to use a *penalty price* to control the transmit power of the base stations. To further understand the idea, we consider a scenario with one macro base station and multiple small base stations. First, the macro base station measures local interference caused by the small base stations. Then, the macro base station sets penalty prices according to interference levels caused by the small base stations. To avoid a high payment, the small base stations locally reduce their transmit power. Such a simple pricing strategy enables the macro base station to mitigate the cross-tier interference while reducing the information exchange and computational complexity. A survey of the interference management schemes based on the economic and pricing models is given in [74].

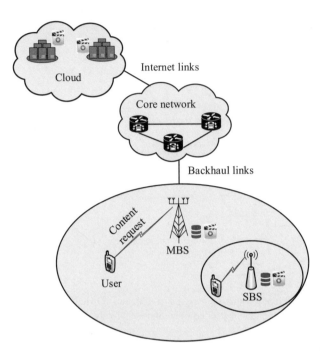

Figure 2.12 A cellular network with wireless caching [122]. MBS and SBS stand for macro base station and small base station, respectively.

2.4.5 Wireless Caching

The exponential growth of mobile data traffic causes a bottleneck issue for backhaul links. In fact, content such as popular films or videos is usually requested by mobile users during a short span of time. To avoid placing a heavy traffic burden on the backhaul links, such popular content can be proactively stored/cached at edge devices (e.g., cellular base stations) close to the mobile users instead of on content servers in the remote cloud. This technique is called *wireless caching*. By leveraging edge nodes for caching the content, wireless caching can (i) reduce the transmission latency of content requests from users, (ii) mitigate the redundant transmission of popular content over backhaul links, (iii) achieve higher energy efficiency, and (iv) significantly improve network capacity as well as coverage.

A general model of a cellular network with wireless caching is shown in Figure 2.12. This model consists of mobile users, also known as clients/viewers; base stations; a core network; and the cloud. In particular, the cloud, thanks to its huge storage capacity, stores a massive amount of original content. The base stations can be macro base stations or small base stations. The base stations can predict mobile user requests, perhaps by observation, and cache popular content in advance. First, each mobile user sends its content request to the base station that is serving the mobile user. If the content is already cached at the base station, the base station directly serves the mobile user's request. Otherwise, the base station sends the content request to the cloud and then delivers the content to the mobile user.

2.4 Data Collection and Resource Management

As such, the wireless caching consists of two phases [122]:

- *Caching placement:* This phase is implemented during off-peak traffic periods to prefetch popular contents. In general, the base stations are constrained in terms of their storage capacities, and the mobile users in different areas may prefer different content. Thus, key issues in the caching placement phase are to decide (i) the popularity of content and (ii) the best base stations at which to cache the content.

  - To determine the popularity of a particular content, the Zipf distribution [123] is often used. The Zipf distribution of the content refers to the request probability for that content. Specifically, the Zipf distribution of content i is expressed by

  $$P_r(i) = \frac{1/i^{r_r}}{\sum_{j=1}^{m} 1/j^{r_r}}, 1 \leq i \leq m \qquad (2.1)$$

  where r_r represents the probability of content reuse, and m is the total amount of content requested [122].

  - To determine the base station for caching the content, *caching placement strategies* need to be adopted. Two well-known strategies, the frequency-based strategy [124] and the recency-based strategy [125], can be used.

- *Content delivery:* The content delivery method decides how to deliver the content to the mobile users upon receiving their content requests. In particular, the content delivery method (i) decides which base stations will deliver the content and (ii) determines the power and channels for transmitting the content.

To improve the QoS experiences for the mobile users and to reduce the cost for the network operators, the design of the caching placement and content delivery phases need, to consider the following performance metrics [122]:

- *Backhaul cost:* Backhaul refers to the communication links among the base stations and the core network. Content such as videos may consume a huge backhaul bandwidth, resulting in a high backhaul cost. The backhaul cost is thus critical in designing the caching placement phase.
- *Network delay:* Network delay refers to the latency from the time that the file is requested until the time that the file is delivered. The network delay affects the QoS experiences of the mobile users, especially delay-sensitive users. Thus, network delay is one of key performance metrics.
- *Cache hit rate:* The cache hit rate of each base station is the probability that the requested content is found in the storage of the base station. A base station with a higher cache hit rate can serve its mobile users better. The cache hit rate is a performance metric for evaluating the caching placement strategies.
- *Power consumption:* The power consumed by the base stations determines the power cost, and power consumed by the mobile users affects their battery life time. Thus, the power consumption is also considered as a performance metric.

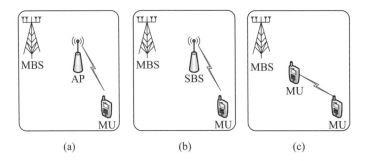

Figure 2.13 Mobile data offloading via (a) Wi-Fi; (b) SBS, such as femtocell; and (c) opportunistic communication, such as D2D communications [126]. MBS, SBS, and MU stand for macro base station, small base station, and mobile user, respectively.

In fact, mobile users themselves can cache the content and then deliver the content to another mobile user. This boosts the revenue and improves network delay and power consumption due to the short-scale communications among the mobile users. However, the mobile users are typically rational, so incentive mechanisms need to be well designed to ensure their resource contributions. For this purpose, economic and pricing models such as auctions can be adopted. A survey of the wireless caching approaches based on economic and pricing models is given in [74].

2.4.6 Mobile Data Offloading

A surge of mobile data traffic can result in overloading and congestion at base stations in cellular networks. Complementary network technologies can be used to offload the data traffic for the cellular networks through a means called *mobile data offloading*. These complementary network technologies can consist of Wi-Fi with access points, small base stations such as femtocells, and even mobile devices. In general, deploying such complementary network technologies is much cheaper and works more quickly than the traditional macrocell deployment. The use of such technologies leads to different mobile data traffic solutions like those shown in Figure 2.13 and described here:

- *Mobile data offloading via Wi-Fi:* In this solution, illustrated in Figure 2.13(a), an access point is used to carry the mobile data. Wi-Fi can be considered to be a natural solution for the data offloading for the following reasons. First, Wi-Fi is able to provide connectivity services with a high data rate. Second, most mobile phones, including smartphones, are equipped with Wi-Fi capabilities. Third, Wi-Fi uses the unlicensed spectrum, which has much cheaper cost than the licensed spectrum used by cellular networks. On the downside, Wi-Fi has limited coverage.
- *Mobile data offloading via small base stations:* In this solution, shown in Figure 2.13(b), a small base station, the femtocell, offloads the data traffic of the macrocell networks. Similar to Wi-Fi, the femtocells are typically deployed in homes or offices and allow for offloading a huge amount of indoor data. However, there are major differences between Wi-Fi and femtocells [127]:

- Wi-Fi operates in unlicensed bands, while femtocells use licensed bands.
- Wi-Fi has a larger free spectrum than femtocells.
- Deploying femtocells is more expensive and requires a more careful plan.
- Femtocells do not require mobile devices to equip or use an extra interface, unlike the Wi-Fi interface.

- *Mobile data offloading via mobile devices*: This solution, shown in Figure 2.13(c), is also known as *mobile data offloading via opportunistic communications*. With the offloading solution via mobile devices, each mobile device can provide data offloading services to nearby mobile devices. This solution can be implemented based on advanced technologies such as D2D and M2M communications. Such solution thus takes advantage of the D2D and M2M communications. In particular, the mobile devices can communicate directly with each other without any need for a network infrastructure. Consequently, offloading solution carries little or no monetary cost. However, it faces issues such as the heterogeneity of data traffic, mobility of the mobile devices, and battery, storage, and communication constraints of the mobile devices. Also, mechanisms need to be adopted that provide an incentive to the mobile devices to carry out the data delivery. The design of the incentive mechanisms for mobile data offloading can be based on the Vickrey–Clarke–Groves (VCG) auction as proposed in [128] and [129], based on the ascending clock auction (ACA) as presented in [130], or based on the Stackelberg game as discussed in [131].

Further discussions of data offloading solutions in cellular networks are presented in [132] and [127].

2.5 Wireless Network Security

Wireless networks are very common, both for organizations and for individuals. The main advantage of wireless networks is that devices are easy to access without any physical connections requirements. However, due to the exposed nature of the wireless medium, wireless networking is prone to some security issues. In particular, attackers can easily break into or use the wireless medium to perform attack actions directed at wired networks. Eavesdropping and DoS attacks, such as jamming and distributed DoS attacks, are especially common security threats. An eavesdropping attack is an incursion in which the attackers try to steal users' information as it is transmitted over wireless networks. This information is often of a, sensitive nature, such as password and credit card information. The eavesdropping attack is a *passive attack*. In contrast, a DoS attack is an *active attack* in which the attackers attempt to interfere with and disrupt network operations. Although these attacks have different strategies and different objectives, they significantly degrade the network's performance, QoS, reputation, and revenue. As a result, it is very important to have wireless security mechanisms.

To efficiently address wireless network attacks, it is important to understand how perpetrators perform attack actions. This section provides background information on attack types as well as emerging security issues in wireless networks. In particular, we first define types of attackers in wireless networks. Then, we describe eavesdropping and DoS attacks. Finally, we introduce and discuss information security issues and malicious behaviors of users in wireless networks.

2.5.1 Users and Attackers in Wireless Networks

The following list presents different types of users and attackers in typical wireless networks [33]:

- *Altruistic users*[1]*:* Altruistic users are users or nodes that behave in such a way as to enhance the overall network performance.
- *Compromised users:* Users within a network can launch insider attacks against a target in the network, with the attack actions beging controlled by an external attacker. Such a user is called *compromised user* or *compromised node* [133], or a *bot*
- *Selfish users:* Selfish users, often referred to as rational users, behave in a way that maximizes their own utilities rather than optimizes the overall network performance. However, they do not intend to cause damage to other users or systems. For example, selfish nodes in MANET may not forward a packet from other nodes because they want to save their resources only for their own transmission. The selfish and altruistic users can be considered to be *regular users* or *legitimate users* in the networks.
- *Malicious users or adversarial users:* In contrast to the regular users, malicious users or adversarial users attempt to damage/harm other users or systems. For example, a Byzantine node is a type of malicious node that purposefully deviates from routing protocols to disrupt the normal network operation [134].
- *Eavesdropper or wiretapper:* An eavesdropper is an unauthorized receiver/party that illegally captures and reads data packets from legitimate transmitters/sources.
- *Jammer:* A jammer is a malicious user that launches a DoS attack by transmitting a high-power noise signal with the aim of degrading or corrupting the signal at the intended receiver, a base station in a wireless network.
- *Friendly jammer:* A friendly jammer is typically used for physical layer security, as described in detail in Section 2.5.2. A friendly jammer assists legitimate sources by transmitting a *friendly jamming* signal to reduce the rate of data leakage from the source to the eavesdropper.
- *Active eavesdropper:* An adversarial user can act as an eavesdropper, a jammer, or both. When the adversarial user performs both jamming and eavesdropping, it is called an *active eavesdropper*.

2.5.2 Eavesdropping Attack

Attackers launch an eavesdropping attack by capturing or stealing information from legitimate communications. In general, an eavesdropping attack is difficult to detect

---

[1] The terms "users" and "nodes" are used interchangeably in this section.

since this attack does not cause any abnormality in the network's operations. Thus, the ability to share secret information reliably is extremely important. To combat eavesdropping attacks, cryptographic techniques such as secret key cryptography [135] and public key cryptography [136] are typically used. These techniques use encryption methods to convert the ordinary information, called plaintext, into an unintelligible form, known as ciphertext. This action is intended to reduce the eavesdropper's ability to decode the information. However, cryptographic techniques require highly complex algorithms for the encryption and decryption. Also, they require additional secure channels for key exchanges that are scarce in mobile environments.

Physical layer security algorithms are proposed as an alternative to cryptographic techniques. Instead of using keys, physical layer security algorithms exploit physical layer properties of wireless channels such as interference, channel gains, and thermal noise to protect the wireless communication against eavesdropping attacks [137], [138], [139]. In particular, physical layer security aims to achieve perfect secrecy data transmission between a legitimate source and its intended destination, while the eavesdropper is unable to decode the information, the eavesdropper obtains zero information [140].

One term commonly used in physical layer security is *secrecy capacity* [141]. Secrecy capacity is the rate at which the legitimate source can transmit secret information to its intended destination. The secrecy capacity is mathematically defined as follows [142]:

$$C_s = \max(C_d - C_e, 0) \tag{2.2}$$

where C_d and C_e are the capacities of the source–destination channel (see Figure 2.14) and the source–eavesdropper channel, meaning the wiretap channel, respectively. C_d and C_e are defined as follows:

$$C_d = W \log\left(1 + \frac{P_s h_d}{\sigma_d^2}\right) \tag{2.3}$$

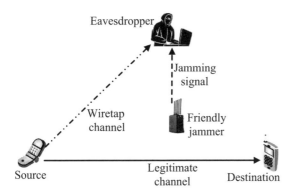

Figure 2.14 A typical eavesdropping attack model with a source, a destination, an eavesdropper, and a friendly jammer [33].

$$C_e = W \log\left(1 + \frac{P_s h_e}{\sigma_e^2}\right) \tag{2.4}$$

where P_s is the transmit power of the source; h_d and h_e are the channel gains from the source to the destination and from the source to the eavesdropper, respectively; W is the bandwidth of the source–destination channel or the source–eavesdropper channel; and σ_d^2 and σ_e^2 are additive white Gaussian noise variances of the source–destination channel and the source–eavesdropper channel, respectively.

To improve the secrecy capacity C_s, friendly jammers are used to degrade the signal quality at the eavesdropper. In particular, the friendly jammers transmit *friendly jamming signals* with high transmit power to introduce extra interference to the eavesdropper. Assume that there is one friendly jammer that transmits the friendly jamming signal with power P. Then, the secrecy capacity with the assistance of the friendly jammer is determined by

$$C_s^J = \max(C_d^J - C_e^J, 0) \tag{2.5}$$

where

$$C_d^J = W \log\left(1 + \frac{P_s h_d}{\sigma_d^2 + P j_d}\right) \tag{2.6}$$

is the capacity of the source-destination channel, and

$$C_e^J = W \log\left(1 + \frac{P_s h_e}{\sigma_e^2 + P j_e}\right) \tag{2.7}$$

is the capacity of the source–eavesdropper channel, where j_d and j_e are the channel gains from the jammer to the destination and from the jammer to the eavesdropper, respectively. It can be seen from (2.6) and (2.7) that employing the friendly jamming power decreases not only C_e, but also C_d. Therefore, the main problem in physical layer security is to determine a proper jamming power to maximize the security capacity, as given in (2.5). This problem can be considered to be a power allocation issue that has been solved by traditional optimization approaches [143], [144].

Recently, auctions have been effectively used to solve the power allocation issue for physical layer security. For example, the share auction is proposed in [145] to improve the secrecy capacity in mobile ad hoc networks; ascending clock auctions are proposed in [146] and [147] to improve the secrecy capacity in cooperative wireless networks; the Vickrey auction is proposed in [148], [149], [150], and [151] to improve the secrecy capacity in cognitive radio networks and mobile ad hoc networks; and the double auction is proposed in [152] and [153] to improve the secrecy capacity in cognitive radio networks. In addition, game theory can be used to develop distributed solutions for physical layer security with low complexity. For example, the noncooperative game is proposed to improve the secrecy capacity in cellular networks in [154].

2.5 Wireless Network Security

2.5.3 Denial-of-Service Attack

The eavesdropping attack is considered to be a passive attack in which eavesdroppers listen to a channel quietly. In contrast, the DoS attack is an active attack in which an adversary attempts to exhaust the resources available to legitimate users. Three common types of the DoS attack in wireless networks are jamming attacks, distributed DoS (DDoS) attacks, and black hole attacks.

Jamming Attack

A jamming attack is a type of DoS attack in which the attackers, either adversaries or malicious nodes, transmit radio-frequency jamming signals with high power to disrupt or cause interference to legitimate communications between sources and destinations. Jamming attacks are mostly launched at the physical layer. The effect of the jamming attack on the legitimate communications depends on the transmit power and the location of the jammer. In particular, the effect is larger if the jammer's transmit power is higher. In general, the jammer may jam the target network using different strategies to make the jamming as effective as possible. Two traditional jamming attacks are discussed here [155]:

- *Proactive jamming [156]:* With proactive jamming, the attacker or proactive jammer continuously transmits interfering signals to a target network regardless of a network activity, the data communication, in the network. The attacker can perform the proactive jamming by sending packets or random bits on a specific channel and does not switch channels until the energy of the attacker is exhausted. This kind of proactive jamming can be divided into constant jamming, deceptive jamming, and random jamming.
- *Reactive jamming [156]:* Different from the proactive jammer, the attacker using reactive transmits the interfering signals to a target network only when the attacker observes a network activity occurring in the network. Reactive jamming thus is more energy efficient than proactive jamming, as well as more difficult to detect. The reactive jammer is thus known to be "smarter" than the proactive jammer.

Apart from proactive and reactive jammers, other advanced jammers include follow-on jammers, channel hopping jammers, pulsed noise jammers, control channel jammers, implicit jammers, and flow jammers [157], [158]. Although these jammers have different strategies and objectives, their attacks lead to serious consequences such as degrading the network performance and QoS as well as losing important data and damaging the network's reputation. To combat such attacks, several jamming detection and countermeasure schemes are proposed. In particular, the jammed-area mapping protocol described in [159] calls for mapping out the jammed area in wireless sensor networks and routing data packets around the affected area. The Ant system–based anti-jamming scheme [160] seeks to detect jamming and redirect data packets to an appropriate destination node. The hybrid system–based anti-jamming scheme proposed in [161] relies on base station replication, base station evasion, and multipath routing

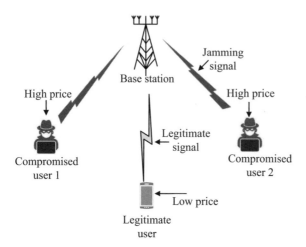

Figure 2.15 A distributed DoS (DDoS) attack [33], [167].

between the base stations. The channel hopping–based anti-jamming scheme described in [162] and [163] requires switching from one jammed channel to an unjammed channel. Since the jammers are subject to the cost of power while attempting to cause the interference, economic and pricing models have been recently used for anti-jamming. In particular, power pricing strategies [164], [165], [166] are applied to discourage jammers from consuming power to carry out their jamming actions.

Distributed Denial-of-Service Attack

A DDoS attack is a large-scale, coordinated attack on the availability of network services or resources [168]. The DDoS attack can be performed at the physical layer or at higher layers such as the network layer or application layer. This attack is typically launched indirectly through a large number of compromised nodes inside the network. The compromised nodes, called *bots*, are illegally tampered with and manipulated to launch the DDoS attack – for example, to transmit radio jamming signals with high power, to a target system, such as a base station. The bots are practically controlled by the external attacker targeting the network. There may be thousands of bots that are distributed over the network. Thus, a DDoS attack is one of the most severe attacks. To address this type of attack, traditional solutions such as the jammed-area mapping protocol and channel hopping can be used. However, these solutions face many challenges or may not work due to the largely jammed area and the boom of the bots.

Recently, pricing models have been effectively used to address the threats posed by DDoS attacks. Figure 2.15 shows an example of using the Bayesian optimal pricing model [167] to prevent bots' attack behaviors. This model consists of two compromised users 1 and 2, one legitimate user (i.e., the target of the attackers), and one base station. Here, the compromised users and the legitimate user are buyers, while the base station acts as a resource seller. Assume that the compromised users are rational, meaning that they behave in such a way as to maximize their own utilities rather than to optimize the overall network performance. In the network, the resources consist of bandwidth and power. To launch attack actions, the compromised users consume many network resources. Since the compromised users are rational, they are subject to the resources' prices [33]. This means that if we set high prices for resources, the compromised users have no incentive to consume more resources to launch their attack actions. This is also the pricing strategy of the Bayesian optimal pricing model. In particular, the Bayesian optimal pricing model sets the resource price for each user in the network proportionally to the probability that the user will launch an attack action. Specifically, high resource prices are set for the compromised users since they have high probabilities of performing malicious actions. Conversely, low prices are set for the legitimate users. As such, the users have no incentive to launch attacks on other users due to the high cost.

Black Hole Attack

The objective of the black hole attack is to disrupt network services such as packet routing or packet forwarding. Indeed, in routing protocols such as Ad hoc On-demand Distance Vector (AODV) [169], a malicious router may claim that it has the lowest cost, referring to the shortest route. Then, the malicious router performs a DoS attack by discarding incoming packets instead of relaying/forwarding them to the destination. Such an attack is called a *black hole attack* or *packet drop attack*. The malicious router, or malicious node in a general case, that performs the back hole attack is called a *black hole node*. Note that the malicious router can drop all packets that come in or perform the attack selectively, by dropping packets only during a specific time period. The former is easy to detect through common networking tools, such as traceroute. However, the latter is often harder to detect since some traffic still flows across the network.

Economic and pricing models can be used to effectively prevent such a black hole attack. For example, a pricing mechanism based on a noncooperative game is proposed in [170] that provides nodes with incentives to forward packets rather than dropping them.

2.5.4 Information Security Issues

Information security is the part of information risk management that involves preventing the unauthorized access, use, damage, leakage, modification, and recording of information [171]. Physical layer security, as mentioned in Section 2.5.2, can be considered to be part of information security since it aims to prevent eavesdroppers from stealing information. Generally, information security seeks to guarantee the confidentiality, integrity, and availability of information.

- *Confidentiality:* Confidentiality is a property of information which guarantees that the information is not available or disclosed to unauthorized users or unauthorized parties [172].
- *Integrity:* Integrity is a property of information which guarantees that the information is not modified or destroyed in an unauthorized manner [173].

- *Availability:* Availability is a property of information which guarantees that the information is always accessible and usable upon being demanded by authorized users [173].

Apart from the aforementioned attributes, information privacy is regarded as part of information security.

- *Information privacy,* also known as *data privacy* or *data protection* [174], refers to the privacy of personal information, such as the personal data stored in computers. Information privacy is considered to be an important aspect of information sharing. Personal information vulnerabilities increase in wireless networks due to the exposed nature of the wireless medium.
- To enhance information privacy, *privacy preserving* mechanisms are required. However, information privacy is a comprehensive concept that covers a wide area of privacy. For example, we have "Internet privacy," "financial privacy," and "medical privacy." Thus, privacy preserving is challenging. In particular, in wireless networks, traditional methods are to employ ciphertext.
- *Ciphertext* or *cyphertext* is the encrypted/encoded text. It is the data or information (*plaintext*) is an encrypted form. The algorithm that performs encryption of the plaintext is called a *cipher* or *cypher* [175]. Thus, the ciphertext contains a form of the original plaintext that is unreadable without the proper cipher to decrypt it. The key used in the encryption algorithm is called a *cryptographic key*.
- Apart from the ciphertext, *anonymity* schemes are used to enhance the information privacy of users. Anonymity refers exclusively to the matters related to the identity of users. More specifically, when anonymity schemes are applied, an attacker cannot identify a particular user within an anonymity set of users. With the technique known as k-anonymity location privacy [176], the user's location information is k-anonymous, meaning that the attacker cannot distinguish the user from the location information of at least $(k-1)$ other users. However, some users may not be sensitive about their own location information, and they may have no incentive to participate in the anonymity set. When there are not enough privacy-sensitive users in the anonymity set, the privacy-sensitive users can act as buyers and invite users that are not sensitive to privacy (i.e., sellers) to participate. To provide the sellers with incentives to take part in the anonymity set, economic and pricing models such as double auction can be used [177].

2.5.5 Illegitimate Behaviors in Wireless Networks

As mentioned earlier, auctions have recently been used to address several resource management and security issues in computer networks, and this trend is increasing. However, the adoption of auctions in computer networks can raise security issues. Consider a general spectrum auction, in which users as spectrum buyers (bidders) submit their bids (bidding prices) for the requested resources (e.g., spectrum) to an auctioneer. To increase utility, the users may perform illegitimate behaviors or misbehaviors as follows:

- *False-name bids:* False-name bids are submitted by a single user using multiple names such as multiple email addresses or identifiers. Such a dishonest action is called *false-name bid cheating*. A bidder that cheats in this way may increase its own utility but reduce the seller's revenue and other bidders' utilities [178]. Moreover, this cheating behavior is very difficult to detect due to the anonymity of bidders in wireless networks. To prevent false-name bid cheating, false-name-proof mechanisms [179], [180] can be applied. A mechanism is *false-name-proof* if no bidder benefits from using multiple identifiers.
- *Collusion:* Collusion is a cheating behavior in which a bidder as a buyer seeks out and persuades other bidders to propose a lower price to the seller. A collection of bidders that collude in an auction is called a *bidding ring*. This approach differs from false-name bid cheating, in which a bidder executes its cheating alone. The collusive behavior significantly reduces the seller's revenue and poses severe threats to the efficiency of spectrum allocation [181]. Similarly, sellers may collude with each other to increase their prices. Additionally, there may be a collusion between an auctioneer and the bidders, a practice called *bid-rigging*.
- *Bid-rigging:* Bid-rigging is a form of collusion in which an untrustworthy auctioneer conspires with greedy bidders to illegally fix the price and manipulate market [182]. To prevent the bid-rigging, the auction models can be integrated with the homomorphic encryption methods.

2.6 Summary

In this chapter, we introduced the background and the fundamentals of modern computer networks. We first provided definitions of IoT and described the general architecture, resources, and services of IoT. Two important components of IoT, wireless sensor networks and mobile crowdsensing networks, were further discussed. Second, we introduced cloud networking models that aim to provide on-demand data storage, computing, and network resources to cloud clients/users. The cloud networking models include cloud data center networking, mobile cloud networking, and edge computing. Furthermore, we presented the cloud-based video-on-demand system that is known to be a new video content delivery model in the development of cloud networking. Third, we provided an overview of key technologies that may potentially be deployed in 5G wireless networks. These technologies consist of massive MIMO, HetNets, mmWave, cognitive radio, D2D communications, and M2M communications. Finally, we discussed several emerging issues in computer networks, including data aggregation, task allocation, user association, interference management, wireless caching, mobile data offloading, and security. In regard to security, we described security issues in wireless networks involving eavesdropping attacks, DoS attacks, information security issues, and malicious behaviors of users. For each issue, we highlighted the motivation for using pricing models as well as auctions.

3 Mechanism Design and Auction Theory in Computer Networks

In this chapter, we introduce mechanism design and required properties of a mechanism. In particular, we first define a mechanism and mechanism design. Then, we discuss one important principle, the revelation principle, when designing a mechanism. After that, we define and discuss required properties of the mechanism, including incentive compatibility, individual rationality, economic efficiency, and budget balance. Finally, we discuss optimal mechanisms.

3.1 Mechanism Design

3.1.1 Mechanism

We consider a trading market as follows:

- The market consists of one seller that has one indivisible item for trading to buyers. For example, in wireless resource trading markets, the seller can be a service provider, and the item is a resource unit, such as a bandwidth unit or a sensing data unit.
- There is a set of N buyers, such as mobile users, that is denoted by $\mathcal{N} = \{1, 2, \ldots, N\}$.
- The buyers have private values of the item. Let v_i denote the value of the item to buyer i. $v_i \in \mathcal{V}_i$, where $\mathcal{V}_i$ is the set of types of buyer i.
- v_i is also known as a *type* of the buyer since it refers to the true value or the preference of the item.
- Let $\mathbf{v} = (v_1, \ldots, v_N)$ denote the set of types reported by the buyers.
- We have $\mathcal{V} = \prod_{i=1}^{N} \mathcal{V}_i$, which is the product of the sets of the buyers' types.

To trade the item to the buyers, the seller can use different schemes/methods. For example, the seller can use auction schemes for trading the item. Accordingly, the buyers, called bidders, are asked to submit their bids to the seller. The bids are the prices that the buyers are willing to pay the seller for the item. The bids of the buyers are known as *messages* of the buyers. Upon receiving the bids, the seller needs to select the best buyer, the winner, for receiving the item. Also, the seller determines the price that the winner needs to pay for winning the item. For example, when the first-price sealed-bid auction is used, the seller selects the best buyer with the highest bid as the winner, and the highest bid is the price that the winner needs to pay. When the second-price sealed-bid auction is used, the seller selects the best buyer with the highest bid as the winner, and the second highest bid is the price that the winner needs to pay. Apart from the two auction schemes, the seller can use other schemes/methods for trading the item. For example, the seller can adopt a scheme in which the seller (i) posts a fixed price for the item, (ii) selects the best buyer that accepts the price first, and (iii) determines the fixed price as the price that the winner pays.

Any scheme or method that the seller adopts for trading the item is called the *mechanism*. For example, the first-price sealed-bid auction and the second-price sealed-bid auction, as just discussed, are mechanisms. Each mechanism contains *rules* selected or set by the seller. The rules of the mechanism consist of (i) the resource allocation or *allocation rule* and (ii) the payment determination or *payment rule*. The allocation rule determines the allocation of the item to the buyers, and the payment rule determines the prices that each buyer needs to pay if it receives the item. The rules are designed by the seller to achieve its objectives or required properties. Designing the allocation and payment rules of the mechanism to meet the desired objectives is called *mechanism design*. Mechanism design is discussed in the next section.

3.1.2 Mechanism Design

To design the mechanism, an input is required. Typically, the input includes messages submitted or reported by the buyers. The messages may be types, meaning the true values or preferences, of the buyers or the bids of the bidders when the auctions are used. A general mechanism is modeled by a tuple $(\mathcal{B}, f^1, f^2)$ where [183]

- $\mathcal{B}$ can be considered to be the input of the mechanism. $\mathcal{B}$ is a product of sets of the buyers' messages that is defined as $\mathcal{B} = \prod_{i=1}^{N} \mathcal{B}_i$, where $\mathcal{B}_i$ is the set of possible messages of buyer i. Again, the messages may be the types or bids of the buyers. Let b_i denote the message submitted by buyer i, and $\mathbf{b} = \{b_1, \ldots, b_N\}$ denote the vector of messages submitted by N buyers. When the message of buyer i is its type, $b_i = v_i$, we say that the buyer reports or submits its type truthfully. Otherwise, when $b_i \neq v_i$, we say that the buyer reports its type untruthfully.
- f^1 is the allocation rule that maps the sets of the buyers' messages to the winning probabilities of the buyers. f^1 is defined as $f^1 \colon \mathcal{B} \to \mathcal{G}$, where $\mathcal{G}$ is the set of the winning probabilities of the buyers. Each element $g_i \in \mathcal{G}$ refers to the probability that buyer i wins the item.
- f^2 is the payment rule that maps the sets of the buyers' messages to the prices that the buyers need to pay the seller. Thus, f^2 is expressed by $f^2 \colon \mathcal{B} \to \mathcal{P}$. $p_i \in \mathcal{P}$ is the price that buyer i is expected to pay the seller if it wins the item. p_i is sometimes called the *expected payment* that buyer i makes.
- f^1 and f^2 are the *social choice functions* of the mechanism. $\mathcal{G}$ and $\mathcal{P}$ are the *outcomes* of the mechanism. For convenience, we can use a common social choice function f to represent the two social choice functions f^1 and f^2.

To further understand the outcomes of the mechanism, we consider two well-known auctions, the first-price sealed-bid auction (described in detail in Chapter 5) and the second-price sealed-bid auction (described in detail in Chapter 6). Assume that the seller receives a vector of messages (i.e., bids) $\mathbf{b} = \{b_1, \ldots, b_N\}$ from the buyers (i.e., the bidders).

With the first-price sealed-bid auction, the seller implements a social choice function $f(\mathbf{b})$ to map the message vector $\mathbf{b}$ to the outcomes as follows:

- The allocation rule for each buyer i is

$$g_i = \begin{cases} 1, & \text{if } b_i > \max_{j \neq i} b_j \\ 0, & \text{if } b_i < \max_{j \neq i} b_j \end{cases} \quad (3.1)$$

- The payment rule for each buyer i is defined as

$$p_i = \begin{cases} b_i, & \text{if } b_i > \max_{j \neq i} b_j \\ 0, & \text{if } b_i < \max_{j \neq i} b_j \end{cases} \quad (3.2)$$

In summary, the first-price sealed-bid auction selects the best buyer with the highest bid as the winner, and the highest bid is the price that the winner needs to pay.

With the second-price sealed-bid auction, the seller implements a social choice function $f(\mathbf{b})$ to map the message vector $\mathbf{b}$ into the outcomes as follows:

- The allocation rule for each buyer i is

$$g_i = \begin{cases} 1, & \text{if } b_i > \max_{j \neq i} b_j \\ 0, & \text{if } b_i < \max_{j \neq i} b_j \end{cases} \quad (3.3)$$

- The payment rule for each buyer i is

$$p_i = \begin{cases} \max_{j \neq i} b_j, & \text{if } b_i > \max_{j \neq i} b_j \\ 0, & \text{if } b_i < \max_{j \neq i} b_j \end{cases} \quad (3.4)$$

In summary, the seller selects the best buyer with the highest bid as the winner, and sets the second highest bid as the price that the winner pays.

These two auctions are the mechanisms that are often used in single-item markets. They are used depending on the seller's objectives and required properties. For example, the seller adopts the first-price sealed-bid auction if it wants to gain revenue or profit. Conversely, the seller uses the second-price sealed-bid auction if it wants to achieve the incentive compatibility property. In fact, the seller can have multiple items for trading with other objectives and required properties. Thus, the seller needs to design an appropriate mechanism to achieve the objectives and required properties.

To design the mechanism, a mechanism designer (the seller) needs to determine or implement the social choice function f. For this, the mechanism designer may need to consider all possible mechanisms and choose the best one that has the objectives equal to or close to the desired objectives. This task is generally a complicated search problem, and the mechanism designer sometimes cannot solve it. Fortunately, the *revelation principle* can be used to reduce the search space and to facilitate the mechanism design task. Before presenting this principle in the next section, we briefly define the equilibrium of the mechanism.

Every mechanism can be defined as a game (e.g., a Bayesian game) of incomplete information among players, referring to the buyers or bidders. The information of each player refers to the type, meaning the true value or preference, of the player that is private. Again, each player i can report its message truthfully ($b_i = v_i$) or untruthfully ($b_i \neq v_i$). b_i is called the *strategy* of player i. The strategy is a complete decision defining an action that the player selects at a stage of the game. We have the following definition:

DEFINITION 3.1 *[184] A vector of strategies* $\mathbf{b} = \{b_1, \ldots, b_N\}$ *of the players is called the equilibrium of the mechanism if given the strategies* $\mathbf{b}_{-i}$ *of other players, strategy* b_i *maximizes the utility or payoff of player i. Here,* $\mathbf{b}_{-i}$ *is the vector of strategies of the players excluding the strategy of player i.*

3.1.3 Revelation Principle

Here, we provide some important concepts related to the revelation principle (and the mechanism design).

- *Truth-revealing strategy:* In the mechanism, the truth-revealing strategy of a player is that the player reports the true information about its type. In particular, the truth-revealing strategy of player i is $b_i(v_i) = v_i$.
- *Direct-revelation mechanism:* A direct-revelation mechanism is the mechanism in which the players are asked/required to perform truth-revealing strategies by reporting directly their true types, consisting of their true values or preferences. This means that we have $\mathcal{B}_i = \mathcal{V}_i, \forall i \in \mathcal{N}$ and $\mathcal{B} = \mathcal{V}$. Thus, a direct-revelation mechanism is modeled as $(\mathcal{V}, f^1, f^2)$. For simplification, the direct-revelation mechanism can be modeled by a tuple with two functions f^1 and f^2 as (f^1, f^2) where $f^1 : \mathcal{V} \to \mathcal{G}$ and $f^2 : \mathcal{V} \to \mathcal{P}$.
- *Incentive-compatible direct-revelation mechanism:* If the truth-revealing strategies in the direct-revelation mechanism constitute an equilibrium, the mechanism is called an incentive-compatible direct-revelation mechanism or strategy-proof direct-revelation mechanism. The equilibrium of the incentive-compatible direct-revelation mechanism is $\mathbf{v} = (v_1, \ldots, v_N)$. For example, the second-price sealed-bid auction or the Vickrey auction is an incentive-compatible direct-revelation mechanism for the single-item allocation problem. The Vickrey–Clarke–Groves (VCG) auction is an incentive-compatible direct-revelation mechanism for the multi-item allocation problem. The first-price sealed-bid auction is a direct-revelation mechanism but not the incentive-compatible

direct-revelation mechanism since each player bids half its true value at the equilibrium.

The revelation principle can be stated as follows:

THEOREM 3.2   *[183] If a social choice function f can be implemented by an arbitrary mechanism, namely the original mechanism, and if this mechanism has an equilibrium corresponding to implementing the social choice function, then*

- *It is the equilibrium for the players to submit their types truthfully.*
- *The social choice function f can be implemented by the incentive-compatible direct-revelation mechanism with the same outcome or the same objective as the original mechanism.*

The proof of Theorem 3.2 can be found in [183] and [185]. The revelation principle means that the social choice function or the outcome of any mechanism can be replicated/implemented by the incentive-compatible direct-revelation mechanism. The revelation principle thus simplifies the task of mechanism design. In particular, if the mechanism designer wants to design the mechanism to achieve a certain objective or a property, the mechanism designer can search for only those incentive-compatible direct-revelation mechanisms that have the same objective or the same property. The social choice function f of the incentive-compatible direct-revelation mechanisms can be used to design the original mechanism. If no such an incentive-compatible direct-revelation mechanism exists, there is no mechanism that can achieve the outcomes, objective, and property. For example, if the mechanism designer wants to design the mechanism to achieve social welfare maximization (the objective) and incentive compatibility (the required property), the mechanism designer can use the social choice function of the well-known VCG auction, which is an incentive-compatible direct-revelation mechanism. By narrowing the search area, the problem of designing as well as finding the mechanism becomes much easier for the mechanism designer to solve.

In the next sections, we discuss important properties of the mechanism, which are summarized here [185]:

- *Incentive compatibility or truthfulness:* This property guarantees that players report their types, or true values, truthfully. This property is important when designing the mechanism to overcome the self-interest or rationality of the players.
- *Individual rationality:* This property guarantees a positive utility/payoff for every player who is participating in the game.
- *Economic efficiency:* This property aims to maximize the total utility or payoff of players.
- *Budget balance:* This property guarantees balanced transfers across players; that is, there are no transfers out of the system or into the system.
- *Revenue maximization:* This objective aims to maximize the revenue of one of the players.
- *Fairness:* This property seeks to guarantee fairness, such an equal winning probability, among players.

3.1.4 Incentive Compatibility

Incentive compatibility is an important property of the mechanism. The reason is that players are rational, meaning that they have an incentive to not report their types truthfully. This behavior reduces the efficiency of the resource allocation. The mechanism holding the incentive compatibility overcomes the rationality or the self-interest of the players, so that the players have an incentive to report their types truthfully. We have the following concepts related to the incentive compatibility:

- *Utility function:* The utility function refers to the welfare or satisfaction of a player when the player receives an item. Let u_i be a utility function of player i. u_i is typically a function of parameters including (i) its type, v_i; (ii) its reported type, b_i; and (iii) the types of other players, $\mathbf{v}_{-i}$, the reported types of other players, $\mathbf{b}_{-i}$, and the social choice function f of the mechanism. Thus, we have $u_i(v_i, b_i, \mathbf{b}_{-i}, f)$. Note that one or some of the parameters in the utility function can be removed for simplification.
- *Dominant strategy:* Strategy b_i of player i is dominant if this strategy maximizes the player's utility given all possible strategies of other players. In other words, the dominant strategy of a player maximizes the player's utility regardless of the strategies of other players. Formally, we have

$$u_i(b_i, \mathbf{b}_{-i}, v_i) \geq u_i(b'_i, \mathbf{b}_{-i}, v_i), \forall i, \text{ and } \forall b'_i \neq b_i \quad (3.5)$$

For example, in the second-price sealed-bid auction, the dominant strategy of each player is the truth-revelation strategy, which entails reporting its type truthfully, $b_i = v_i$. The reason is that the bid of the player is the price that the player accepts, but it is not the actual price that the player pays. The price that the player pays is completely independent of its bid; it is the second highest bid. This means that the auction guarantees a gain utility for the player when it becomes the winner. This is further explained and discussed in Chapter 6.

- *Nash equilibrium:* In a Nash equilibrium, every player selects a utility-maximizing strategy given the strategies of other players.

DEFINITION 3.3 *A strategy profile* $\mathbf{b} = (b_1, \ldots, b_N)$ *is the Nash equilibrium if*

$$u_i(b_i, \mathbf{b}_{-i}, v_i) \geq u_i(b'_i, \mathbf{b}_{-i}, v_i), \forall i \text{ and } \forall b'_i \neq b_i \quad (3.6)$$

- *Bayesian Nash equilibrium:* In a Bayesian game, each player has incomplete information about the types of other players, but the player is assumed to know a common prior about the distribution of other players' types. The Bayesian Nash equilibrium is defined as follows:

DEFINITION 3.4 *A strategy profile* $\mathbf{b} = (b_1, \ldots, b_N)$ *is the Bayesian Nash equilibrium if*

$$E_{\mathbf{v}_{-i}}[u_i(b_i, \mathbf{b}_{-i}, v_i)] \geq E_{\mathbf{v}_{-i}}[u_i(b'_i, \mathbf{b}_{-i}, v_i)], \forall i \text{ and } \forall b'_i \neq b_i \quad (3.7)$$

Definition 3.4 means that each player selects a strategy to maximize the expected utility in conjunction with the expected-utility maximizing strategies of other players.

Now we can define the incentive compatibility property as follows. This mechanism is called incentive compatible or truthful or strategy-proof if every player can achieve the highest utility by reporting their types truthfully. There are two common types of incentive compatibility: *Bayesian–Nash incentive-compatibility* and *dominant-strategy incentive-compatibility*.

DEFINITION 3.5   The mechanism is said to hold the Bayesian–Nash incentive-compatibility if

$$E_{\mathbf{v}_{-i}}[u_i(v_i, \mathbf{v}_{-i}, v_i)] \geq E_{\mathbf{v}_{-i}}[u_i(v_i, (\mathbf{v}_{-i}, b_i))], \forall i, \forall v_i, \text{and } \forall b_i \qquad (3.8)$$

Definition 3.5 means that each player that reports its type truthfully achieves the highest expected utility given that other players report their types truthfully.

DEFINITION 3.6   The mechanism is said to hold the dominant-strategy incentive-compatibility if

$$u_i(v_i, \mathbf{b}_{-i}, v_i) \geq u_i(v_i, (\mathbf{b}_{-i}, b_i)), \forall i, \forall v_i, \forall b_i, \text{and } \forall \mathbf{b}_{-i}. \qquad (3.9)$$

Definition 3.6 means that the player achieves the maximum utility by reporting its type truthfully no matter what other players submit.

It can be seen from Definition 3.5 and Definition 3.6 that every dominant-strategy incentive-compatibility mechanism is the Bayesian–Nash incentive-compatibility mechanism, but the Bayesian–Nash incentive-compatibility mechanism may exist even if no dominant-strategy incentive-compatibility mechanism exists. Therefore, the Bayesian–Nash incentive-compatibility is said to be "weaker" and the dominant-strategy incentive-compatibility is said to be "stronger."

3.1.5   Individual Rationality

Individual rationality is known as a "voluntary participation" constraint. The reason is that the mechanism, such as a game or an auction, may not attract a player to participate in it if the player's expected utility is negative. For example, a bidder may not have an incentive to participate in an auction if its expected utility is negative due to a high price that the bidder needs to pay. Let $u_i(f(\mathbf{v}))$ denote the expected utility of player i at the equilibrium of the outcome when the player participates in the mechanism. Also, let $u_i^0(v_i)$ be the expected utility achieved by the player for non-participation – that is, the player is outside of the mechanism. We have the following definition.

DEFINITION 3.7   [186] A mechanism is said to hold individual rationality (i.e., the mechanism is individually rational) if the mechanism implements a social choice function $f(\mathbf{v})$ such that

$$u_i(f(\mathbf{v})) \geq u_i^0(v_i), \forall i \text{ and } \forall v_i \qquad (3.10)$$

In practice, we often assume that the expected utility of the player when it does not participate in the mechanism is zero, $u_i^0(v_i) = 0$. Thus, we can define that the mechanism holds individual rationality if for every player i, $u_i(f(\mathbf{v})) \geq 0$, or if every player achieves a non-negative utility when participating in the mechanism.

3.1.6 Economic Efficiency and Budget Balance

Economic Efficiency

Economic efficiency refers to an economic state at which items are optimally allocated to buyers in the best way while minimizing waste of resources and inefficiency. Thus, the mechanism that holds the economic efficiency ensures that the items are allocated to those buyers that value them the most. We also say that the mechanism is allocatively efficient.

DEFINITION 3.8 *The mechanism is said to hold the economic efficiency property if it can implement a social choice function f that maximizes the total value over all buyers.*

For example, the double auction, as presented in Chapter 8, is a mechanism that holds the economic efficiency property since it guarantees that the items of sellers are allocated to buyers that value them the most.

It is worth mentioning the concept of "social welfare." In general, social welfare is defined as the sum of utilities of all players. Since the utility of each player is proportional to the value of the item to the player, social welfare is closely related to economic efficiency. In particular, the mechanism that maximizes the total utility over players also maximizes the total value of all the players. Thus, we can define the economic efficiency as follows: the mechanism is efficient if it maximizes social welfare.

Budget Balance

Consider a mechanism with N players in which there are $N - K$ buyers and K sellers. Also, there is a broker that conducts the trading. Let p_i denote the price that buyer i pays for receiving items, and let p'_j denote the price that seller j receives for selling its items. In general, the budget balance introduces constraints over the total monetary transfer made from the players to the broker. Depending on the total monetary transfer, there are two different degrees of the budget balance.

- The mechanism is said to hold the strong budget balance if

$$\sum_{i=1}^{N-K} p_i + \sum_{j=1}^{K} p'_j = 0 \qquad (3.11)$$

    Equation (3.11) means that there are no monetary transfers to the broker; that is, the monetary transfers are done only between the buyers and the sellers.
- The mechanism is said to hold the weak budget balance if

$$\sum_{i=1}^{N-K} p_i + \sum_{j=1}^{K} p'_j \geq 0. \qquad (3.12)$$

    Equation (3.12) means that there can be some monetary transfers to the broker.

Double auction with the average payment rule [187] is a mechanism that assures a strong budget balance since all monetary transfers are among buyers and sellers, and the auctioneer (i.e., the broker), does not gain money.

3.2 Optimal Mechanisms

In this section, we introduce two most fundamental objectives of an optimal mechanism design: social surplus or total welfare maximization and profit or revenue maximization. We first define the social surplus and profit of the mechanism. Then, we discuss the problem formulations corresponding to the two objectives. Further details of the optimal mechanism design can be found in [188].

3.2.1 Social Surplus and Profit

We consider again the market model in Section 3.1.1. In particular, there is one seller that has one indivisible item for trading to buyers. There are N buyers that are willing to buy the item. The value of the item to buyer i is v_i, and let $\mathbf{v} = (v_1, \ldots, v_N)$ be the value profile of the buyers. We assume that the seller designs a mechanism that has an allocation $\mathbf{g} = (g_1, \ldots, g_N)$, where g_i indicates whether player i receives the item, and the payment $\mathbf{p} = (p_1, \ldots, p_N)$, where p_i is the payment made by buyer i given the allocation g_i. Note that given the allocation $\mathbf{g}$, the seller may need to pay the cost $c(\mathbf{g})$. In particular, for computer network environments, the cost can be the resource maintenance cost. For example, when delivering streaming live videos to viewers (i.e., buyers), a content provider needs to pay the cost for leasing network links. Also, to provide cloud network resources to cloud tenants, a cloud provider needs to pay the bandwidth cost to network providers. The cost $c(\mathbf{g})$ is sometimes called service cost. We have the following definitions [188].

- *Buyer surplus:* The surplus of buyer i, denoted by S_i^b, is defined as the difference between the value of the item to the buyer and the price that the buyer pays. Thus, S_i^b is defined as $S_i^b = g_i v_i - p_i$.
- *Seller surplus:* The surplus of the seller, denoted by S^s, is defined as the difference between the price that the seller receives from selling the item and the cost of the item. Thus, S^s is defined as $S^s = \sum_i^N p_i - c(\mathbf{g})$.
- *Social surplus:* The social surplus or social welfare of the mechanism, denoted by S, is the sum of the buyer surplus and the seller surplus:

$$S = \sum_i^N S_i^b + S^s$$

$$= \sum_i^N g_i v_i - c(\mathbf{g}) \qquad (3.13)$$

From (3.13), the social surplus is actually the difference between the total value of buyers and the service cost. The service cost $c(\mathbf{x})$ is typically fixed. Thus, the mechanism that maximizes the social surplus also maximizes the total value of the buyers. According to Definition 3.8, such a mechanism is allocatively efficient.

- *Profit:* The profit of the mechanism is defined as the seller surplus, the difference between the total payment made by the buyers and the service cost:

$$\pi = \sum_i^N p_i - c(\mathbf{x}) \qquad (3.14)$$

Note that $\sum_i^N p_i$ is defined as the revenue of the seller. Since the service cost $c(\mathbf{x})$ is typically fixed, the mechanism that maximizes the profit also maximizes the revenue of the seller.

In general, designing the optimal mechanism in terms of social surplus maximization seems to be simpler than designing the optimal mechanism in terms of profit maximization. For example, to design the optimal mechanism in terms of social surplus maximization, we can simply adopt the VCG auction, a generalization of the second-price sealed-bid auction described in Section 6.2. For the optimal mechanism in terms of profit or revenue maximization, such an optimal mechanism does not exist. The mechanism designer needs to know the distribution of the players' types to derive the optimal mechanism.

3.2.2 Social Surplus Maximization Problem

In this section, we derive the optimal mechanism for the social surplus. The optimization problem of maximizing social surplus is to find an allocation rule $\mathbf{g}$ to maximize the surplus $S(\mathbf{v}, \mathbf{g})$:

$$\arg\max_{\mathbf{g}} S(\mathbf{v}, \mathbf{g}) \qquad (3.15)$$

The allocation rule is implemented as follows:

- Assume that the optimal social surplus obtained by solving Equation (3.15) is $S^*(\mathbf{v})$. This means that

$$S^*(\mathbf{v}) = \max_{\mathbf{g}} S(\mathbf{v}, \mathbf{g}) \qquad (3.16)$$

- Consider a particular buyer i. There are two possible cases for the buyer [188]:

  - Case I: The item is assigned to buyer i, $g_i = 1$, so $S^*(\mathbf{v})$ can be expressed as follows:

  $$S^*(\mathbf{v}) = v_i + \max_{\mathbf{g}_{-i}} S\big((\mathbf{v}_{-i}, 0), (\mathbf{g}_{-i}, 1)\big) \qquad (3.17)$$

  We define $S^*_{-i}(\mathbf{v}) = \max_{\mathbf{g}_{-i}} S((\mathbf{v}_{-i}, 0)(\mathbf{g}_{-i}, 1))$, and we have $S^*(\mathbf{v}) = v_i + S^*_{-i}(\mathbf{v})$.

- Case II: The item is not allocated to buyer i, $g_i = 0$, so $S^*(\mathbf{v})$ can be expressed as follows:

$$S^*(\mathbf{v}) = \max_{\mathbf{g}_{-i}} S\big((\mathbf{v}_{-i},0),(\mathbf{g}_{-i},0)\big) \tag{3.18}$$

We define $S^*(\mathbf{v}_{-i}) = \max_{\mathbf{g}_{-i}} S((\mathbf{v}_{-i},0),(\mathbf{g}_{-i},0))$, and we have $S^*(\mathbf{v}) = S^*(\mathbf{v}_{-i})$.

- To maximize the social surplus, the item is allocated to buyer i whenever the surplus in case I is greater than or equal to the social surplus in case II [188]:

$$S^*(\mathbf{v}) = v_i + S^*_{-i}(\mathbf{v}) \geq S^*(\mathbf{v}) = S^*(\mathbf{v}_{-i}) \tag{3.19}$$

- Let $S_i^0 = S^*(\mathbf{v}_{-i}) - S^*_{-i}(\mathbf{v})$. Then, we can say that the item is allocated to buyer i whenever its value is greater than or equal to S_i^0. Note that S_i^0 does not depend on v_i of buyer i, and thus S^0 is considered to be a *critical value* [188].

The critical value $S^0 = S^*(\mathbf{v}_{-i}) - S^*_{-i}(\mathbf{v})$ is known as the externality that buyer i imposes on the other buyers due to receiving the item [188]. In other words, since buyer i receives the item, the social surplus of the other buyers is $S^*_{-i}(\mathbf{v})$ instead of $S^*(\mathbf{v}_{-i})$. Buyer i needs to pay a price that is equal to the externality that it imposes on the other buyers.

One well-known optimal mechanism in term of social surplus maximization is the VCG auction described in Section 6.2. The allocation and payment rules of the VCG auction aim to maximize the social welfare while guaranteeing the incentive compatibility. Thus, the VCG auction is applicable in several scenarios of computer networks that aim to satisfy buyers' QoS or the fairness among the buyers. For example, the VCG auction is adopted for allocating data rates to M2M applications as proposed in [189], for spectrum allocation in the 4G LTE network as presented in [190], and for bandwidth reservation in cloud networking as proposed in [191]. In particular, for the bandwidth reservation, the considered model consists of a cloud provider, or seller, which owns a number of distributed data centers, and cloud tenants, or buyers, which act as application and service providers. Cloud tenants rent bandwidth from the cloud provider to serve their subscribers. To avoid the high bandwidth reservation payment, the cloud tenants can lie about their revenues obtained by serving subscribers. The VCG auction is adopted for the bandwidth reservation to achieve both optimal social welfare and incentive compatibility such that the cloud tenants have no incentive to lie about their revenue information. Specifically, the cloud tenants are required to submit their bids to compete for bandwidth to the cloud provider. Each bid consists of bandwidth demands and the price per unit of bandwidth for which the cloud tenant is willing to pay. To achieve the highest social welfare for the allocation, the winners are determined through a linear programming model that can be solved in polynomial time. The VCG mechanism is then applied to calculate the charge for each winner. The charge is the difference between the social welfare when the winner does not participate and that when the winner participates in the auction. Further details are presented in [191]. Since the proposed approach has an optimal allocation and calculates the charge based on the VCG auction, it is concluded to be the optimal auction mechanism in terms of social welfare maximization for the bandwidth reservation.

3.2.3 Profit Maximization Problem

Profit here refers to the profit of the seller as defined in (3.14). Since the service cost is typically fixed, the profit maximization problem is equivalent to the revenue maximization problem. In general, designing the optimal mechanism in terms of profit maximization is more difficult than designing the optimal mechanism in terms of social surplus maximization. The main reason is that improving the profit of the seller also means reducing the utility/payoff/benefit of the buyers. Thus, maximizing the profit of the seller may make the utility of the buyers negative, and the mechanism may not attract the desired buyers. Therefore, the mechanism designer needs to optimize the trade-off between the profit of the seller and the utility of the buyers. In other words, the mechanism designer needs to solve an optimization problem that maximizes the expected profit or expected revenue of the seller while guaranteeing the rational individuality property. In addition, incentive compatibility is an important property that needs to be introduced in the optimization problem. For this, the mechanism designer needs to know the distribution of the values of the items to the buyers. Based on the values drawn from the distribution, the mechanism designer determines the allocation and payment rules to maximize the expected profit while guaranteeing the required properties – that is, the rational individuality and incentive compatibility. The mechanism with the allocation and payment rules is called *Bayesian optimal mechanism*. Here, "Bayesian" means that the probability distribution of a particular buyer's value is known to other buyers and even the mechanism designer. A simple distribution function is $F(v) = v$, as the buyer's value follows the uniform distribution $U[0, 1]$.

The optimization problem for the optimal mechanism in terms of profit maximization is defined as follows [192]:

$$\max_{\mathbf{g}, \mathbf{p}} \left[\sum_i p_i(\mathbf{v}) \right] \quad (3.20)$$

s.t. [IC], [IR]

where **g** is the allocation rule, **p** is the payment rule, and [IC] and [IR] are the incentive compatibility and individual rationality constraints, respectively.

The problem in (3.20) can be found in different scenarios of computer network environments. For example, it can be found in fog resource trading markets [193], where

- The market consists of one service provider, the seller, that has M computing resource units for trading.
- There are N users, and each user i has a value, a type, v_i of the computing resource unit. The valuation profile of the users is $\mathbf{v} = (v_1, \ldots, v_N)$. v_i is drawn independently from distribution F_i over a possible valuation profile $\mathbf{V}_i$.

- The service provider may not know type v_i of each user i, but the service provider can know the distribution functions, for example through the observation.
- The users are required to submit the prices that they are willing to pay to the service provider. Let $b_i \in V_i$ denote the price submitted by user i, and $\mathbf{b} = (b_1, \ldots, b_N)$ denote the price profile of the users.

Upon receiving the price profile $\mathbf{b}$ from the users, the service provider determines an allocation rule and a pricing rule. The allocation rule includes the winning probabilities $g_i, i = 1, \ldots, N$, of the users, and the pricing rule includes the conditioned prices $p_i, i = 1, \ldots, N$, for the users. The service provider needs to determine the allocation and pricing rules to maximize its revenue. Moreover, to provide an incentive to the users to participate in the market, the utility of the users must be non-negative. For this, the service provider can formulate its problem as shown in (3.20). The problem in (3.20) is the constrained optimization. In general, solving such a constrained optimization problem to derive the optimal mechanism is difficult [192]. The Myerson's optimal mechanism [194] can be adopted by using the concept of *virtual values* and *monotone transform functions*. However, the Myerson's optimal mechanism is limited to a single item.

In recent years, machine learning technique that has the ability to automatically identify relevant features of data has gained considerable attention. Recent theoretical results in [195] show that machine learning using stochastic gradient descent can successfully find globally optimal solutions for complex problems. Thus, machine learning can be used to solve the constrained optimization problem in (3.20). The use of machine learning for the optimal mechanism design is proposed in [192] and described in detail in Sections 10.3 and 10.4.

There are various properties for an auction. First, allocative efficiency means that in all such auctions the highest bidder always wins (i.e., there are no reserve prices). Second, it is desirable for an auction to be computationally efficient. Finally, to study the revenue (expected selling price) of different auctions, we have one of the major findings of auction theory: the celebrated revenue equivalence theorem. The revenue equivalence theorem is used to predict the strategy of each bidder in the auctions, and determine the equilibrium in the auctions. The revenue equivalence theorem is presented in Section 4.3.2.

3.3 Auction Theory in Computer Networks

Auctions are known as mechanisms that are widely used in computer networks. The history of auction theory in computer networks dates from the use of spectrum license distribution in wireless systems. Prior to the application of pricing and auction theory, static resource management approaches were used. In these approaches, the spectrum licenses are assigned to users in a static manner. One example of the static resource management approaches is the first-come-first-serve approach [196]. However, such a static approach is inefficient since the demand and supply of the resources do not always match. In particular, the resources may not be assigned to the users that value the resources most. To enhance the efficiency of the resource allocation, pricing and auction theory can be adopted. In the auction, the users submit their bids for the spectrum licenses. The bids are the prices that the users are willing to pay for the spectrum licenses. These prices reflect the demands or the values of the spectrum licenses to the users. The users with the highest bids are the winners of the spectrum licenses. As such, the adoption of the auction methodology increases the market competition and enhances the efficiency of the resource allocation. This further improves the revenue of the seller. For example, by applying the auction approach for spectrum license distribution, the Federal Communications Commission (FCC), an independent agency of the US government, gained $40 billion from 1994 to 2001.

In this section, we first present the basics of auction. Then, we discuss the motivations for and significance of applying the auctions to computer networks. Finally, we define basic terminologies in auction theory.

3.3.1 Auction Basics

As mentioned earlier, auctions are regarded as market mechanisms in which the item allocation and pricing and payment determination are performed by a *bidding process* [197]. Various auctions are designed with different objectives and economic properties. In general, auctions can take many forms, but they share two major characteristics. First, auctions are universal since they can be used in anywhere to sell any item. Second, auctions are anonymous since the outcome of the auction does not depend on the identity of participants, referring to the buyers or the bidders. Moreover, in most auctions, participants are required to submit their bids. Here, the bids refer to the prices, or amounts of money, that the participants are willing to pay for the items. The winner of the auction is the participant with the highest bid. The auction can be defined as follows.

DEFINITION 3.9 *[197] An auction is a market mechanism that includes an explicit set of rules for determining item allocation and the corresponding prices on the basis of bids from the market participants, the bidders. The traditional mechanisms include English and Dutch auctions, and the first-price and second-price sealed-bid auctions.*

From the perspective of game theory, an auction can be considered to be an incomplete information game. The game is represented by a set of players, a set of strategies available to each player, and a payoff vector that corresponds to each combination of the strategies, the strategy profile, of the players. Here, the players can be the buyers (i.e., the bidders) or the sellers. In the case that the players are the buyers, the strategy of the player is a bid function that maps its value of the item to a bid, a bidding price. The payoff obtained by the player corresponding to each strategy profile is the expected utility or profit of the player given the strategy profile. The strategy profile constitutes the equilibrium of the auction if each strategy b_i in the profile maximizes the payoff of the corresponding player, player i, given the strategies of the other players.

There are two game-theoretic models of auctions: *common value auctions* and *private value auctions*.

- *Common value auctions:* In the common value auctions, bidders have equal values of the item; that is, the value of the item is identical among the bidders. However, the bidders do not have perfectly accurate information about this value, so they do not know an exact value of the item. Each bidder can assume that any other bidder obtains a random signal, which is used to estimate the true value, from a probability distribution common to all bidders. Examples of common value auctions are Treasury Bill auctions, auctions of timber, spectrum auctions, and auctions of oil and gas leases. In each case, the value of the item is the same to all the bidders, but different bidders have different information about what that value actually is.
- *Private value auctions:* In contrast to common value auctions, in private value auctions, each bidder knows the value, or private value, of the item, but the bidder does not know the values of the other bidders. We say that the values are *independent* across bidders since the value that a particular bidder assigns to the item is independent of the values assigned by the other bidders.

Various types of auctions exist, and there are also many ways to categorize auctions. Some simple examples of categorization follow.

- *Forward and reverse auctions:* The forward auction typically consists of multiple bidders (i.e., buyers) and one seller (see Figure 3.1(a)). The bidders bid for the items by offering increasingly higher prices. The reverse auction has multiple sellers and one buyer (see Figure 3.1(b)). The sellers compete for the buyer's attraction by submitting their asks to the buyer or the auctioneer. Here, each ask refers to the price that the seller is willing to receive for trading the items.

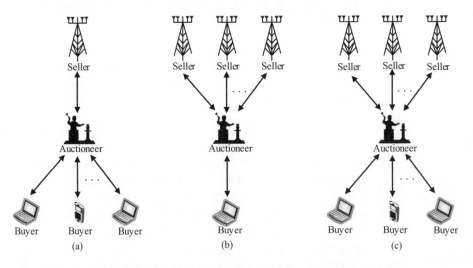

Figure 3.1 Different types of auctions: (a) forward auction, (b) reverse auction, and (c) double auction. The arrows indicate transactions of items and money among auction players [86].

The price in the reverse auction typically decreases to the lowest price that the buyer can accept.
- *Single-sided and double-sided auctions:* In the single-sided auction, there are only buyers or sellers that submit bids or asks. In the double-sided auction or double auction (see Figure 3.1(c)), both buyers and sellers submit their bids and asks, respectively.
- *Open-cry and sealed-bid auctions:* In the open-cry auction, bids of bidders are made out in the open market. Since the bids of the bidders are disclosed to the other bidders during the auction, the open-cry auction is regarded as a public auction that gives the bidders a chance to compete for the item with the best price. In the sealed-bid auction, all bidders simultaneously submit their sealed bids to the seller (or the auctioneer). Thus, no bidder knows how much the other bidders bid.
- *Single-item and multi-item auctions:* In the single-item auction, there is a single item for sale. In the multi-item auction, there are multiple items for sale.

Apart from these categorizations, another typical approach is based on the auction rules. This approach facilitates the selection of proper auctions for the mechanism designers or the sellers. There are four traditional auctions:

- *English auction:* The English auction or open ascending-bid auction is the oldest and perhaps the most popular auction. Here, *open* refers to the fact that the bids of all bidders are disclosed to the other bidders during the auction. The general idea of the English auction is that the auctioneer initially sets a low price for the item and raises the price gradually until only one bidder expresses its willingness to buy the item at the price. The last bidder that is willing to buy the item is the winner of the auction. The winner receives the item and pays the price at which the last-second bidder dropped out.
- *Dutch auction:* In contrast to the English auction, the Dutch auction is a descending-bid auction in which the auctioneer initially sets a high asking price, a ceiling price, for the item and then decreases the price until one of the bidders accepts the price. The winning bidder pays the final price and receives the item.
- *First-price sealed-bid auction:* In the first-price sealed-bid auction, the bidders submit their bids in sealed envelopes to the auctioneer. Upon receiving the bids, the auctioneer selects the bidder with the highest bid as the winner. The winning bidder receives the item and pays the highest bid.
- *Second-price sealed-bid auction:* This auction is similar to the first-price sealed-bid auction. However, the winning bidder pays the second-highest bid.

Apart from these auctions, there are other auctions such as the VCG auction, double auction, and combinatorial auction. They have different objectives, such as social welfare maximization and revenue maximization, and advance various desired properties, such as truthfulness, economic efficiency, and individual rationality. These auctions are discussed in the next chapters of this book.

3.3.2 Auction Theory for Computer Networks

This section explains the motivations and significance of applying the auctions for the resource management in computer networks.

To fully support various emerging multimedia applications, modern computer networks such as IoT, 5G wireless networks, cognitive radio, and cloud networking have been advancing rapidly in recent years. However, the adoption of emerging technologies introduces new challenges in the design and optimization of the network resource management.

- The computer network may consist of billion of devices, such as IoT devices. These devices are required to make optimal decisions without or with minimal human intervention given their constrained resources and the dynamic nature of the network environment. This leads to many challenges in efficiently controlling and managing the devices. Thus, new approaches with higher efficiency and more flexibility to adapt to dynamic networks need to be developed.
- The computer network has become more decentralized and ad hoc in nature. The traditional network resource management methods, such as system optimization, face many challenges or even may not work since they usually require a centralized entity. Thus, it is crucial to develop and adopt new resource allocation and control schemes that are suitable for distributed autonomous decision making.
- The decentralization of the computer network further increases the need for online resource sharing and resource reallocation among the network entities. To cope with the dynamic and unpredicable resource demand as well as to match resource supply and demand profiles in time and space in the network, dynamic, flexible, and scalable resource management schemes need to be considered.
- The computer network is a large-scale entity with a high density of network devices. Perfect global network state information may be too costly, impractical, and impossible to obtain. Thus, control decisions on resource management have to be made with partial or no knowledge of the parameters of the optimization problem.
- The computer network may include a number of rational and selfish entities. These entities may seek to maximize their own utilities by misreporting local parameters that may reduce the socially optimal and global resource allocation within the network. For example, they can misreport information related to channel demands and channel values that increases their own utilities. This can reduce the spectrum utilization as well as the revenue of the seller. Such behaviors need to be understood through game theoretic models and prevented through mechanisms that promote truthfulness and cooperation.
- The entities and stakeholders in the computer network are diverse and heterogeneous. They have different objectives, such as high data rates, low latency, utility maximization, cost minimization, and profit maximization, which may conflict with each other. The traditional methods merely focus on the system performance metrics given system parameters and constraints rather than economic factors, including profit, cost, and revenue. Thus, resource management methods that incorporate economic implications into the solution need to be adopted.

These complexity characteristics make modern computer networks analogous to real markets [86]. In particular, both have various participants in the system, and those participants perform transacting items or commodities, including sharing information and network resources, under certain regulations. Therefore, economics and business management approaches [198] can be employed to dynamically and efficiently manage the resources of the computer networks. Auction [199] is one kind of interdisciplinary method used to solve the resource management issues. A major advantage of using the auction mechanisms is the ability to guarantee the efficiency of the resource management by allocating the resources to those buyers that value the resources most. Moreover, the auction mechanisms can address the challenges of resource management in the computer networks in the following ways:.

- Auctions as game models can model and analyze complex interactions among the network entities and stakeholders [74]. Through these interactions, each entity can observe, learn, and predict the status/actions of other entities, and then make the best decisions based on the equilibrium analysis. Therefore, auctions are inherently suitable for distributed autonomous decision making and can cope with the diverse and conflicting interests of autonomous network entities in the computer networks.
- Auctions can support different objectives, ranging from revenue maximization for the auctioneer to social welfare maximization for the entities, and own various desired properties, including truthfulness, economic efficiency, and individual rationality. Thus auctions can be used to design incentive mechanisms that cope with the rationality, selfishness, and even maliciousness of the network entities.
- Auctions offer the flexibility of setting prices for items or commodities dynamically and efficiently based on current supply and demand in a market. They meet the requirement for matching the dynamic spatiotemporal patterns of demand and supply in the computer networks.
- By adopting auctions, desired resource allocation schemes can be achieved without knowledge of the utility functions of the network entities [200]. Thus auctions are able to provide control decisions for resource management under conditions of limited or no network state and node utility information.

3.3.3 Basic Terminology in Auction Theory

This section provides the basic terminology that is used through this book. These terms are essential to understand the auction approaches discussed in subsequent chapters.

- *Seller:* A seller offers its items for sale. The items are things that the seller sells in the auction. In computer networks, the items can be network resource, (e.g., bandwidth, spectrum, energy, cloud, and storage) or network services (e.g., relay, caching, and offloading services). Sellers can be wireless service providers, cloud providers, or even users.
- *Bidder:* A bidder is a buyer that wants to buy the items from the seller. In computer networks, bidders can be end-users, mobile users, mobile devices, or even

service providers. Bidders want to buy network resources and have to compete with each other for the resources.

- *Auctioneer:* An auctioneer is as an intermediate agent that conducts the auction. In particular, the auctioneer initializes the auction, determines the winners, and identifies the prices that the bidders need to pay or that the sellers receive. In many cases, the auctioneer is the seller itself. In computer networks, the auctioneer can be a base station or an access point that can conduct resource auctions using its auction controller.
- *Player:* The auction can be considered to be a game model in which the players are the sellers or the bidders.
- *Bidding price* and *asking price:* The bidding price is the price that the bidder is willing to pay for a requested item. The asking price is the price of the item that the seller is willing to sell/offer, and the price that the seller accepts. In computer networks, the asking price can be the cost for maintaining the network resources or network services.
- *Bid:* The bid is typically the bidding price, the price that the bidder is willing to pay for the item. However, in computer networks, the bid can refer to the resource demand of the bidder. For example, a bid can be a power demand, the number of power units, as presented in [145] and [201].
- *Ask:* The ask is typically the asking price, the price of the item that the seller accepts for trading. The ask is determined by the seller.
- *Price:* The price in the auction can refer to the bid or the ask. Also, it can be the price that winning bidders need to pay for winning the item or the price that the winning sellers receive for trading the items.
- *Strategy:* In the auction, the bidders and the sellers are players that have their strategies. The strategy of the bidder is to determine its bid, and that of the seller is to determine its ask. For example, the strategy of a service provider, the seller, is to determine the resource price, and the strategy of a user, the bidder, is to determine the resource price that it is willing to buy. The objective is to achieve a desired outcome or payoff. The payoff of a player depends on not only the player's own strategy, but also on the strategies of others. The payoff can be the revenue, profit, or utility. In particular, the utility is related to the values of the items to the buyers and the sellers. In particular, the value of the item to a particular bidder can be of the following types:
    - *Private value:* The value that a particular bidder assigns to the item is independent of the values of the other bidders. The private value of the bidder is typically unknown to other bidders.
    - *Interdependent value:* The value that a particular bidder assigns to the item depends on or is a function of the other bidders' values.
    - *Common value:* The bidders assign the same interdependent value to the item.

Other terms used in auction theory are as follows:

- *Efficiency:* An auction is considered to be efficient if the item is sold to the bidder that has the highest value.
- *Signal:* Each bidder has only an estimate/private information of the value, and the estimate or the private information is called a *signal* of the bidder. For example, in fog computing, users in different locations may have different estimates or signals of the same computing unit [193]. This is because the users have different latency, which results in different experiences.
- *Winner's curse:* A bidding strategy of each bidder entails increasing in its signal, and the bidder with the highest signal wins the item in the auction. If the bidder discovers that the value of the resource is less than its bid, then this case is considered to be *winner's curse*. To avoid the winner's curse issue, each bidder has to shade its bid. Sealed-bid auctions such as first-price sealed-bid and second-price sealed-bid auctions can guarantee this requirement.

We often use the following assumptions when analyzing an auction [202].

- *A1:* Bidders in the auction are risk neutral; that is, the bidders seek to maximize their expected utilities. Here, the utility of the bidder is the difference between its value and the price that the bidder needs to pay. For example, in the first-price sealed-bid auction, risk-averse bidders are willing to bid more to increase their chances of winning, which increases their expected utility. This allows the first-price sealed-bid auction to generate higher expected revenue than the English auction.
- *A2:* Bidders in the auction have independent private values; that is the values of the items to different bidders are independently distributed.
- *A3:* Bidders in the auction are symmetric; that is the values of the bidders are distributed according to same distribution function F. We also say that the bidders possess symmetric information.
- *A4:* Payment is a function of bids alone. In particular, let $\mathbf{b} = (b_1, \ldots, b_N)$ denote the bid profile of the bidders. Then, the payment p_i made by bidder i is a function of the bid profile, $p_i(\mathbf{b})$.

3.4 Summary

In this chapter, we introduce mechanism design and auction theory. In particular, we first define the mechanism and the allocation rule and the payment rule of the mechanism design. The mechanism design task is generally a complicated search problem. Thus, we introduce the revelation principle that can be used for facilitating the mechanism design task. We also present the required properties of the mechanism, which include incentive compatibility, individual rationality, economic efficiency, and budget balance. We further define and discuss optimal mechanisms in terms of social surplus maximization and profit maximization. After that, we introduce basics of auction theory and present the motivations as well as the significance of applying auctions to computer networks.

4 Open-Cry Auction

In this chapter, we present two types of open-cry auction, the English and Dutch auctions, and their applications in computer networks. Specifically, we first introduce the theory of the English auction from computer networks' perspective, and then discuss how to design this auction for spectrum sharing in computer networks. Then, we present the theory of the Dutch auction and its application to deal with emerging issues in computer networks – namely, network security, relay selection, and channel allocation. Finally, the combination of the English and Dutch auctions is also discussed.

4.1 English Auction

The English auction is a type of open-outcry ascending dynamic auction that is commonly used for selling items such as antiques and artwork. In computer networks, the items can be network resources such as computing units, storage, power, bandwidth, and wireless spectrum. The term *ascending* implies that bids submitted by bidders increase monotonically in multiple iterations of the auction. Here, bids are equivalent to bidding prices. The term *open* refers to the fact that the bids and even the identities of all bidders are disclosed publicly during the auction. The English auction is thus considered to be *open* or fully transparent.

Bids in an English auction can be indicated to the auctioneer in many ways, such as raising a paddle, raising a hand, or following the form of prearranged signals. In computer networks, a bid can be sent as an additional protocol signaling [203]. Also, bidders can submit their bids for the resource sequentially or simultaneously to the auctioneer. The auction terminates if no new higher bids are submitted, and the bidder with the highest bid wins the resource and pays the final price. The final price reached at auction is sometimes called the *hammer price*. The process of the English auction is implemented as follows.

4.1.1 English Auction Process

There are various ways to raise the bidding price. Two common approaches are that either the auctioneer sets price of resource in an increasing manner, typically in small increments, and announces the price to all bidders in iterations, or the bidders set their bids themselves and submit the bids to the auctioneer in iterations. These approaches are essentially equivalent and have one common characteristic: at any iteration, all bidders know the current highest bid. In the second approach, the auctioneer first sets and announces an initial price for the resource. Then, the auctioneer accepts increasingly higher bids from the initial price, submitted by bidders with an interest in the resource. The highest bid of a bidder at any iteration is considered to be a new price for the resource. The new price can be replaced by a higher bid from a competing bidder in the next iterations. If no competing bidder submits a higher bid than the new price, the bidder with the new price becomes the winner, and the resource is sold to the winner. The winner pays the auctioneer a final price that is typically higher than the lowest price that the auctioneer can accept and lower than the winner's budget. If the final price is lower than the lowest price deemed acceptable by the auctioneer, the auction will terminate without resource trading.

Example 4.1 Consider a resource trading market in cloud networking. The model for this market, shown in Figure 4.1, consists of three mobile users numbered from 1 to 3 and one service provider. The service provider has a unit of bandwidth for sale, and the service provider adopts the English auction for auctioning the bandwidth unit. Therefore, the service provider acts as the seller or auctioneer, the mobile users are potential buyers or bidders, and the item is the bandwidth unit. User 1 is willing to pay $1.5, user 2 is willing to pay $1.2, and user 3 is willing to pay $1.0 for the bandwidth unit. The prices $1.5, $1.2, and $1.0 can be considered the values of the bandwidth unit to the users. The auction process is implemented in multiple iterations as follows.

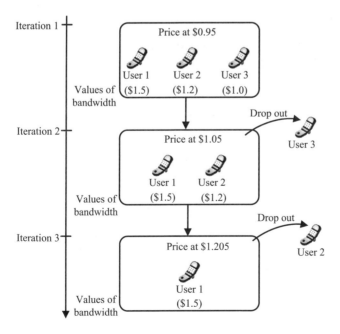

Figure 4.1 English auction for bandwidth resource trading market.

Initially, the service provider sets a price of $0.95 for the bandwidth unit. Since the price is less than $1.0, all three users will bid for the bandwidth unit in the first iteration. In the second iteration, the service provider increases a new price, $1.05. User 3 drops out of the auction because it is not willing to pay more than $1.0, and thus only users 1 and 2 are willing to buy the resource. In the third iteration, when the price of bandwidth unit exceeds $1.2, in this case $1.205, user 2 drops out of the auction, leaving user 1 as the only bidder. Thus, user 1 becomes the winner for the bandwidth unit, and the minimum price that user 1 needs to pay the service provider for the bandwidth unit is $1.205.

In general, the winner of the English auction has to bid just a little more than the individual who places the second highest value on the resource. In Example 4.1, user 1 agrees to buy the bandwidth unit at the price of $1.205, which is slightly greater than $1.2, the valuation of user 2. Therefore, when the auction involves a single unit of resource for sale, and each participant has an independent private value for the resource auctioned, the expected payment and expected revenue of an English auction is theoretically equal to that of the second-price sealed-bid auction. However, in the English auction, when one user drops out, the other users can increase their estimations of valuation. Such a valuation inference is impossible in the second-price sealed-bid auction, and thus the English auction and the sealed-bid second-price auction are not equivalent in terms of bidders' strategies and revenue.

4.1.2 Equilibrium Strategies

This section describes equilibrium strategies of bidders in an English auction. This is important in auction theory in which bidders are naturally selfish, and equilibrium strategies are their best responses. We first introduce basic terms commonly used in the English auction.

- *Signal:* Each bidder has only one estimate of, or private information about, the value of the item. The estimate or private information is called a *signal*, denoted by s_i, of the bidder. For example, in wireless networks, mobile users may have different values and thus different signals of the same spectrum and bandwidth unit. In an English auction, signals of the bidders are affiliated or correlated, meaning that a high signal of a bidder makes signals of other bidders high as well. In other words, the value of an item to a bidder is a function of not only its own signal but also the signals of other bidders. In general, the value is an increasing function of the signals of all bidders. Consider again Example 4.1, and let s_1, s_2, s_3 denote the signals of users 1, 2, and 3, respectively. Then, the value of a bandwidth unit to bidder 1 can be expressed as $v_1(s_1, s_2, s_3)$. Note that v_i is a realization of a random variable variable V_i. In general, the value of an item can be random for bidders. For example, mobile users have different and random values for a bandwidth unit depending on their channel quality. Similarly, s_i is a realization of a random variable S_i.

- *Symmetric model:* In the symmetric model, values of an item to bidders are drawn from the same distribution. Moreover, when signals of the bidders are affiliated, there is the symmetry of the distribution of signals, as signals of the bidders are drawn from the same probability distribution. Thus, for a particular bidder, the signals of other bidders can be interchanged without affecting the bidder's value. In Example 4.1, the value of a bandwidth unit to user 1 is $v_1(s_1, s_2, s_3) = v_1(s_1, s_3, s_2)$. When signals are affiliated, the value can be expressed by its expectation as $v_1(s_1, s_2, s_3) = \mathbb{E}[V_1 | S_1 = s_1, S_2 = s_2, S_3 = s_3]$. This can be interpreted as the conditional expectation of user 1 when it receives signals s_2 and s_3 from users 2 and 3.
- *Active and non-active bidders:* In an English auction, the auctioneer sets a low price and gradually raises it. This price at any iteration is observed by all bidders. At a certain price, bidders that are willing to buy the item are called *active bidders*, and bidders that drop out the auction are called *non-active bidders*. The non-active bidders will not participate in the next iterations of the auction. The information of the non-active bidders and the prices at which they drop out are commonly known to all bidders. A bidder can show its willingness to buy the item by various ways. In computer networks, a user as a bidder can send bit "1" to show that it is willing to buy the bandwidth unit and bit "0" to show that it will drop out from the auction. Then, the service provider, acting as the auctioneer, uses broadcast channels to send the information, including the prices at which the bidders drop out as well as the number of active and non-active bidders, to all bidders.
- *Bidding strategy:* To make the decisions (i.e., 1 or 0), bidders use their own bidding strategies. The strategy of the bidder is used to determine the price at which the bidder will drop out. This price is also considered to be a bid of the bidder or the expectation of the value of an item to the bidder. In general, the strategy of the bidder depends on its signal and the prices at which non-active bidders drop out. In this section, we denote a bid of the bidder as $b^K(s_i)$, referring to the price at which the bidder i drops out when there are K active bidders. Note that the bidding strategy is also an increasing function of its signal. Since signals of bidders may be different, the prices at which the bidders drop out are different. The following discussion presents how bidders determine their equilibrium strategies.

Consider again the bandwidth resource trading market in Example 4.1 with N users as bidders. Let $s_1 \geq s_2 \geq, \ldots, \geq s_N$ denote the signals of users $1, 2, \ldots, N$, respectively. In other words, user 1 has the highest signal, followed by user 2, and so on. Also, let $p_1 \geq p_2 \geq, \ldots, \geq p_N$ denote the prices of the bandwidth unit set by the service provider at which users $1, 2, \ldots, N$ drop out, respectively. Due to the symmetric model, we consider the strategy of user 1 during the auction to understand how the users determine their equilibrium strategies at each iteration.

Before the auction starts, assume that the price set by the service provider is low enough that all N active users participate in the auction. In the first iteration, the strategy of user 1 is to determine the price at which it will drop out. The price at which user

1 drops out matches its expectation of the value. Thus, user 1 determines its expectation of the value based on the signals of the other users. However, in this iteration, the signals of the other users are not available to user 1, so user 1 assumes that the other users have the same signal s_1. Let $S_1, S_2, \ldots, S_N$ denote the highest, second highest, and so on, signals of N users, respectively. Then, the strategy of user 1 is determined as follows:

$$b^N(s_1) = \mathbb{E}[V_1 | S_1 = s_1, S_2 = s_1, \ldots, S_N = s_1] \tag{4.1}$$

where V_1 is the value to user 1 of the bandwidth resource, and $\mathbb{E}[V_1 | S_1 = s_1, S_2 = s_1, \ldots, S_N = s_1]$ is user 1's conditional expectation of the value given the signals of the other users.

Suppose that in this iteration, the service provider sets the price of the bandwidth resource at p_N, at which point user N with a signal s_N drops out. User N drops out because its bid $b^N(s_N)$ reaches the price p_N set by the service provider, $b^N(s_N) = p_N$. Since $b^N(s_N)$ is a continuous and increasing function of s_N, user 1 can infer a unique value from signal s_N of user N by inverting the bidding function $b^N(s_N)$. Since the users are affiliated, user 1 updates its signal and calculates its new strategy, meaning the new price at which it will drop out, as follows:

$$b^{N-1}(s_1) = \mathbb{E}[V_1 | S_1 = s_1, S_2 = s_1, \ldots, S_{N-2} = s_1, S_N = s_N] \tag{4.2}$$

Similarly, in the second iteration, assume that the service provider sets the price $p_{N-1} \geq p_N$, at which point user $N-1$ with signal s_{N-1} drops out. Then, user 1 can infer the signal of user $N-1$, and its strategy changes to

$$b^{N-2}(s_1) = \mathbb{E}[V_1 | S_1 = s_1, S_2 = s_1, \ldots, S_{N-2} = s_1, S_{N-1} = s_{N-1}, S_N = s_N] \tag{4.3}$$

The auction is repeated in the same way. After $N-2$ iterations, there are $N-2$ users (i.e., users $N, \ldots, 3$), which drop out at prices $p_N \leq \cdots \leq p_3$, respectively. There are two remaining active users, users 1 and 2, with the highest signals. At this iteration, user 1 updates its strategy as follows:

$$b^2(s_1) = \mathbb{E}[V_1 | S_1 = s_1, S_2 = s_1, S_3 = s_3, \ldots, S_N = s_N] \tag{4.4}$$

When the service provider sets the price p_2, at which point user 2 drops out, user 1 becomes the winner since the signal of user 1 is higher than that of user 2. User 1 thus infers the signal s_2 of user 2 and updates its strategy as follows:

$$b^2(s_1) = \mathbb{E}[V_1 | S_1 = s_1, S_2 = s_2, \ldots, S_N = s_N] \tag{4.5}$$

Now, we consider the price that user 1, the winner, needs to pay the service provider. Note that user 2 drops out at price p_2, meaning that p_2 reaches user 2's strategy $b^2(s_2)$, which is given by

$$b^2(s_2) = \mathbb{E}[V_2 | S_1 = s_2, S_2 = s_2, \ldots, S_N = s_N] \tag{4.6}$$

User 1 then pays exactly the service provider the price $b^2(s_2)$, and the expected profit of user 1 is given by

$$\pi_1(s_1) = \mathbb{E}[V_1 | S_1 = s_1, S_2 = s_2, \ldots, S_N = s_N]$$
$$- \mathbb{E}[V_2 | S_1 = s_2, S_2 = s_2, \ldots, S_N = s_N] \quad (4.7)$$

Although these strategies, $(b^N(s_1), \ldots, b^2(s_1))$, are considered for user 1, the other users also follow the same strategies in the symmetric model. Therefore, we have the following proposition.

PROPOSITION 4.1 *[183]A strategy profile* $b = (b^N, \ldots, b^2)$ *is said to be a symmetric equilibrium strategy in the English auction.*

Proposition 4.1 implies that the users in the English auction cannot do better than to follow strategy **b**. This is verified by the following example.

Example 4.2 Consider again the bandwidth resource trading market with three users 1, 2, and 3. Three users as bidders compete for a bandwidth unit of the service provider, which acts as the auctioneer. The considered model is symmetric, such that the values V_i and signals S_i of the users are symmetric. Again, this means that the signals of users 2 and 3 can be interchanged without affecting the value of user 1, and similarly for other users. Specifically, we assume that the values assigned by users 1, 2, and 3 to the bandwidth resource have the following forms:

$$V_1(S_1, S_2, S_3) = S_1 + \frac{1}{3}S_2 + \frac{1}{3}S_3$$
$$V_2(S_1, S_2, S_3) = S_2 + \frac{1}{3}S_1 + \frac{1}{3}S_3 \quad (4.8)$$
$$V_3(S_1, S_2, S_3) = S_3 + \frac{1}{3}S_1 + \frac{1}{3}S_2$$

Since the users are symmetric, we can suppose that the signal of user 1 is the highest, followed by users 2 and 3, and the realizations of the signals of users 1, 2, and 3 are $(s_1, s_2, s_3) = (0.7, 0.4, 0.1)$, respectively. The equilibrium strategies of users are to determine the prices at which they drop out. At the beginning of the auction, each user only knows its signal and assumes that other users have the same signals. At this stage, the strategies of users 1, 2, and 3 are, respectively, as follows:

$$b^3(s_1) = v_1(s_1, s_1, s_1) = s_1 + \frac{1}{3}s_1 + \frac{1}{3}s_1 = \frac{5}{3}s_1 = 1.66$$
$$b^3(s_2) = v_2(s_2, s_2, s_2) = s_2 + \frac{1}{3}s_2 + \frac{1}{3}s_2 = \frac{5}{3}s_2 = 0.66 \quad (4.9)$$
$$b^3(s_3) = v_3(s_3, s_3, s_3) = s_3 + \frac{1}{3}s_3 + \frac{1}{3}s_3 = \frac{5}{3}s_3 = 0.166$$

If the service provider sets a price of the bandwidth resource of $p_3 = 0.166$, user 3 drops out from the auction since the price, p_3, reaches its value, $v_3(s_3, s_3, s_3)$. In the next iteration, there are two active users, 1 and 2. After inferring the signal s_3 of user 3, the equilibrium strategies of users 1 and 2 change to

$$b^2(s_1) = v_1(s_1, s_1, s_3) = s_1 + \frac{1}{3}s_1 + \frac{1}{3}s_3 = \frac{4}{3}s_1 + \frac{1}{3}s_3 = 0.966 \quad (4.10)$$

$$b^2(s_2) = v_2(s_2, s_2, s_3) = s_2 + \frac{1}{3}s_2 + \frac{1}{3}s_3 = \frac{4}{3}s_2 + \frac{1}{3}s_3 = 0.566$$

Clearly, the value of the bandwidth resource to users 1 and 2 decreases after they know that the signal of user 3 is low. Thus, the prices at which users 1 and 2 will drop out decrease since the bidding functions $b^3(.)$ and $b^2(.)$ are strictly increasing in their signals. User 1 pays the service provider a price of $b^2(s_2) = 0.566$. Thus, the profit that user 1 receives is $\pi_1(s_1) = b^2(s_1) - b^2(s_2) = 0.966 - 0.566 = 0.4 > 0$. It is clear that by following the strategy $\mathbf{b} = (b^3, b^2)$, user 1 becomes the winner and winning the resource results in a positive profit. Therefore, user 1 cannot do better than to follow strategy $\mathbf{b} = (b^3, b^2)$. This is similar to the general case with N users, where $\mathbf{b} = (b^N, \ldots, b^2)$ is the best response strategy, the equilibrium strategy, of user 1 given the signals of the other users.

The equilibrium strategy $\mathbf{b}$ is achieved when the signals and values of all users are commonly known. In such an auction, known as a complete information game, the equilibrium strategies of the users form an *ex-post* equilibrium or Nash equilibrium in the game. Since the user determines its strategy after it infers all signals of users, the equilibrium strategy in the English auction has a strong *no-regret* property. In particular, users that drop out the auction do not regret losing, and the user that wins the auction does not regret winning.

In summary, bidders in an English auction drop out in order of their signals. The bidder with the lowest signal drops out first, followed by the bidder with the second lowest signal, and so on. Importantly, signal interference may occur in an English auction, meaning that the bidding strategy of a user depends not only on the user's signal but also on the number of dropped-out bidders. The equilibrium strategy of the English auction has the *no-regret* property.

4.2 Development of English Auction for Computer Networks

Modern computer networks such as the 5G wireless networks are expected to support a high density of mobile broadband users. To achieve this goal, maximizing the spectrum efficiency is critical, as the available spectrum resources are limited. In this section, we approach the problem of maximizing the spectrum efficiency in the context of cognitive radio for 5G systems.

There have been a number of works related to spectrum sharing that modeled dynamic spectrum leasing using economic approaches. In the scenarios with one primary user and multiple secondary users [204], [205], the primary user has a set of unused channels, and the secondary users are willing to buy the channels. Given the limited number of channels, the scenario is a competitive market, also referred to as one-sided matching markets. The prices of the channels regulate the quantities bought by the secondary users and are updated depending on the demand and supply of the channels. The desired

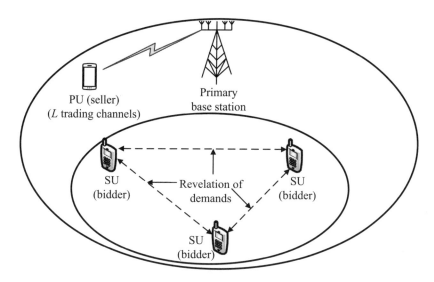

Figure 4.2 English auction for channel assignment in cognitive radio networks [205] where PU and SU stand for primary user and secondary user, respectively.

outcome is the state at which the demand equals the channels' supply, which is called *Walrasian equilibrium* or *competitive equilibrium*. To reach the Walrasian equilibrium, a price adjustment process is required. This process can be implemented using a well-known economic model, the demand and supply model.

Apart from the demand and supply model, the English auction has recently been used to reach the Walrasian equilibrium [205]. In particular, the adoption of the English auction for spectrum leasing in cognitive radio networks has clear advantages. First, this auction allows autonomous and rational buyers to locally decide their resource demands, and this is favorable for distributed operation of the secondary users in cognitive radio networks. Second, the auction relies only on binary decisions that reflect demands or interests in competitive markets, so it has low communication overhead.

In this section, we present channel assignment using the English auction in the scenario with one primary user and multiple secondary users [205]. Specifically, we first describe the system model and present the formulation of the channel assignment problem in the cognitive radio networks. We then define the Walrasian equilibrium and conditions of existence of the Walrasian equilibrium. We discuss how the English auction is used to reach the Walrasian equilibrium as well as to solve the distributed channel assignment.

4.2.1 System Model and Problem Formulation

The considered model is a cognitive radio network as shown in Figure 4.2. The model is a competitive market including a set of N secondary users, denoted by $\mathcal{N} = \{1, \ldots, N\}$, and one primary user. N secondary users compete for a set of L orthogonal channels, denoted by $\mathcal{L} = \{1, \ldots, L\}$, which are licensed to the primary user. A *coordination mechanism* can be used in which a coordinator performs the assignment of primary channels and the secondary users. To avoid the payment requirement for the coordinator, this section considers a *cooperative mechanism*.

The cooperative mechanism relies on direct communications among secondary users and does not require a coordinator. The objective is to maximize the utilities of both secondary users and the primary user. This is a multi-objective optimization problem, and one common approach is to employ the optimization of the weighted sum of the objectives. Let $\mathcal{B} \in \mathcal{L}$ denote a set of channels that is assigned to secondary user k. Then the optimization problem formulated for secondary user k is given by

$$W(\mathcal{B}, k) = \omega u_k^{SU}(\mathcal{B}) + (1 - \omega) \sum_{l \in \mathcal{B}} u_l^{PU}(k)$$

where $u_k^{SU}(\mathcal{B})$ is the utility, the aggregate transmission rate, of secondary user k when having the set $\mathcal{B}$ of channels, and $u_l^{PU}(k)$ is the utility of the primary user when channel l is assigned to secondary user k. Here, ω is the weight, which refers to the priority of one objective to the other objectives. The weight is typically predefined, for example, by network operators. In particular, if ω is close to zero, the utility optimization of the primary user is prioritized. On the contrary, if ω is close to 1, the utility optimization of secondary users is more important. The overall optimization problem is given by

$$\sum_{k \in \mathcal{N}} \sum_{\mathcal{B} \in \mathcal{L}} W(\mathcal{B}, k) x(\mathcal{B}, k) \qquad (4.11)$$

where $x(\mathcal{B}, k)$ is an assignment variable. $x(\mathcal{B}, k) = 1$ means that set $\mathcal{B}$ is assigned to SU k, and $x(\mathcal{B}, k) = 0$ indicates that there is no channel assigned to secondary user k. Note that a channel is assigned to only one secondary user, but the secondary user can be assigned more than one channel. However, to guarantee the fairness in the channel allocation for secondary users, the number of channels for the secondary user is restricted by a *quota* q_k, $|\mathcal{B}| \leq q_k$. Given these constraints, problem (4.11) is an integer optimization problem, and its solution is a *Walrasian equilibrium* of the competitive market [205]. Thus, problem (4.11) can be solved by finding the Walrasian equilibrium of the presented model. Further details of the Walrasian equilibrium and how to reach it are given in the next subsections. The general idea is that the secondary users reveal their demands to each other and identically set the channel prices. Then, each secondary user locally decides its demand set, in this case the set of demand channels, given the channel prices so as to maximize the net utility of the secondary user. In particular, the net utilities of the secondary users are determined as follows.

Assume that secondary user k knows the utility of the primary user, the weight ω, and the quota q_k. Then, the weighted sum-performance or the unit-less utility of secondary user k is given by

$$U_k(\mathcal{A}) = \max_{\mathcal{B} \subseteq \mathcal{A}} \omega u_k^{SU}(\mathcal{B}) + (1 - \omega) \sum_{l \in \mathcal{B}} u_l^{PU}(k) \qquad (4.12)$$

s.t. $|\mathcal{B}| \leq q_k$

Let $\mathbf{p} = (p_1, \ldots, p_L)$, where $p_l \geq 0$, denote the set of initial prices of the channels. The initial prices of the channels can be set by the primary user. The net utility of each secondary user k can be defined as follows:

$$u_k^{\text{SU-NET}}(\mathcal{A}, \mathbf{p}) = U_k(\mathcal{A}, \mathbf{p}) - \sum_{l \in \mathcal{A}} p_l \quad (4.13)$$

The net utility of secondary user k is used for determining its demand set. Secondary user k determines the demand set by finding a collection of channels that maximize the secondary user's net utility. We can consider the demand set of the secondary user as its strategy. Let $\mathcal{A}_k$ denote the demand set of secondary user k. Then $\mathcal{A}_k$ is defined as follows:

$$\mathcal{A}_k = \{\mathcal{A} \subseteq \mathcal{L} | u_k^{\text{SU-NET}}(\mathcal{A}, \mathbf{p}) \geq u_k^{\text{SU-NET}}(\mathcal{B}, \mathbf{p}), \forall \mathcal{B} \subseteq \mathcal{L}\} \quad (4.14)$$

We further discuss how the secondary user determines its demand set in Section 4.2.3.

4.2.2 Walrasian Equilibrium

Walrasian equilibrium, also known as competitive equilibrium, is the traditional concept of economic equilibrium used for the analysis of commodity markets with flexible prices and multiple sellers. This equilibrium is the desired outcome of the market and is defined as a state at which the demand equals the supply for the commodity in the market. The Walrasian equilibrium is considered since it is the solution of the integer optimization problem (4.11). It is essentially composed of equilibrium strategies, representing the best responses of secondary users. This is similar to the Nash equilibrium in which the strategies of bidders are the demand sets of secondary users. However, apart from the demand sets, the secondary users set channel prices. Thus, the Walrasian equilibrium additionally includes the channel prices. We have the following definition.

DEFINITION 4.2 *[205] A Walrasian equilibrium is defined as a tuple* $(\mathbf{p}, \mathcal{A}_1, \ldots, \mathcal{A}_N)$, *where* $\mathbf{p}$ *is a vector of channel prices,* $\mathcal{A}_k \subseteq \mathcal{L}$ *is a set of channels assigned to secondary user* k, $\mathcal{A}_k \cap \mathcal{A}_j = \emptyset$ *for* $k \neq j$, *and* $\bigcup_{k=0}^{N} \mathcal{A}_k = \mathcal{L}$, *such that for each secondary user* $k \in \mathcal{N}, u_k^{\text{SU-NET}}(\mathcal{A}_k, \mathbf{p}) \geq u_k^{\text{SU-NET}}(\mathcal{B}, \mathbf{p})$ *for all* $\mathcal{B} \subseteq \mathcal{L}$.

According to Definition 4.2, the Walrasian equilibrium consists of two elements. The first element is a set of prices at which each secondary user buys channels, and the second element is the sets of channels allocated to secondary users. It is important to know what the conditions for guaranteeing the existence of the Walrasian equilibrium are. The Walrasian equilibrium exists if and only if the utility function U_k in (4.12) satisfies two conditions, the monotonicity property and the gross substitutes condition [206]. In particular for the presented model, U_k upholds the monotonicity property since it increases as more channels are allocated to the secondary user. In particular, the aggregate transmission rate keeps increasing as there are more channels. For the gross substitute condition, this implies that if a secondary user demands a set of channels and the prices of some channels in the set increase, then the secondary user would still demand channels for which the prices did not change. In our model, U_k preserves the gross substitutes property since it is additively separable, meaning that $U_k(\mathcal{B}) = \sum_{l \in \mathcal{B}} U_k(\{l\})$ [206].

The existence of the Walrasian equilibrium ensures that the solution of problem (4.11) is identical to the solution of its linear programming relaxation. In the next subsection, we discuss how the English auction is used to achieve the Walrasian equilibrium.

4.2.3 English Auction for Walrasian Equilibrium

The English auction used in this section is essentially similar to the single-item English auction described in Section 4.1. Specifically, the auction is implemented in multiple iterations, and the strategies, or demand sets, of other secondary users are known to a particular secondary user. Also, if a channel is simultaneously demanded by more than one secondary user, the price of that channel is increased. When only one secondary user demands the channel, the channel will be assigned to the secondary user. However, there are some differences between the English auction used here and the single-item English auction. First, there are multiple items, the channels in the presented model, and thus more iterations may be required to finish trading all the channels. Second, instead of the primary user (i.e., the seller), secondary users identically update the channel prices knowing the demand sets of all secondary users. For these reasons, the English auction used in this section can be called a modified English auction [205]. The process of the auction is presented in the following discussion. Without loss of the generality, the auction begins with the ith iteration.

At the ith iteration, given the prices of the channels, each secondary user k determines its demand set $\mathcal{A}_k^{(i)}(\mathbf{p})$ that maximizes secondary user k's net utility. Accordingly, the secondary user calculates a set of its net utilities when associated with L channels, denoted by $\mathcal{U}_k^{\text{SU-NET}} = \{u_k^{\text{SU-NET}}(\{1\}, \mathbf{p}), \ldots, u_k^{\text{SU-NET}}(\{L\}, \mathbf{p})\}$. The secondary user sorts $\mathcal{U}_k^{\text{SU-NET}}$ in a descending order of the net utilities. It then takes the first q_k elements in the sorted $\mathcal{U}_k^{\text{SU-NET}}$ as its demand set $\mathcal{A}_k^{(i)}$. We have the following results.

LEMMA 4.3 *[205] If prices of the channels are greater than zero, then the demand set of secondary user k at iteration i satisfies $|\mathcal{A}_k^{(i)}| \leq q_k$, $\forall k \in \mathcal{N}$.*

The proof of Lemma 4.3 can be found in [205]. In particular, Lemma 4.3 means that if the prices of channels are strictly greater than zero, the maximum number of channels that a secondary user demands is equal to its quota q_k. After determining the demand set, the secondary user broadcasts its demand set to the other secondary users. Note that the secondary user broadcasts this update to the other secondary users only in the case that the demand set of the secondary user has changed. The secondary user can broadcast a bit message including the indices of the channels that it demands. For example, given L channels, the bid message includes bits "1" and "0" to indicate the channels that the secondary user demands. When a secondary user knows all the other secondary users' demands, the secondary user calculates an excess demand set $\mathcal{Z}^{(i)}(\mathbf{p})$. The elements of $\mathcal{Z}^{(i)}(\mathbf{p})$ are the channels that are simultaneously demanded by more than one secondary user. For example, if channels 1, 4, and 5, are demanded by more than one secondary user, then $\mathcal{Z}^{(i)}(\mathbf{p}) = \{1, 4, 5\}$. $\mathcal{Z}^{(i)}(\mathbf{p})$ is a function of channel prices $\mathbf{p}$ since the number of channels demanded by each secondary user decreases as the prices of the channels increase, and vice versa.

At this iteration, a price step Δp is synchronized for all secondary users in the network, and the secondary users identically update prices for channels in the next iteration as follows:

$$p_l^{(i+1)} = \begin{cases} p_l^{(i)} + \Delta p, & \text{if } l \in \mathcal{Z}(\mathbf{p}^{(i)}) \\ p_l^{(i)}, & \text{otherwise} \end{cases} \quad (4.15)$$

Equation (4.15) means that the secondary users update the prices for the channels only in the excess demand set. Note that the step Δp influences the convergence speed of the algorithm. A large value of Δp enables the algorithm to converge faster. However, some channels may have suddenly high prices. These channels become unattractive to any secondary users, and they are not demanded by any secondary users. For a sufficiently small Δp, $\Delta p \to 0$, the algorithm converges to the Walrasian equilibrium as stated in Definition 4.2. Algorithm 4 is guaranteed to converge since prices of the channels can only be incremented and the secondary users' utilities have finite values.

The simulation results for the English auction–based channel assignment are given in [205]. As shown, for different signal-to-noise ratios (SNR), the secondary user's sum transmission rate using the English auction approach is very close to that of the optimal approach. Here, the optimal approach is the Hungarian method [207], which determines the optimal channel assignment by maximizing the weighted sum rate of the secondary users and primary user in a centralized fashion.

4.3 Dutch Auction

This section introduces the other type of open-cry auction, Dutch auction. The Dutch auction is similar to the English auction, meaning that the bid of a particular bidder is disclosed to the other bidders when the bidder submits its bid to the auctioneer. However, the English auction is a ascending-bid auction while the Dutch auction is a descending-bid auction. Thus, the Dutch auction is also called an *open-outcry descending-price auction*. The name "Dutch auction" reflects that this type of auction was commonly used in flower markets in the Netherlands.

4.3.1 Dutch Auction Process

The Dutch auction is known as a descending-bid auction. Thus, the process of the Dutch auction is essentially opposite of that of the English auction. Accordingly, the auctioneer initially sets a high asking price for the resource, and then it decreases the price until one of the buyers accepts the price or until the price reaches a predetermined reserve price. The winning buyer pays the final price and receives the resource. We provide the following example to clarify how the Dutch auction works.

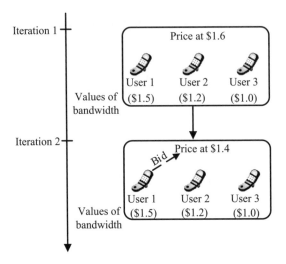

Figure 4.3 Dutch auction for bandwidth resource trading market.

Example 4.3 We consider again the cloud networking model described in Example 4.1. This model consists of one service provider, one bandwidth unit, and three mobile users. The service provider is the seller or auctioneer, the mobile users are potential buyers (i.e., bidders), and the bandwidth unit is the item. The private values of the bandwidth unit to users 1, 2, and 3 are $1.5, $1.2, and $1.0, respectively. Assume that the service provider sets a price decrement step as $0.2, and the users submit bids only when the price of the bandwidth unit is less than their values. Here, the bid refers to the willingness-to-buy of the corresponding user. The user shows its bid by sending a signal, such as bit "1," to the service provider. The auction process is shown in Figure 4.3 and implemented in multiple iterations as follows. Initially, the service provider sets an initial price of $1.6 for the bandwidth unit. At the first iteration, none of users submits its bid since the initial price is greater than the values of all three users. The service provider reduces the price to $1.6 − $0.2 = $1.4. This price is less than the value of user 1, and thus in the second iteration, user 1 bids and the auction terminates. User 1 wins the bandwidth unit and pays the service provider a price of $1.4. As such, user 1 obtains a surplus or a positive profit of $1.6 − $1.4 = $0.2.

From Example 4.3, we can see that $1.4 is the first price that the winning user bids and pays. Therefore, the Dutch auction is strategically similar to the first-price sealed-bid auction, which is presented in Chapter 5. Also, in the Dutch auction, each bidder makes only a single decision, by sending bit "1" if it is willing to buy the bandwidth unit. Otherwise, no bit needs to be sent to the service provider. This is different from the English auction in which each user sends "1" or "0" to show its willingness-to-buy or dropping-out, respectively. The price of the bandwidth unit at which the user should submit its bid is called the *bidding* or *equilibrium strategy* of the bidder. Note that since there is no dropping-out decision in the Dutch auction, a particular user cannot infer the values or signals for the bandwidth unit of the other users when the user determines its equilibrium strategy. Alternatively, the user determines its equilibrium strategy by assuming that the value distributions of all users are publicly known. This assumption is reasonable, and can be realized by observing historical bids. The Nash equilibrium in the Dutch auction, if it exists, is thus called *Bayesian–Nash equilibrium*. We determine the Nash equilibrium in the Dutch auction by using the revenue equivalence theorem presented in the next section. Note that the revenue equivalence theorem is also used to determine the Nash equilibrium in the first- and second-price sealed-bid auctions.

4.3.2 Revenue Equivalence Theorem

Revenue equivalence is a concept in auction theory that given certain conditions, any mechanism that has the same outcomes in the sense of allocations of items to the same bidders also has the same expected revenue. We can use revenue equivalence to predict the bidding strategy of a bidder in an auction. In particular, for the Dutch auction, the bidding strategy, denoted by b_i, of a bidder is the price at which the bidder should submit its bid. We first recall basic concepts of distribution and density functions of a continuous random variable.

- The distribution function or cumulative distribution function (CDF) of a continuous random variable X evaluated at x is defined as $F(x) = P(X \leq x)$, and represents the probability that X will take a value less than or equal to x.
- When $F(x)$ is differentiable, such that its derivative exists at each point in its domain, we have a density function defined as $f(x) = \frac{F(x)}{dx} = F'(x)$. For example, if X is a variable that is uniformly distributed in $[0, 1]$, then the density function of $X = x$ is $f(x) = \frac{1}{1-0}$ if $0 \leq x \leq 1$ and $f(x) = 0$, otherwise.

We now consider again the bandwidth resource trading market mentioned in Section 4.1.2. In this model, the service provider as an auctioneer trades a bandwidth unit to N mobile users as the bidders. Here, we assume that the values of the bandwidth unit to the users are identically and independently distributed according to a known distribution F. The term "independently" expresses that the information about each bidder's private value of the item is independent of that for every other bidder's private value. We also assume that the user with the highest value purchases the bandwidth unit if and only if its value is at least r. When the user decides to bid to buy the bandwidth unit, it expects that it will obtain the maximum payoff. The payoff of the user is given by

$$\pi_i = Q_i(v)v - p_i(v) \tag{4.16}$$

where Q_i is the probability that user i wins, and p_i is the expected payment that the user pays the service provider. We need to determine Q_i and p_i. Note that in the auction, the user with the highest value will win the auction and receive the bandwidth unit. Thus, when user 1 decides to bid for the bandwidth unit, user 1 expects that $(N-1)$ remaining users have values less than the value of buyer 1. Since the values of all users follow the same distribution functions, $F_1(v) = \cdots = F_N(v) = F(v)$, and are independent with each other, then the probability that user 1 wins is defined as $Q_i(v) = F^{N-1}(v)$. By substituting $Q_i(v)$ into (4.16), taking the derivative of π_i and letting it be zero, we have

$$\frac{dF^{N-1}(v)v}{dv} = \frac{dp_i(v)}{dv} \quad (4.17)$$

Suppose that at a value of r, the user gets an expected payoff equal to zero. Then, taking integration by parts over the interval $[r, v]$, we determine the expected payment p_i as follows:

$$p_i(v) = F^{N-1}(v)v - \int_r^v F^{N-1}(t)dt \quad (4.18)$$

Let $b_i(v)$ denote the bidding strategy of user i. This means that the user purchases the bandwidth unit when the price of the bandwidth unit is $b_i(v)$. Thus, the expected payment is equal to the price multiplied by the probability of winning, $p_i(v) = Q_i(v)b_i(v) = F^{N-1}(v)b_i(v)$. As such, the payoff of the user with value v can be rewritten as follows:

$$\pi_i = F^{N-1}(v)(v - b_i) \quad (4.19)$$

Substituting $p_i(v)$ into (4.18), we obtain the bidding strategy of the user:

$$b_i(v) = v - \frac{\int_r^v F^{N-1}(t)dt}{F^{N-1}(v)} \quad (4.20)$$

We discuss further the bidding strategy in the next section.

4.3.3 Equilibrium in Dutch Auction

Based on (4.20), we have the following definition of the equilibrium of the Dutch auction:

DEFINITION 4.4 *[208] If N users have independent private values that are drawn from the common distribution F, then the equilibrium strategies of the users given by*

$$b(v) = v - \frac{\int_r^v F^{N-1}(t)dt}{F^{N-1}(v)} \quad (4.21)$$

constitute a symmetric Nash equilibrium of a Dutch auction.

The equilibrium strategy $b(v)$ as defined in (4.21) can be interpreted as users' best response. The equilibrium strategy is determined by the user knowing the distribution of values of other users, and thus the Nash equilibrium is also called Bayesian–Nash equilibrium. From (4.21), it can be seen that in the Dutch auction, the equilibrium strategy of the Dutch auction is well defined regardless of the values of the bidder's rivals. In contrast, in the English auction the equilibrium strategy, represented by (4.2) and (4.4), for each user is determined based on knowing the values and signals of the bidder's rivals.

To more easily understand the physical meaning of (4.21), we assume that the distribution function of the value of the users is $F(v) = v$, and the minimum value is zero, $r = 0$. Then, we can calculate the equilibrium strategy of each user as follows:

$$b(v) = v - \frac{v}{N} \tag{4.22}$$

If there are two users, $N = 2$, we can obtain the equilibrium strategy of each user as $b(v) = \frac{v}{2}$. This means that the equilibrium strategy for each user is to bid half of its valuation. Also, the bidding strategy of the user is an increasing function of the user's value. For example, if the private values of the bandwidth unit to users 1 and 2 are $v_1 = \$1.5$ and $v_2 = \$1.2$, then the corresponding equilibrium strategies of users 1 and 2 are 0.75 and 0.6, respectively. User 1 also has a higher probability of winning the bandwidth unit since it has a higher value.

4.4 Development of Dutch Auction for Computer Networks

Future-generation wireless networks will become more decentralized and ad hoc in nature, and traditional approaches, such as system optimization, for resource management may not be used due to the fact of that there are no central nodes. The Dutch auction, which allows bidders or users to locally make their decisions, may be used to provide dynamic and distributed approaches. This section discusses the application of the Dutch auction to emerging issues in modern computer networks. Specifically, we first present the use of the Dutch auction as an incentive mechanism for preventing black hole attacks in mobile ad hoc networks (MANETs). Second, we discuss the application of the Dutch auction for the quick relay selection in the Internet of Things (IoT). Third, we introduce the use of the Dutch auction as a distributed solution for the channel allocation in 5G heterogeneous networks (HetNets).

4.4.1 Prevention of Black Hole Attacks in Mobile Ad Hoc Networks

As presented in Chapter 2, in a black hole attack, a malicious node, known as the black hole node, attempts to disrupt network services such as packet routing or forwarding. To detect and prevent the malicious node from taking action, this section presents a security approach, namely *key-revocation scheme* [208], for MANETs using the Dutch auction. Here, the term "key-revocation" refers to removing a key or an identity of a node in the network. This makes the node unavailable in the network and not able to communicate with other nodes in the network. The Dutch auction is used since (i) it is based on the local decisions of bidders (i.e., nodes) without requiring any permanently available central authority (CA); (ii) it guarantees that the malicious node is quickly revoked because of its simplicity – that is, the node makes only a single decision and the auction terminates when the single decision of any node is submitted; and (iii) it incentivizes rational nodes to revoke the malicious node due to its payment policy.

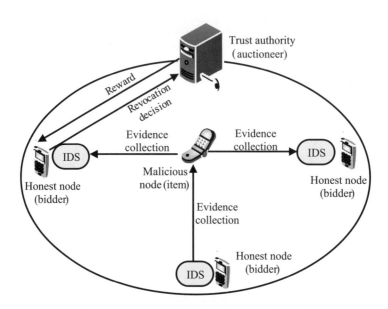

Figure 4.4 Dutch auction for revocation model, where IDS stands for intrusion detection system.

This section presents the application of the Dutch auction for the revocation model [208] in a MANET. Specifically, we first discuss how to map the traditional Dutch auction to the revocation model and then analyze the equilibrium strategies of nodes.

Dutch Auction and Revocation Model Mapping

Consider the MANET shown in Figure 4.4, which includes N honest nodes, one malicious node, and one trust authority. The node that revokes the malicious node earliest receives a reward from the trust authority. Therefore, the honest nodes act as the bidders (i.e., buyers) that compete for the reward, the trust authority is the auctioneer, and the malicious node is the item. In fact, the honest nodes do not know which nodes are malicious, so they use their intrusion detection systems (IDSs). The IDS collects evidence of any malicious behavior, such as the non-forwarding or the non-routing of packets, of malicious neighbor nodes. In the example network, there is a malicious node, and the IDSs of N honest nodes collect evidence of the malicious node. In general, the accuracy of the IDS in detecting a malicious node becomes higher as the IDS collects more evidence. However, this requires more time or more observed rounds.

Let $0 \leq P_i^c \leq 1$ denote the probability that the IDS of node i correctly identifies the malicious node. P_i^c depends on the number of rounds k observed by the IDS. A simple form of P_i^c is $P_i^c(k) = k/k_{perfect}$, where $k_{perfect}$ is the number of observed rounds after which $P_i^c = 1$. For example, given $k = 60$ and $k_{perfect} = 100$, $P_i^c(k) = 0.6$. This means that the probability that the IDS correctly identifies the malicious node is 0.6.

Assume that all honest nodes join the auction at the same time and gather the same evidence about the malicious node. Then P_i^c are equal for all honest nodes, $P_i^c = P^c, i = 1, \ldots, N$. Node i needs to select the value of P^c at which it should revoke the

4.4 Development of Dutch Auction for Computer Networks

Table 4.1 Linkage between the traditional Dutch auction and the revocation model [208].

Dutch auction	Revocation model
Bidders	Honest nodes
Item	Malicious node
Auctioneer	Trust authority
Value v_i	Risk appetite r_i^a
Value distribution $F(v_i)$	Risk appetite distribution $F(r_i^a)$
Bidding strategy b	Determination of risk taken r^t
Payoff $v_i - b$	Payoff $r_i^a - r^t$

malicious node. Node i can choose a high value of P^c to make its revocation decision. However, this requires more time, which translates into more rounds. Consequently, node i may miss the chance to revoke the malicious node and to receive the reward since another node revokes the malicious node first. Otherwise, if node i selects a low value of P^c for the revocation decision, it can make an incorrect assessment and gets a penalty/loss from the trust authority. This can be considered to be a *risk* of the node. Thus, the node with the highest risk makes a revocation decision early and has a high probability of winning the auction. Two types of risks are discussed: *risk taken*, or *actual risk*, and *risk appetite*, or *risk tolerance*. Based on these risks, Table 4.1 highlights the relationship between the traditional Dutch auction and the revocation model. The relationship and the risks are further explained here.

- *Risk taken:* Risk taken or actual risk is the risk that node i takes when the node revokes the malicious node. In the traditional Dutch auction, the risk taken is equivalent to the price that the bidder pays the auctioneer for winning the item. Assume that all nodes collect the same amount of evidence at a given time. Then the risk taken by all nodes is the same, $r_i^t = r^t, i = 1, \ldots, N$. r^t is determined as follows [208]:

$$r^t = 1 - a((1+d)P^c - 1) \quad (4.23)$$

  where a is the scaling factor, and $d > 0$ is the fixed reward that gives the honest nodes an incentive to revoke. Equation (4.23) means that the taken risk r^t is inversely proportional to the probability P^c that the IDS of the node correctly identifies the malicious node. Note that P^c is a function of the number of rounds observed by the IDS.
- *Bidding strategy:* The bidding strategy or equilibrium strategy of the node is to determine the value of risk taken r^t for its revocation decision. Then, based on (4.23), in which r^t is a linear function of the probability P^c, the node can determine the value of P^c of the IDS on which the node makes its revocation decision. In the traditional Dutch auction, r^t is equivalent to the price b at which bidder i should submit its bid.
- *Risk appetite:* Risk appetite, also known as risk tolerance, is the risk that the node accepts/tolerates the decision to revoke the malicious node. Let r_i^a denote the

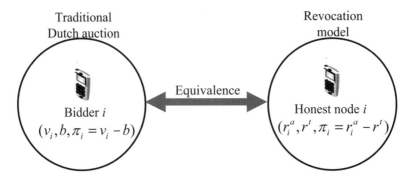

Figure 4.5 Equivalence of traditional Dutch auction and revocation model.

risk appetite of node i. In the traditional Dutch auction, the risk appetite r_i^a is equivalent to the value v_i of the item to the bidder.
- *Payoff*: The payoff of node i is defined as the difference between the risk appetite r_i^a and the risk taken r^t: $\pi_i = r_i^a - r^t$.
- *Risk appetite distribution:* In the traditional Dutch auction, we have the distribution function of value, $F(v_i)$, of bidders. Correspondingly, in the revocation model, we have the distribution function of risk appetite. Consider the symmetric model, in which the nodes have the same distribution function, denoted by $F(r_i^a)$. $F(r_i^a)$ refers to the probability that the risk taken is less than or equal to the node's risk appetite, $r^t \leq r_i^a$. Assume that P^c is defined as $P^c(k) = k/100$, where k is the number of observed rounds of the IDS, which is uniformly distributed on $[50, 100] \subset \mathbb{N}$. Then, as proved in [208], $F(r_i^a)$ is determined as $F(r_i^a) = r_i^a$. As seen, the formulation of $F(r_i^a)$ is similar to that of $F(v) = v$ in the traditional Dutch auction.

To help visualize these concepts, Figure 4.5 shows the equivalence of the traditional Dutch auction and the revocation model. In particular, when honest node i with risk appetite r_i^a makes its revocation decision at r^t, honest node i receives a payoff of $\pi_i = r_i^a - r^t$. The following discusses the details involved in determining equilibrium strategy r^t as well as its physical meaning.

Equilibrium Strategy

Based on distribution function $F(r_i^a) = r_i^a$, we can determine the equilibrium strategies for the honest nodes by directly using (4.22), in which $b = r^t$ and $v = r_i^a$. Here, we add the index i to refer to a particular node i. The equilibrium strategy of the node is given by [208]

$$r^t(r_i^a) = r_i^a - \frac{r_i^a}{N} \qquad (4.24)$$

Equation (4.24) illustrates that the equilibrium strategy of node i is an increasing function of its risk appetite r_i^a. This means that $r^t(r_i^a)$ increases as r_i^a increases. Recall that $r^t(r_i^a)$ is the risk taken of node i. The large value of $r^t(r_i^a)$ means that the number of observed rounds that the node uses as a basis to make its revocation decision is small. In other words, the node revokes the malicious node early and has a high probability of winning the auction. For example, consider a scenario with two honest nodes, $N = 2$, in which the risk appetite values of nodes 1 and 2 are $r_1^a = 0.8$ and $r_2^a = 0.4$, respectively. Using (4.24), the equilibrium strategies of nodes 1 and 2 are $r^t(r_1^a) = 0.4$ and $r^t(r_2^a) = 0.2$, respectively. The equilibrium strategy of node 1 is higher than that of node 2, so node 1 is the first one to revoke the malicious node. Consequently, node 1 is the winner of the auction and receives a payoff of $0.8 - 0.4 = 0.4$. It is also seen from (4.24) that the equilibrium strategy of the node increases as the number of honest nodes increases. This is reasonable: given a higher number of competitors, the node should take a higher risk to make its revocation decision earlier.

The simulation results of the Dutch auction-based revocation scheme are given in [208]. The scenarios with $N = 2, 5$, and 10 honest nodes are considered. As shown, given $N = 2, 5$, and 10 honest nodes, the number of rounds required to revoke the malicious node with a probability of 100% is 100, 83, and 70, respectively. This means that as the number of honest nodes (i.e., competitors) increases, the malicious node is revoked earlier. This shows that the Dutch auction-based revocation scheme is a suitable solution for future-generation wireless networks in which there will be a high density of honest nodes to quickly revoke malicious nodes.

4.4.2 Relay Selection in the Internet of Things

An important requirement of IoT is to guarantee the freshness of the sensing data. Thus, approaches for fast routing and relay selection for the data communication in IoT need to be investigated. The Dutch auction can be used for the selection decision since it is able to quickly decide the winner.

This section discusses the application of the Dutch auction for the relay selection in single-hop wireless sensor networks (WSNs) in IoT [209]. WSNs are the main component of IoT in which sensors collect data from an environment and transmit the sensing data to the sink for further processing. Due to the coverage limit, the sensor, known as the source node or source for short, may ask other sensors, known as relay nodes, to forward the source's data to its sink, representing its destination. To meet QoS requirements such as freshness of the sensing data, the source needs to choose optimal routes, such as the shortest route and the least-energy-consumption route, for its data forwarding. However, the relay nodes on the different routes are typically rational and selfish, so pricing strategies can be employed as incentive mechanisms to persuade the relay nodes to forward the sensing data.

The model shown in Figure 4.6 includes multiple relay nodes, one source, and its sink (the destination). The source needs to choose the best relay node, comprising the relay node closest to the sink, for the source's data forwarding. To incentivize the relay nodes, the source pays the selected relay node for its relay service. Since the model has one buyer (the source) and multiple sellers (the relay nodes), the auction is considered a *Dutch reverse auction* [209]. The relay node selection process using the Dutch reverse auction is implemented as follows.

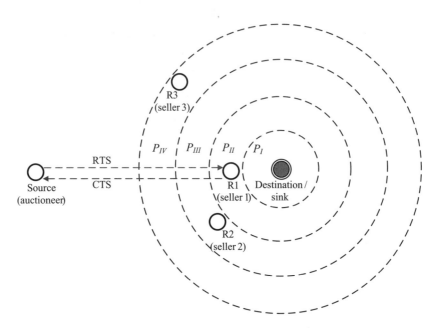

Figure 4.6 Dutch auction–based relay selection in IoT [209], where R stands for relay node, and RTS and CTS stand for request-to-send and clear-to-send frames, respectively.

First, the source divides the network into *forwarding regions* with different priorities. The priority of the forwarding region is defined according to the advance increment by which the relay nodes in the region forward the source's data to the sink. This means that those regions closer to the sink have the higher priorities. As illustrated in Figure 4.6, the source divides the network into four forwarding regions with four different priorities. The forwarding regions are distinguished from each other by circles. Forwarding region I is closest to the sink and has the highest priority, followed by forwarding regions II, III, and IV. The relay nodes in region I have the highest probability of winning the auction. Similarly, the relay nodes in forwarding region II have a higher probability of winning than the relay nodes in forwarding regions III and IV. Let $p_I > p_{II} > p_{III} > p_{IV}$ denote the priorities of forwarding regions I, II, III, and IV, respectively. Assume that there are three relay nodes, denoted by 1, 2, and 3, which are located in the forwarding regions as shown in Figure 4.6.

After dividing the network into the forwarding regions, the source initializes the request-to-send/clear-to-send handshake. The source sends the information including the forwarding regions and the corresponding priorities through the request-to-send frame to the relay nodes. Based on its location information, each relay node locally determines its forwarding region and priority. For example, in Figure 4.6, relay node 1 in forwarding region II determines its priority as p_{II}, and relay nodes 2 and 3 determine their priorities as P_{III} and P_{IV}, respectively. The relay node will claim that it can forward the data if its priority is greater than or equal to the priority requested by the source. Recall that the traditional Dutch auction, the priority of the relay node is the price at which the bidder submits its bid to buy the item. The relay node selection process can be implemented in multiple iterations as follows.

At iteration 1, the source sends p_I through the request-to-send frame, which has the highest priority for all relay nodes. The objective is to discover the relay node that is closest to the sink. Each relay node compares its priority with the priority requested by the source. Since there is no relay node in forwarding region I, then no relay node holds p_I. If the source does not receive any reply from the relay nodes in a certain time, the source implements iteration 2.

At iteration 2, the source sends p_{II}, which has the second highest priority for all relay nodes. Similarly, the relay nodes compare their priorities with the new priority requested by the source. Since relay node 1 holds priority p_{II}, the relay node replies to the source through the clear-to-send frame. Relay node 1 is the first one that replies to the source. Therefore, relay node 1 is selected as the winner for the data forwarding, and the auction terminates.

In fact, there may be more than one potential relay node in the same forwarding region, and thus these relay nodes have the same priority. In this case, a collision will occur since at the same iteration, two clear-to-send frames are sent to the source. The source can solve this problem by redividing the forwarding region into small forwarding regions with different priorities. This information is then sent back to the relay nodes such that they update the new priorities.

The simulation results obtained in [209] show that the relay selection based on the Dutch auction can achieve a stable packet delivery success ratio up to 90% while this ratio of the scheduled relay selection is around 70%. Here, the scheduled relay selection chooses the relay that is the closest to the source.

4.4.3 Channel Allocation in 5G Heterogeneous Networks

5G HetNets are expected to deploy a massive number of small cell base stations (SBSs), which introduces challenges for radio resource management. A major requirement of 5G HetNets is to improve the spectrum efficiency. To do so, one solution is to allow the SBSs to reuse the same channels when the SBSs are far enough from each other. Typically, a macrocell base station (MBS) performs the channel allocation in a central manner since it has full knowledge of locations of the SBSs. However, the future-generation wireless networks will become more decentralized and ad hoc in nature so that network infrastructure such as MBSs may not exist. In such scenarios, the SBSs need to make allocation decisions locally. The Dutch auction can be an appropriate solution for the channel allocation. The first reason is that the auction allows the SBSs to make local decisions without a central authority. The second reason is that the auction enables the SBSs to reveal information, in the form of the requested channels and allocated channels, based on which the SBSs can decide to reuse the same channels. The following discussion focuses on the use of the Dutch auction for channel allocation in 5G HetNets as proposed in [210].

System Model

The model includes N transmitting SBSs, or transmitters, and N receiving SBSs, or receivers. They are distributed as shown in Figure 4.7. The transmitters as bidders compete for a single radio channel, the item, to connect to their desired receivers. Each

Open-Cry Auction

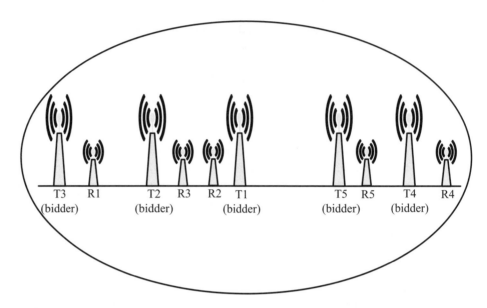

Figure 4.7 Dutch auction for channel allocation in 5G HetNets [210], where T and R stand for transmitter and receiver, respectively.

transmitter has its own value v_i of the radio channel and has its bidding strategy b_i as given in (4.22). Recall that bidding strategy b_i is the price at which the transmitter submits its bid, signified by bit "1," to indicate that it is willing to use the channel. Note that b_i may not be the same as v_i. There is no network infrastructure, and thus the information related to the auction, such as the starting time, initial price, price decrement step, and clock, needs to be synchronized among all the transmitters and the receivers. The synchronization task is generally challenging, but it can be achieved and maintained through, for example, global positioning systems. Each terminal, comprising the transmitter or receiver, has its own index.

The main idea behind channel allocation using the Dutch auction is as follows [210]. Whenever the price of the channel is less than or equal to the bidding strategy of the transmitter, this transmitter bids the channel by broadcasting bit "1" and the index of its target receiver. If the receiver is available and the channel is available, the receiver replies to the transmitter with a *confirmation message*. The transmitter broadcasts this confirmation message, which implies that it wins the channel. The process of the scheme is further explained through a specific scenario as follows.

Specific Scenario

Figure 4.7 illustrates a specific scenario including five transmitters, denoted by T1, T2, T3, T4, and T5, and five receivers, denoted by R1, R2, R3, R4, and R5. Table 4.2 shows the indices of the transmitters, the indices of the target receivers, and the bidding strategies of the transmitters. For example, in the second row of the table, the index of T1 is 1, that of its target receiver is 1, and its bidding strategy is $1.0. The bidding

Table 4.2 Index of transmitters and receivers, and bidding strategies [210].

Transmitter (bidder)	Receiver	Bidding strategy
1	1	$1.0
2	2	$0.7
3	3	$0.6
4	4	$0.5
5	5	$0.4

strategy is $1.0, meaning that T1 submits its bid, by sending bit "1", when the price of the channel is less than or equal to $1.0. The following assumptions are made, some of which may be different from the traditional Dutch auction:

- The transmitters have different values of the channel and thus have different bidding strategies. For example, in Table 4.2, the bidding strategies of T1 and T2 are $1.0 and $0.7, respectively. This means that T1 submits its bid when the price of the channel is less than or equal to $1.0, and T2 submits its bid when the price of the channel is less than or equal to $0.7.
- The auction starts at t_0 and stops at $t_0 + \tau$, where τ is the auction duration.
- The initial price of the channel is $1.1.
- In auction duration τ, the price of the channel decreases with a price decrement step of $0.01 at every "clock."
- The duration between two consecutive clocks is much smaller than the auction duration τ, such that multiple transmitters may become the winners in the Dutch auction as long as the channel is available and its price is less than the values of the transmitters. This aims to improve the efficiency of the channel allocation.

The whole process of the Dutch auction is shown in Figure 4.8. Note that the figure shows the process from clock = 1 at which the terminals start with their activities. The detail is explained next [210].

At clock = 0, which equates to t_0, all the transmitters and the receivers know that the Dutch auction starts. However, the initial price of $1.1 is too high for every transmitter, so no transmitter submits a bid. At clock = 1, the price reaches $1.0, and transmitter T1 immediately broadcasts the information including bit "1" and the index of its target receiver "1". This information implies that T1 is willing to access the channel to connect to receiver R1. However, as shown in Figure 4.7, receiver R1 is out of coverage of T1, so R1 does not respond. T1 drops out of the auction, and there is no winner at this time.

At clock = 2 and 3, the price drops to $0.9 and $0.8, respectively, which are too high for the remaining transmitters. At clock = 4, the price reaches $0.7, and transmitter T2 broadcasts the information including bit "1" and the index of its target receiver "2". This information is heard by T1, and by receivers R1, R2, and R3 since they are within the coverage of T2. Receiver R2 replies to T2 with a confirmation message, and then T2

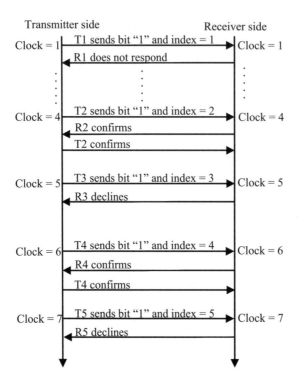

Figure 4.8 Process of Dutch auction for channel allocation in 5G HetNets, where T and R stand for transmitter and receiver, respectively.

broadcasts this message to announce that it wins and occupies the channel. Note that the confirmation message from T2 is heard by transmitter T1 and receivers R1, R2, and R3.

Since T3 has not heard the messages, it "thinks" that there is no winner and continues the auction. At clock = 5, the price of the channel reaches $0.6, so T3 broadcasts the information including bit "1" and the index of its target receiver "3", R3. However, R3 knows that the pair T2–R2 currently occupies the channel, and therefore the channel is not available, since it heard the confirmation message of T2. R3 declines the request of T3, and T3 drops out of the auction.

The process continues with T4. Similar to T3, T4 has not heard any previous messages, and it also "thinks" that there is no winner. At clock = 6, when the price reaches $0.5, T4 broadcasts its information including bit "1" and the index of its target receiver "4," R4. Since R4 is out of coverage range of T2, R4 "thinks" that the channel is available and replies to T4 with a confirmation message. T4 broadcasts this message to indicate that it wins the channel. Note that this message is heard by R5, and thus this receiver will decline T5's request for the channel to connect to R5 at clock = 7.

Assume that the auction finishes at clock = 7. T2 and T4 are the winners of the Dutch auction. They simultaneously win the channel in the auction since both (i) have the highest values for the channel and (ii) are located far enough away to not cause interference with each other. This helps improve the resource efficiency. Since T2 and T4 occupy the channel for a certain period, the whole process of the Dutch auction is repeated such that other transmitters can use the channel when it becomes available again.

4.5 English–Dutch Auction

The English–Dutch auction can be considered to be a subtype of the Dutch auction that essentially employs the ascending-bid auction, or English auction, to set an initial price for the Dutch auction [211]. Indeed, as presented in Example 4.3, the service provider sets the initial price, $1.6, of the bandwidth unit, and then decreases the price until user 1 submits its bid. The service provider can establish the initial price, perhaps based on the preceding day's prices of the bandwidth unit. However, in some cases, such as the first time that the service provider trades the bandwidth unit, the initial price can be set much lower than the highest value of the users. Thus, the service provider receives less revenue for trading the bandwidth unit. For example, if the initial price is set as $1.2 and only user 1 submits its bid to show its willingness to buy the bandwidth unit, then the service provider receives a profit of $1.2 rather than $1.4.

Such an example shows that setting the initial price is very important; ideally, this price should exceed the highest value of the users in the auction such that the service provider receives the highest possible profit. Otherwise, as discussed in Section 4.2, the values of the users can be deduced through the English auction. The English auction can thus be used for setting the initial price of the Dutch auction. This combination is known as an English–Dutch auction. The English–Dutch auction consists of two phases. Phase I is implemented by an English auction, in which the service provider starts the auction with a low price and then increases it until no user is interested in buying the bandwidth unit. The final price of the first phase is then set as the opening bid of the second phase, which is performed as a Dutch auction. Note that the user with the highest value will not win the bandwidth unit unless the user keeps the highest value in the second phase, the Dutch auction. We provide the following example to explain the process of the English–Dutch auction.

Example 4.4 Consider again the cloud networking model in Example 4.1, which consists of one service provider as a seller, one bandwidth unit as an item, and three mobile users as bidders 1, 2, and 3. We assume that the private values of the bandwidth unit to users 1, 2, and 3 are $v_1 = \$1.5$, $v_2 = \$1.2$, and $v_3 = \$1.0$, respectively. Also, the users are interested in buying the bandwidth unit when the price of the bandwidth unit is less than their values. The service provider implements an English–Dutch auction in two phases as shown in Figure 4.9. In phase I, an English auction is used in which the service provider sets the initial price as $p^{(I,0)} = \$0.9$ for the bandwidth unit. At iteration 1, all three users want to buy the bandwidth unit. Suppose that the price increment is set to $0.2, and then the service provider increases the initial price to $0.9+\$0.2 = \$1.1$. At iteration 2, since the price $1.1 is greater than $v_3 = \$1.0$ of user 3, this user is not willing to buy the bandwidth unit. User 3 can show its non-willingness-to-buy by sending bit "0" to the service provider. This is different from the traditional English auction as discussed in Section 4.2, in which the user shows its non-willingness-to-buy by dropping out/leaving the auction. Now, there are users 1 and 2, both of which are willing to buy the bandwidth unit. The service provider increases the price to $1.1 + \$0.2 = \$1.3$. At iteration 3, user 2 is not willing to buy the bandwidth unit, leaving only user 1 in the

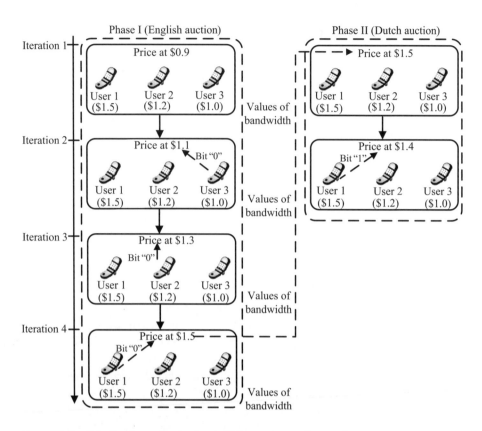

Figure 4.9 English–Dutch auction for bandwidth resource trading market.

auction. The service provider increases the price to $1.3 + $0.2 = $1.5, and user 1 is not willing to buy the bandwidth unit at iteration 4. The service provider now infers that $1.5 is the price that may exceed the values of all users. The service provider terminates phase I and triggers phase II.

Phase II is the Dutch auction with an initial price that is the final price of phase I, $p^{(II,0)} = \$1.5$. Assume that in phase II, the service provider sets a price decrement of $0.1, and the users will not bid until the price of the bandwidth unit is less than their values. At iteration 1, the price $1.5 is greater than the values of all three users, and hence the users do not bid. Thus, the service provider reduces the price to $1.5 − $0.1 = $1.4. This price is greater than the values of users 2 and 3 but less than that of user 1. Thus, at iteration 2, only user 1 bids for the bandwidth unit by sending bit "1" to the service provider. User 1 wins the bandwidth unit, and the price that it pays the service provider is $1.4. User 1 receives a positive surplus as $1.5 − $1.4 = $0.1.

In general, by using the English auction in phase I for establishing the initial price of the Dutch auction, the English–Dutch auction enables the service provider to receive the highest possible profit. However, this may not be guaranteed if in the users are untruthful in phase I; that is, users send their non-willingness-to-buy messages even if the price of the bandwidth unit is less than their values. The users have an incentive to do this since they may gain the high surplus. For example, in phase I in Example 4.4, user 1 might signal its non-willingness-to-buy at iteration 3 instead of iteration 4. Thus, the service provider sets the final price in phase I as $1.3 instead of $1.5. Consequently, user 1 receives the high surplus, $1.5 − $1.2 = $0.3, but the profit of the service provider decreases. Another shortcoming compared to the traditional English or Dutch auction is that the English–Dutch auction is more complex and requires more time to finish.

Despite the aforementioned disadvantages, there are some research directions that can be investigated. One research direction is to apply the English–Dutch auction for sensing data trading in crowdsensing networks in IoT. Indeed, consider a crowdsensing network including one mobile user and multiple service providers. The mobile user as a seller trades its sensing data to the service providers as buyers. When the mobile user first joins the sensing trading market, it is often not aware of the market price or the value of the sensing data. In this case, the mobile user can conduct an English–Dutch auction for the sensing data trading to exploit the value of the sensing data and to achieve the highest revenue.

4.6 Summary

In this chapter, we introduce two types of open-cry auction, the English and Dutch auctions, and discuss their applications in computer networks. Specifically, we first describe the theory of the English auction. A specific example in the context of computer networks is provided to show how to determine the equilibrium strategies in the English auction.

Second, we discuss the application of the English auction for spectrum leasing for cognitive radio in 5G wireless networks. In particular, we present the system model, problem formulation, and Walrasian equilibrium, as well as discuss the use of the English auction to reach the Walrasian equilibrium.

Third, we provide the definition and process of the Dutch auction in the context of computer networks. In particular, we introduce the revenue equivalence theorem, which is used to determine the Nash equilibrium in the Dutch auction. Unlike in the English auction, the Nash equilibrium in the Dutch auction is the Bayesian–Nash equilibrium since a particular bidder determines its equilibrium strategy by knowing the distribution of values of other bidders rather than knowing their actual values. This is a common situation in computer networks where there is no centralized controller to maintain information on all users. This is one of the reasons that the Dutch auction is more widely used in computer networks compared with the English auction.

Fourth, we discuss the applications of the Dutch auction for prevention of black hole attacks in MANETs, relay selection in IoT, and channel allocation in 5G HetNets. Finally, we introduce the combination of the English and Dutch auctions as a solution for some situations in which the auctioneer has no information about the market price of its item.

5 First-Price Sealed-Bid Auction

This chapter describes the first-price sealed-bid auction, which is a common type of sealed-bid auction. We first introduce the definition of the first-price sealed-bid auction and give a specific example of this auction in computer networks. Through the example, we provide the strategic analysis of bidders in the first-price sealed-bid auction. Then, we show that the revenue equivalence theorem can be used to determine the Bayesian–Nash equilibrium in the auction. After that, we introduce the first-price sealed-bid reverse auction, which is a variant of the first-price sealed-bid auction. Finally, we discuss related works that help to explain how the first-price sealed-bid auction and its variant are used to address emerging issues in computer networks – namely, data aggregation, task allocation, relay selection, and denial-of-service attack prevention.

5.1 Definition

The first-price sealed-bid auction is a commonly used type of auction. Unlike an opencry auction (i.e., the Dutch and English auctions), in which bids of all bidders are disclosed to each other during the auction, the first-price sealed-bid auction is a blind auction where bidders simultaneously submit sealed bids. As such, in the first-price sealed-bid auction, each bidder does not know the other buyers' bids (i.e., bidding strategies) and thus cannot change its bid. Also, the first-price sealed-bid auction is a single-round auction. The bidder with the highest bid is the winner of the auction, and the winner pays the seller the price that the winner submits. This pricing rule is also called *pay-what-you-bid*. In general, the first-price sealed-bid auction is easy to understand and implement. However, under its pricing rule, bidders receive a zero payoff when they submit bids equal to their values of the item or network resource. This can either discourage the bidders from participating in the auction or motivate them to decrease their bids. This can be further explained by the following example.

Example 5.1 Consider the cloud networking model as presented in Example 4.1, in which three mobile users as bidders compete for a single bandwidth unit owned by the service provider. The private values of the bandwidth unit to users 1, 2, and 3 are $v_1 = \$1.5$, $v_2 = \$1.2$, and $v_3 = \$1.0$, respectively. Assume that the users would truthfully submit their bids, $b_i = v_i, i = 1, \ldots, 3$ to the service provider. Then, the

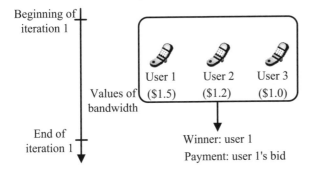

Figure 5.1 First-price sealed-bid auction for bandwidth resource trading market.

auction has a single round, as shown in Figure 5.1. In this round, the service provider receives bids from the users and determines the winner as well as the price that the winner pays. Since user 1 has the highest bid, user 1 is the winner of the bandwidth unit. The price that user 1 pays the service provider is $p = b_1 = \$1.5$, and the payoff that user 1 receives is $\pi_1 = p - b_1 = \$0$. As such, when the user submits a bid equal to its value, the user receives a zero payoff even if the user is the winner. Therefore, in practice, the users tend to bid below their values, $b_i < v_i$. However, the bid should not be too low since the user may not win the auction. As such, the user faces a trade-off. An increase in the bid will increase the winning probability of the user, but at the same time reduce its payoff gain from winning. Therefore, the key issue in the first-price sealed-bid auction is how the user determines its optimal bid. This issue can be solved by finding the equilibrium of the auction, as discussed in the following section.

5.2 Equilibrium

In this section, we discuss how to determine the equilibrium in the first-price sealed-bid auction. For convenience, we present the strategic analysis of users in the auction first.

5.2.1 Strategic Analysis

As presented in Example 5.1, when the user submits a bid equal to its value of the bandwidth unit, the user receives a zero payoff even if it wins the auction. Therefore, the user would like to bid less than its value so that it may receive a positive payoff. Specifically, users 1, 2, and 3 actually submit bids $b_1 < v_1$, $b_2 < v_2$, and $b_3 < v_3$, respectively. However, the exact payoff gain of the user depends on the bids of the other users. In particular, user 1 needs to bid greater than both bids b_2 and b_3. Assume that $b_2 > b_3$, and user 1 can observe these bids. Then, user 1 needs to submit a bid that is greater than b_2. b_1 can be $b_1 = b_2 + \epsilon$, where ϵ is any positive price, such as $\$0.2$, that can be added such that user 1 wins.

In fact, it is challenging for a particular to determine its optimal bid, or equilibrium strategy, since the bids of other users are not disclosed. The user cannot also infer the values of other users as in the English auction since no dropping-out decision is revealed. However, similar to the Dutch auction, the user can determine its equilibrium strategy assuming that the value distributions of all users are publicly known. Also, the auction is considered to be a Bayesian game in which players, the users, have incomplete information, such as the strategies or payoffs, for the other players, but they have the beliefs with a known probability distribution. The Nash equilibrium in the first-price sealed-bid auction is thus a Bayesian–Nash equilibrium.

5.2.2 Bayesian–Nash Equilibrium

In this section, we discuss how to find the Bayesian–Nash equilibrium or equilibrium strategies, representing the optimal bids, of users in the first-price sealed-bid auction. In particular, we show that the equilibrium strategies can be determined using the revenue equivalence theorem as presented in Section 4.3.2. Thus, we can directly use (4.20) of the revenue equivalence theorem as the equilibrium strategies in the first-price sealed-bid auction. Recall that (4.20) refers to the equilibrium strategy of the user with value v as given by

$$b_i(v) = v - \frac{\int_r^v F^{N-1}(t)dt}{F^{N-1}(v)} \tag{5.1}$$

where N is the number of users, $F(v)$ is the distribution function of v, and r is the minimum value of v.

We next consider the bandwidth resource trading market in Section 4.3.2. This market has N mobile users, the bidders, which compete for a single bandwidth unit from the service provider, the seller. Each user has its own value v_i of the bandwidth unit. Similar to the Dutch auction, for simplification, we consider the symmetric auction model in which the values v_i of the users are identically and independently distributed according to some known cumulative distribution function $F(\cdot)$. Here, the term "independently" implies that the information about a particular user's value is independent of that for the other users' values.

User i submits a sealed bid b_i to the service provider. b_i is the price that user i pays the service provider if the user is the winner. As stated in Section 5.2.1, the user tends to bid b_i less than its value v_i such that it can receive a positive payoff. Thus, the payoff of the user can be expressed by

$$\pi_i = \begin{cases} v_i - b_i, & \text{if } b_i > \max_{j \neq i} b_j \\ 0, & \text{if } b_i < \max_{j \neq i} b_j \end{cases} \tag{5.2}$$

The expression in (5.2) means that user i wins the auction if its bid is greater than the maximum bid of all users except user i. Also, the user receives a zero payoff; that is, the user loses in the auction if its bid is less than the maximum of the bids submitted by the other users. We can ignore the case $b_i = b_j$ since it occurs with a low probability. In general, the bid is a monotonically increasing function of the user's value, meaning that the user with the highest value will produce the highest bid and win the auction. Thus, we can say that user i wins the auction if $v_i > \max_{j \neq i} v_j$. Let $Q_i(v_i)$ denote the winning probability of user i with value v_i. Then, when user i submits bid b_i, the user expects that $N-1$ other users will have values less than its own value. Since the values are identically and independently from the same distribution $F(\cdot)$, the winning probability of user i is $Q_i(v_i) = F^{N-1}(v_i)$. Consequently, the expected payoff of user i is $Q_i(v_i)(v_i - b_i)$. In general, if we replace $v_i = v$, the expected payoff of the user with value v is expressed by

$$\pi_i = F^{N-1}(v)(v - b_i) \tag{5.3}$$

We can observe that the expression in (5.3) is exactly the same as (4.19), which presents the expected payoff of a user with value v. Therefore, we can directly use the equilibrium strategies as given in (4.20) of the revenue equivalence theorem in Chapter 4 for determining the equilibrium strategies of the users in the first-price sealed-bid auction. Specifically, we have the following proposition.

PROPOSITION 5.1 *Bidding strategies given by*

$$b_i(v) = v - \frac{\int_r^v F^{N-1}(t)dt}{F^{N-1}(v)} \tag{5.4}$$

where r is the minimum value of the bandwidth unit, are equilibrium strategies of the users in the first-price sealed-bid auction.

It can be seen that the equilibrium strategies in the first-price sealed-bid auction are similar to those in the Dutch auction (see (4.21)). Thus, we can say that the first-price sealed-bid auction and the Dutch auction are strategically equivalent. In particular, we assume that the distribution function of the value of the users is $F(v) = v$ and $r = 0$, and we also have the equilibrium strategies of the users:

$$b_i(v) = \frac{N-1}{N}v = v - \frac{v}{N} \tag{5.5}$$

Similar to the Dutch auction, (5.5) shows that the equilibrium strategies of the users are increasing functions of their values. Moreover, each user bids lower than its value, $b_i < v$, by an amount $\frac{v}{N}$. For example, in the case that $N = 3$, the user with value v bids $b = \frac{2v}{3}$, which is its equilibrium strategy. As such, the user bids lower than its value by an amount $\frac{v}{3}$. The amount of $\frac{v}{3}$, or $\frac{v}{N}$ for the case of N users, is called *bid shading*. Bid shading is used to compensate for *winner's curse*. Winner's curse is a situation in which the user discovers that the value of the resource is less than its bid when the user wins. This situation often occurs in sealed-bid auctions, in which bids and values of the users are not disclosed, and the users may overestimate the value of the resource. To avoid the winner's curse issue, the user has to reduce its bid with bid shading. However, the amount of bid shading depends on the number of users in the auction. When the number of users increases, such that N is large, the competition among the users is high. Thus, the user reduces its bid by only a small bid shading so that the user has a high chance of winning the auction. In this case, winner's curse may more readily occur, and we say that the severity of the winner's curse increases.

5.3 First-Price Sealed-Bid Reverse Auction

The first-price sealed-bid auction described in Section 5.1 can be considered to be a forward auction. In the forward auction, buyers (i.e., the users) compete for the resource from the seller (i.e., the service provider) by submitting their bids to the seller. The service provider thus receives the highest price among the users. In practice, the resource trading market may have multiple service providers, and the service providers compete with each other on the resource price to attract users/buyers. As a result, the price of the resource typically decreases to the lowest price that the users can accept. Such an auction is called a *reverse auction*. When the reverse auction is combined with the first-price sealed-bid auction, we have the first-price sealed-bid reverse auction. With this model, compared with the first-price sealed-bid auction, the roles of the buyers and the sellers in the first-price sealed-bid reverse auction are reversed.

As an example, we consider again the resource trading market in Section 5.2.2, but the model is assumed to have multiple service providers/sellers, and one user/buyer. The service providers compete to attract the user's demand by submitting *asks* or *asking prices* to the user. The ask, denoted by a_i, is the price that the service provider is willing to receive from the user. Upon receiving the asks, the user selects the service provider with the lowest ask as the winner:

$$\hat{i} = \arg\min_i a_i \qquad (5.6)$$

The winning service provider assigns the bandwidth to the user and receives the price from the user. The price that the winning service provider receives is determined as follows:

$$p = a_{\hat{i}} \qquad (5.7)$$

Equation (5.7) means that the price that the service provider receives is equal to the ask that it submits. Thus, the service provider naturally submits an ask greater than its true value to receive a positive payoff. The true value is the minimum price at which the service provider is willing to sell its bandwidth. Therefore, submitting truthful asks is not dominant strategy among the service providers, and we can say that the first-price sealed-bid reverse auction is not incentive-compatible.

Despite a non-truthful auction, the first-price sealed-bid reverse auction has been widely used in computer networks since it allows participants to readily discover the lowest ask. Specifically, in computer networks, the asks can be differently defined to achieve different objectives. For example, to minimize the payment or incentive cost of data aggregation in IoT, the ask is defined as the price of the sensing task [95], [212]. Also, to maximize the energy balancing among sensors and hence improve the overall network lifetime of IoT, the ask is defined as a function of the residual energy of sensors in the networks [213]. To minimize the latency of forwarding sensing data from sensors to sink nodes, the ask is defined as a function of the hop count of the relays to the sink nodes [214]. Such approaches are further presented in the following sections.

5.4 Development of First-Price Sealed-Bid Auction for Computer Networks

In this section, we present applications of the first-price sealed-bid auction in computer networks. More specifically, we discuss how to apply the auction to address emerging issues including data aggregation, task allocation, relay selection, and denial-of-service attack prevention in IoT.

5.4.1 Incentive Mechanism for Data Aggregation

IoT has employed emerging sensing paradigms such as participatory sensing and crowdsensing networks to gather data from portable smart devices. A common data aggregation model in IoT is shown in Figure 5.2. In this model, the participants, which are users, use mobile phones to provide their sensing data to the service providers via their servers or platforms. Since collecting and transmitting sensing data require the users to consume resources, including energy and bandwidth, a challenge in data

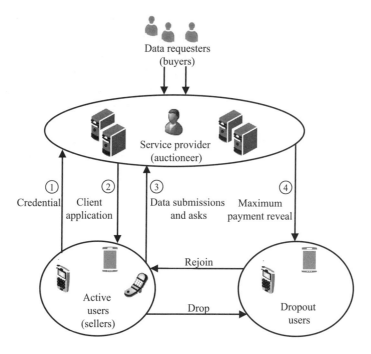

Figure 5.2 Data aggregation in IoT using the first-price sealed-bid reverse auction [212].

aggregation is to provide an appropriate incentive to the users. Minimizing the incentive cost, which comprises the price paid to the users to provide data, for the service provider and maintaining high data quality are also the challenges. The first-price sealed-bid reverse auction can be used to address these issues. This type of auction allows the service provider to simply select a single user with the lowest price, which reduces incentive cost, removes data redundancy, saves energy, and guarantees timeliness of the sensing results. The following discussion presents the use of the first-price sealed-bid reverse auction for data aggregation in participatory sensing applications as proposed in [212].

The model shown in Figure 5.2 includes multiple mobile users, one service provider, and data requesters. The data requesters, as buyers, require sensing data such as temperature and traffic speed. The users, as sellers, own mobile devices that can perform sensing tasks required by the data requesters. The users can communicate with the service provider by using mobile client applications. The service provider acts as the auctioneer, conducting the auction and facilitating the interactions between the data requesters and the users. In general, data aggregation using the auction is implemented as follows:

- The service provider first receives the data requests from the data requesters.
- The service provider then broadcasts a sensing task description to all users.
- The interested users, also called *active users*, perform the sensing tasks.
- After completing the sensing tasks, the users submit their asks a_i to the service provider. In particular, the asks are the prices that the users are willing to receive from the service provider.
- Upon receiving the asks, the service provider selects the user with the lowest ask as the winner for providing the sensing data as $\arg\min_i a_i$. The service provider then makes the payment to the winner. The payment paid to the winner is the ask that the winner submits to the service provider.
- Finally, the winner sends the sensing data to the data requesters through the service provider.

Note that the sensing data has a time-sensitive perishable property, so it may be unusable if it is not used at the current time. After collecting the sensing data for a certain time period, the service provider requires new sensing data for the next periods so as to provide real-time services continuously. Thus, the execution of the first-price sealed-bid reverse auction needs to be repeated in the next periods or next rounds. Naturally, the losers, meaning the users that lose the auction in the current round, have no incentive to participate in the next rounds, which reduces the number of participants. As a result, the price competition level is decreased, and the winners increase their asking prices to gain better payoffs, which may cause an incentive cost explosion. To address this problem, virtual participation credits are introduced. In particular, user i, which lost in the previous round $k-1$ of the auction and participates in the current round k of the auction, receives virtual participation credit $d_i^{(k)}$ as a reward for its last participation. $d_i^{(k)}$ is defined as follows:

5.4 Development of First-Price Sealed-Bid Auction for Computer Networks

$$d_i^{(k)} = \begin{cases} d_i^{(k-1)} + \alpha, & \text{if user } i \text{ lost in round } k-1 \\ 0, & \text{otherwise} \end{cases} \quad (5.8)$$

where α is the amount of virtual participation credit. Equation (5.8) means that when the user loses multiple rounds of the auction consecutively, the amount of α is added to the user's virtual participation credit. Also, the virtual participation credit is reset to zero whenever the user won or dropped out in the previous round of the auction. The user can use the virtual participation credit only for reducing its ask, and the low ask increases the winning probability of the user in the current round of the auction. An ask that has been reduced by the virtual participation credit is called a *competitive* ask. It differs from the actual ask claimed by the user. The relationship between the competitive ask and the actual ask is given by

$$a_i^{(k)'} = a_i^{(k)} - d_i^{(k)} \quad (5.9)$$

where $a_i^{(k)}$ is the actual ask claimed by the user, and $a_i^{(k)'}$ is its competitive ask. The service provider uses $a_i^{(k)'}$ to select the winners rather than $a_i^{(k)}$. With the reduced value of the competitive ask, the loser has more opportunities to win in the current round. Hence, the loser is encouraged to continue to participate in the sensing application.

In fact, even if the winners receive rewards from the service provider, they may still drop out of the participatory sensing application. This scenario often occurs when the received rewards do not meet the expectations of the winners. The dropping-out by participants also decreases the price competition, which in turn causes an explosion of the incentive cost. Therefore, recruiting the dropout users is as important as retaining the losers to ensure that a high competition level is still maintained and the selling price is low. To recruit the dropout users, the service provider also designs an incentive mechanism for the users. In the proposed scheme, this incentive mechanism is designed based on a *return on investment* of the users. The return on investment is one of the major metrics that the users use to decide whether they should drop out from the auction.

In general, the return on investment for the user is defined as a ratio of the profit to the cost of investment. Here, the profit is the total reward, or return, that the user has received, and the cost of investment is the total cost that the user has invested for performing sensing data until the current round of the auction. In particular, the cost of investment includes costs of battery power consumption, device resources, and privacy. As such, the return on investment of the user is given by

$$RoI_i^{(k)} = \frac{\tilde{d}_i^{(k)} + \beta_i}{C_i^{(k)} + \beta_i} \quad (5.10)$$

where $RoI_i^{(k)}$ denotes the return on investment of user i until round k of the auction, $\tilde{d}_i^{(k)}$ is the total reward that the user has received until round k, $C_i^{(k)}$ is the total cost that the user invests in its data sensing until round k, and β_i represents the sensitivity of the user i to its winning and losing in the auction. In particular, the user with a large β_i slowly responds to its volume of losses or winnings, and the user has a smaller value of its return on investment.

The user drops out in round k when the value of its return on investment $RoI_i^{(k)}$ goes below its satisfaction threshold (e.g., 0.5). To provide the dropout users with an incentive to participate to reveal next rounds, the service provider needs to increase the value of the return of investment. One solution is for the service provider to reveal the maximum price paid to the winner in the previous rounds to the dropout users. Like an invitation, this price information helps the dropout users to reevaluate their return on investment in the next round $k+1$:

$$RoI_i^{(k+1)} = \frac{\tilde{d}_i^{(k)} + p_{max}^{(k)} + \beta_i}{C_i^{(k+1)} + \beta_i} \quad (5.11)$$

where $p_{max}^{(k)}$ is the maximum price paid to the winner in the previous rounds. As such, the return on investment of the user is expected to increase in the next round. This provides the user with an incentive to rejoin in the next round of the auction.

In simulation results shown in [212], performance is evaluated in terms of incentive cost and incentive mechanism stabilization. The baseline scheme is the random selection–based fixed-pricing mechanism [215]. As shown in [212], to maintain 20 active users, the proposed scheme can reduce the incentive cost up to 63% compared with the baseline scheme. The reason is that the proposed scheme selects those users with low selling prices, whereas the baseline scheme randomly selects the winners and pays them the same price, the fixed payment. Compared with the baseline scheme, the proposed scheme is also able to better stabilize the low incentive cost over auction rounds. This is due to the fact that the proposed scheme introduces the virtual participation credit and dropout user recruiting mechanisms. Such mechanisms maintain the high number of participants, thereby preventing an incentive cost explosion. This is not guaranteed by the baseline scheme since some winners with a high return on investment may not be satisfied with the fixed payments and drop out. Since the number of participants is stabilized, the sensing service quality provided by the proposed scheme does not deteriorate significantly relative to that provided by the baseline scheme.

In [212], the service provider expects that revealing the maximum price $p_{max}^{(k)}$ can meet the reward expectations of the dropout users. In fact, the service provider does not know the reward expectations of the dropout users, and thus the price may be higher than the reward expectations. This may create a high incentive cost for the service provider. The study in [95] suggests that performance, as measured by incentive cost, can be improved if the satisfactory levels of the reward expectations for the dropout users can be quantified and incorporated in the auction. However, determining the expectations of the users in advance is a challenge to the service provider, as users may voice their satisfaction/dissatisfaction only after getting the reward. The exponential smooth method can be adopted to deal with this problem [216]. The exponential smooth method employs the most recent payment history to predict the next expected reward from the users. The simulation results in [216] show that the total allocated reward is close enough to the total required reward for those users with a small tolerance of ±0.05.

In summary, by using the first-price sealed-bid reverse auction and the virtual participant credit, the schemes proposed in [212] and [95] tackle the issues of incentive and cost explosion. The proposed schemes are thus more economically feasible for data aggregation. However, ways to determine the optimal virtual participant credits for the users are not discussed. Moreover, negotiating based only on the data price is not enough to meet the data quality requirements of the service provider as well as the data requesters. Additional attributions of the sensing data, such as the location accuracy and the sensing time, need to be considered for the ask evaluation.

5.4.2 Market-Based Adaptive Task Allocation

This section presents the application of the first-price sealed-bid reverse auction for sensing task allocation in wireless sensor networks (WSNs). WSNs are the main components of IoT that collect data from the environment and transmit the data to the sink nodes, or sinks for short. However, the sensors and wireless links in WSNs have both energy and capacity constraints. Therefore, removing the data redundancy and improving energy efficiency and network lifetime are crucial in WSNs. The first-price sealed-bid reverse auction allows for selecting a unique seller with the lowest ask as the winner. By defining appropriate asks, the auction can be used to reduce sensing data redundancy and maximize energy balancing among the sensors as well as the overall network lifetime of the WSNs. Such an approach is proposed in [213] and is described here.

The model, as shown in Figure 5.3, comprises one server and multiple sensors performing the sensing tasks. The sensors act as sellers, and the server is the buyer.

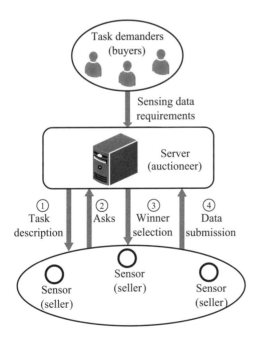

Figure 5.3 Sensing task allocation in IoT using the first-price sealed-bid reverse auction [16].

Accordingly, after receiving a data requirement from a customer, such as data monitoring of a specific area, the server broadcasts the task message, consisting of the task description, to all the sensors in the area. The task message includes the task, task size, and task deadline. The task size is quantified by the expected energy necessary to perform sensing and communicate the data sensing to the server, and the task deadline is the time required to complete the task. Upon receiving the task message, sensors that are interested in the task calculate their asks for completing the task. In general, the ask of the sensor is determined based on its current status as measured by available energy, the communication cost C_i^{com}, the task deadline t^D, and the resource release time t_i^R. In particular, the release time is the time that the sensor can start executing the sensing task, the time that the sensor is readily available to execute the sensing task. The release time is introduced so that the sensing task can be assigned to sensors with high energy requirements even if these sensors are executing other tasks. The proposed scheme aims to balance energy among the sensors in the network. Thus, the ask a_i of sensor is defined by

$$a_i = (C_i^{com} + a_i^B)(1 + f(t^D, t_i^R)) \qquad (5.12)$$

where C_i^{com}, a_i^B, and $f(t^D, t_i^R)$ are defined as follows:

- C_i^{com} is the bandwidth cost of submitting the sensing data to the server. In general, C_i^{com} is a function of (i) the size of the data packet and (ii) the number of hops between the sensor and the server.
- a_i^B denotes the base price, which is actually the energy cost for completing the task. The base price needs to be designed to guarantee that the sensor with higher remaining energy is selected. Since we are using a reverse auction, the base price should be inversely proportional to the remaining energy level of the sensors. The formulation of the base price is thus expressed by

$$a_i^B = \frac{E_{task}\alpha}{1 - e^{-E_i/\beta}} \qquad (5.13)$$

  where E_{task} is the task size, or the expected energy for completing the sensing task; E_i is the remaining energy of sensor i; and α and β are scaling parameters.
- $f(t^D, t_i^R)$ is a function of the task deadline and the resource release time. We expect that if the current time is close to the release time of a sensor with a high remaining energy level, then this sensor should be selected for the sensing task. Since this sensor can be requested by multiple sensing tasks, the task with the most urgent deadline should be assigned to the sensor. Since we are using a reverse auction, function $f(t^D, t_i^R)$ should be defined such that the sensor with a low value of $f(t^D, t_i^R)$ has a high possibility of being selected by the server. Therefore, $f(t^D, t_i^R)$ is defined as follows:

$$f(t^D, t_i^R) = e^{\frac{\epsilon(t^D - t)}{-(t_i^R - t)}} \qquad (5.14)$$

where ϵ is the scaling parameter and t denotes the current time. Note that we only consider $t < t^D$ and $t < t_i^R$, so that $t^D - t$ and $t_i^R - t$ are non-negative. The expression in (5.14) shows that the value of $f(t^D, t_i^R)$ is small when the task is urgent (i.e., $t^D - t$ is small) and the sensor can start the task early (i.e., $t_i^R - t$ is small).

After calculating their asks, the sensors simultaneously submit them to the server. Based on the asks, the server determines the winner of the auction as:

$$\hat{i} = \arg \min_i a_i \tag{5.15}$$

Thus, the server selects the sensor with the lowest ask as the winner for executing the sensing task. The winner selection process aims to achieve two objectives. The first objective is to achieve fair energy balancing among the sensors and hence maximize the lifetime of the network. The ask of the sensor is inversely proportional to its remaining energy E_i, as shown in (5.13). Thus, selecting the sensor with the low ask means that the sensor with high energy is selected for the task execution. The second objective is to reduce the latency when executing the sensing task and transmitting the sensing data to the server. The low ask indicates a low value of the function $f(t^D, t_i^R)$ and a small communication cost C_i^{com}. The low value of $f(t^D, t_i^R)$ means that the release time of the sensor is small, so the sensor will be available early to execute the sensing task, and the low cost of C_i^{com} implies that the distance between the sensor and the server is short. The winning sensor then performs sensing and sends the sensing data to the server.

The aforementioned scheme is a centralized approach in which all the sensors are required to submit their asks to the server. This may increase the communication overhead and the energy consumption cost. To address this disadvantage, a distributed approach can be used. In particular, after calculating the ask, each sensor locally determines its *waiting time*. The waiting time is proportional to its ask, $T_i^w = \gamma a_i$, where γ is a linear coefficient. Note that γ is commonly used for all sensors to guarantee the fairness among them. Then, instead of submitting the ask to the server immediately, the sensor switches to a listen mode. If the waiting time is completed and the sensor does not receive the winner information, the sensor submits its ask to the server. Thus, the sensor with the lowest ask will submit its ask first, and the sensor is selected as the winner for executing the sensing task. The server broadcasts the winner information to all the remaining sensors. The remaining sensors will leave the auction without sending their asks to the server. In this way, the communication overhead and the energy consumption for sending non-winning messages to the server are reduced. Since only one sensor, the winner, will reply, the advantage is still guaranteed even if the number of sensors in the network increases.

To evaluate the performance of the proposed centralized and distributed schemes, simulation results are provided in [213]. The baseline scheme comprises static task allocation [217], which is based on the energy balance–critical node path tree. From the results, after one auction round of task scheduling, the remaining energy for all sensors under the proposed scheme appears more balanced compared with that under the baseline scheme. The reason is that the proposed schemes continuously adapt to the changes in available energy of the sensors, a factor that is not considered in the baseline scheme. Moreover, the overall energy consumption of the proposed schemes is significantly lower than that of the baseline scheme since the price formulation takes the communication cost into account. In particular, the distributed scheme has lower energy consumption than does the centralized scheme due to its lower communication overhead. Especially since only one sensor replies to the server, the distributed scheme can maintain the low energy consumption even if the number of sensors increases. In future work, assigning multiple simultaneous tasks to multiple sensors needs to be considered for emerging applications such as target tracking. In this case, a combinatorial reverse auction can be used.

5.4.3 Market-Based Relay Selection

Due to their energy constraints, source sensors, which own the sensing data in IoT, may need to communicate with the server through other sensors known as relays. In general, the sensing data is sensitive to latency, as the utility assigned to the data will decrease over time due to the reduction of its freshness. Thus, it is important that the source sensor selects relays in the network for forwarding the sensing data with the lowest latency. In this regard, the first-price sealed-bid reverse auction allows for quickly determining the seller with the lowest price, and thus it can be used as an efficient solution to meet this requirement. The following discussion presents the application of the first-price sealed-bid reverse auction for relay selection in WSNs for IoT applications as proposed in [214].

The model illustrated in Figure 5.4 consists of one source sensor, its destination (i.e., the sink), and multiple intermediate sensors. The relay selection process is implemented in multiple stages. In every stage, the first-price sealed-bid reverse auction is used with

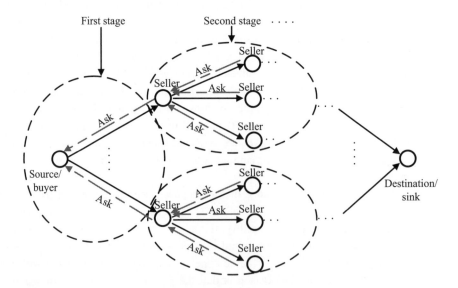

Figure 5.4 Relay selection in IoT using the first-price sealed-bid auction [214].

5.4 Development of First-Price Sealed-Bid Auction for Computer Networks

Table 5.1 Information of neighboring sensors of the source sensor [214].

Neighboring sensor(i)	Count hop(N_i^h)	Residual energy(E_i)
1	1	2.5 joule
2	2	1.5 joule
3	2	1.3 joule

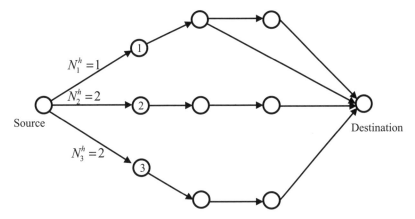

Figure 5.5 Network topology [214].

one sensor as the buyer (i.e., the auctioneer) and its neighboring sensors as sellers. The buyer selects one of the sellers as a relay for forwarding sensing data. The stages have the same relay selection process, and thus we consider only the relay selection in the first stage.

In the first stage, there is one source sensor that has N neighboring sensors. Before conducting the auction, the source sensor builds its own table containing information about the residual energy of neighboring sensors and the hop counts of the neighboring sensors to the sink. Figure 5.5 shows the scenario in which the source sensor has $N = 3$ neighboring sensors, which are labeled as "1", "2", and "3". Let N_i^h denote the minimum hop count of neighboring sensor i to the sink and E_i denote the residual energy of the neighboring sensor. Table 5.1 shows the details. The neighboring sensor 1 has the smallest hop count, $N_1^h = 1$, and the highest residual energy, $E_1 = 2.5$. To reduce the latency and to achieve high reliability, the source sensor should consider both the two parameters for relay selection.

The source sensor calculates the *link quality* of each neighboring sensor. The link quality is a function of the residual energy and the hop count of the neighboring sensor to the sink. In particular, link quality a_i of neighboring sensor i is defined as follows:

$$a_i = \frac{N_i^h}{E_i} \quad (5.16)$$

Here, we denote "a_i" as the link quality since it is used as the "ask" for the winner determination in the reverse auction. In general, a low value for a_i implies that the forwarding service provided by neighboring sensor i has low latency and high reliability.

Since the source sensor has knowledge of the link quality of all its neighboring sensors, the source sensor can determine an average link quality $\bar{a}$. The average link quality $\bar{a}$ is a threshold for the winner determination in the auction. The average link quality is calculated as follows:

$$\bar{a} = \sum_{i=1}^{N} \frac{a_i}{N} \qquad (5.17)$$

After calculating the average link quality, the source sensor conducts the first-price sealed-bid reverse auction for the relay selection, in which the neighboring sensors are required to submit their a_i as asks to the source sensor. Upon receiving the asks, the source sensor selects neighboring sensor $\hat{i}$ as the winner for forwarding sensing data if that neighboring sensor has a link quality lower than the average link quality, $a_{\hat{i}} < \bar{a}$. In the case where multiple neighboring sensors have link quality a_j lower than the average link quality, the source sensor chooses the one with the lowest link quality as the winner:

$$\hat{i} = \arg\min_{j} a_j \qquad (5.18)$$

The winner receives a price as its reward for forwarding the sensing data. The price for the winner can be simply determined as $p = Ba_{\hat{i}}$ as proposed in [214], where B is the budget provided by the sink. However, it can be seen that such a payment determination is not fair. In particular, even if the winner is able to provide the forwarding service with low latency and high reliability, the winner receives a low price since the value of $a_{\hat{i}}$ is small. This may not provide neighboring sensors with a sufficient incentive to participate. After the relay selection in the first stage finishes, the process is implemented similarly in the next stages. Note that the winner in the previous stage becomes the buyer in the current stage. The buyer selects its neighboring sensor with the lowest link quality as a relay for forwarding sensing data. Eventually, when the relay selection process in all stages finishes, we have a route between the source sensor and the sink that includes multiple relays. Since these relays have good link quality, the route is able to provide the forwarding service with low latency and high reliability.

5.4.4 Denial-of-Service Attack Prevention

This section discusses the application of the first-price sealed-bid auction for security issues in WSNs. WSNs are vulnerable to security threats due to the absence of infrastructure and dynamically changing topologies. One common security issue in these networks is a black hole attack, a type of denial-of-service attack in which some sensors in the network attempt to interrupt or suspend network services such as packet forwarding and packet routing. Sensors that perform the attack are considered malicious nodes and called *blackhole nodes*. Various security protocols for WSNs are classified and evaluated in [218]. However, the traditional protocols, including the CONFIDANT

5.4 Development of First-Price Sealed-Bid Auction for Computer Networks

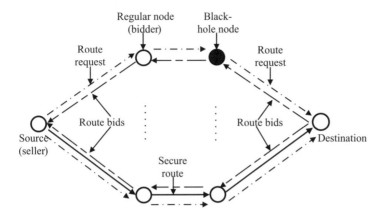

Figure 5.6 Secure routing protocol based on first-price sealed-bid auction [33].

protocol [219], lack interactions among the strategies of rational decision makers, the malicious nodes and regular nodes, the might strengthen the network security. Auctions can be employed to address the security issue by providing sensors with incentives to forward packets rather than dropping them.

In the following discussion, we explain how to apply the economic and pricing model to address black hole attacks in WSNs. More specifically, we describe how to use the first-price sealed-bid auction to obtain a secure routing protocol in WSNs as proposed in [170]. Here, the term "secure route" refers to a route that does not include any sensor that acts maliciously by dropping incoming packets. The first-price sealed-bid auction can be used since it is able to rapidly discover the most secure route for packets by quickly deciding the winner and the payment.

The model shown in Figure 5.6 includes one source sensor, one destination (i.e., the sink), and N intermediate sensors. In this model, the source sensor acts as the seller, and the intermediate sensors in the network are bidders, or buyers. The intermediate sensors can form multiple routes between the source sensor and its destination. Among the available routes, the source sensor needs to select the most secure one for the data forwarding operation. The selected route consists of those intermediate sensors with the highest reputation. These intermediate sensors are regarded as the winning sensors, and they are charged prices based on their power consumption during the packet forwarding. The winning sensors receive a good reputation from the source sensor as their reward. We now present the details of the use of the first-price sealed-bid auction for the secure routing protocol.

Initially, the source sensor broadcasts a *route request* message to its neighboring sensors. The route request message contains information about the destination and the number of data packets. Upon receiving the message, if the neighboring sensor is not the destination, that sensor calculates its bid or its bidding strategy. Note that the source sensor chooses a route including sensors with both a high reputation and energy efficiency. Thus, the bid of the sensor should represent of the sensor and the energy that the sensor consumes during the packet forwarding. In particular, bid b_i of the sensor is determined using the theory of equity, reciprocity, and competition [220] as follows:

$$b_i = \alpha_i u(\pi_i) + \beta_i \quad (5.19)$$

where α_i and β_i are positive constants, π_i is the payoff of sensor i, and $u(\pi_i)$ is a differentiable, strictly increasing, and concave function of π_i. In particular, the payoff of the sensor is calculated based on of the sensor and the energy consumed by the sensor as follows:

- Energy consumed by the sensor includes energy used for the communication and computation of the sensor. Typically, the energy used for the communication is much higher that that used for the computation. Thus, the energy consumed by the sensor is the energy consumption for the communication. Since the sensor performs the packet forwarding, the sensor can act as a receiver to receive the packet and then as a transmitter to forward the packet. Let P_i^f and P_i^r denote, respectively, the transmit power and the receive power of sensor i. Then, the energy E_i consumed by the sensor can be determined as follows:

$$E_i = P_i^f + P_i^r \quad (5.20)$$

- The reputation of sensor i, denoted as Φ_i, is defined as the ratio of the number of forwarded packets to the total number of received packets. Let M_i^f and M_i^r be, respectively, the number of forwarded packets and the number of received packets. Then, the reputation of the sensor is calculated as

$$\Phi_i = \frac{M_i^f}{M_i^r} \quad (5.21)$$

The expression in (5.21) means that the reputation value of sensor i decreases when the sensor misbehaves – for example, by dropping received packets – since M_i^f decreases. Thus, to receive a high reputation as the reward, the sensor needs to successfully forward received packets to increase M_i^f. However, this action requires energy consumption E_i. Therefore, the payoff of the sensor is defined by

$$\pi_i = \Phi_i - E_i \quad (5.22)$$

From (5.19) and (5.22), and by noting that $u(\pi_i)$ is an increasing function of π_i, it can be seen that bid b_i of the sensor is proportional to the reputation of the sensor and inversely proportional to its energy consumption. As such, the route that includes sensors with high bids has a high reputation and needs less energy for the packet forwarding. After determining its bid, the sensor adds the bid to the route request message and forwards the message to the others. When the destination receives this request message, it sends back a *reply message*, which includes a *route bid*. The route bid is the sum of bids of the sensors on the route. Let $\bar{b}^k$ denote the route bid of route k. Then $\bar{b}^k$ is expressed by

$$\bar{b}^k = \sum_{i=0}^{N^k} b_i^k \qquad (5.23)$$

where N^k is the number of sensors belonging to route k, and b_i^k is the bid of sensor i in route k, which is determined according to (5.19). As shown in Figure 5.6, multiple routes are available between the source sensor and the destination. Correspondingly, the destination may receive different request messages, and then reply to the source sensor with different route bids. When receiving the route bids $\bar{b}^k$, the source sensor selects the route with the highest route bid as the winning route for the packet forwarding:

$$\hat{k} = \arg\max_{k} \bar{b}^k \qquad (5.24)$$

The sensors in the winning route $\hat{k}$ receive a good reputation from the source sensor as rewards. These sensors pay the source sensor prices based on their power consumption for the packet forwarding. Since the sensors on a route desire to have a good reputation from the source sensor, they are forced to cooperate to forward received packets rather than dropping the packets. In fact, some malicious sensors might potentially agree to the auction but later subvert the route. A watch list can be used to recognize such malicious sensors. The watch list records misbehavior of the sensors and warns all sensors not to communicate with them.

The simulation results of the proposed scheme are given in [170]. The total number of sensors in the considered scenario is 100, of which one third (approximately 33) are malicious nodes. The average number of dropped packets in the proposed scheme is less than half the number with the CONFIDANT protocol [219]. The reason is that the reputation of sensors is not considered in the CONFIDANT protocol, but is regarded as a major metric for the route selection in the proposed scheme. Moreover, as the percentage of malicious nodes changes from 10% to 90%, the proposed scheme still keeps the number of dropped packets lower and more stable than occurs with the CONFIDANT protocol. This is due to the fact that the proposed scheme dynamically updates the reputation of the sensors and guarantees that low-reputation sensors are ignored by the majority of sensors. Although the proposed scheme achieves a highly secure route protocol, other QoS factors related to the network, such as the delay or latency of the packet forwarding, are not considered. Therefore, the information, such as the number of hops, needs to be incorporated into each route bid.

5.5 Summary

In this chapter, we introduce the first-price sealed-bid auction and its variant, the first-price sealed-bid reverse auction, and discuss their applications in computer networks. Specifically, we first provide the definition of the first-price sealed-bid auction and give a specific example in the context of computer networks. Through an example, we present the strategic analysis of bidders in the first-price sealed-bid auction. We then discuss how to find the Bayesian–Nash equilibrium in the first-price sealed-bid auction. In particular, we show that the equilibrium strategies of bidders in the first-price sealed-bid auction can be determined using the revenue equivalence theorem as in the Dutch auction. Therefore, the first-price sealed-bid auction and the Dutch auction are strategically equivalent. Next, we introduce the first-price sealed-bid reverse auction, which is a variant of the first-price sealed-bid auction. Finally, we discuss the applications of the first-price sealed-bid auction and its variant to address emerging issues in IoT including data aggregation, task allocation, relay selection, and denial-of-service attack prevention.

6 Second-Price Sealed-Bid Auction

This chapter introduces one of the most common types of sealed-bid auction, the second-price sealed-bid auction or Vickrey auction. In general, the chapter consists of two main parts. In the first part, we introduce the theory of the second-price sealed-bid auction and its applications in computer networks. In particular, we first provide the definition of the second-price sealed-bid auction. Then, we prove that truthful bidding is a dominant strategy of the auction. Furthermore, we compare the dominant strategy in the second-price sealed-bid auction and the equilibrium strategy in the English auction. After that, we present a variant of the second-price sealed-bid auction, the second-price sealed-bid reverse auction. At the end of the first part, we discuss approaches that use the second-price sealed-bid auction to address emerging issues in computer networks.

The second part presents the Vickrey–Clarke–Groves (VCG) auction, which is known as a generalization of the second-price sealed-bid auction with multiple items. In particular, we first provide the formal description of the VCG auction and discuss the dominant strategy in the auction. Then, we present important virtues of the VCG auction. Finally, we discuss applications of the VCG auction in computer networks such as mobile device clouds, massive MIMO, video streaming systems, and cognitive radio networks.

6.1 Second-Price Sealed-Bid Auction

6.1.1 Definition

The second-price sealed-bid auction is a type of sealed-bid auction. It is also known as the Vickrey auction for a single item since the auction was first introduced academically by William Vickrey [221]. Thus, in this book, the terms "second-price sealed-bid auction" and "Vickrey auction" are equivalent and used interchangeably. This auction is essentially similar to the first-price sealed-bid auction. That is, bidders simultaneously submit their sealed bids to the seller in one round. Upon receiving the bids, the seller selects the bidder with the highest bid as the winner of the auction. However, different from the first-price sealed-bid auction, the winner in the second-price sealed-bid auction pays the seller the price equal to the second highest bid. The pricing rule is thus called a *second-price rule*. Since the price that the winner pays the seller is lower than the winner's bid, the second-price rule guarantees a positive payoff for any bidder if the

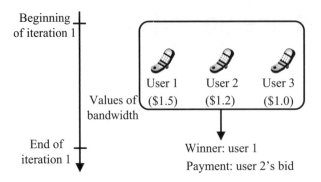

Figure 6.1 Second-price sealed-bid auction for bandwidth resource trading market.

bidder wins. This provides bidders with an incentive to participate in the auction. By using the second-price rule, in the next section we show that the second-price sealed-bid auction achieves incentive compatibility or truthfulness. In general, truthfulness is a desired economic property in designing an auction in which the bidders are incentivized to bid their true values of an item. We can say that bidding in the second-price sealed-bid auction is *sincere bidding* or *truthful bidding*. Since the item is allocated to the bidder that truly values the item the most, we also say that the second-price sealed-bid auction achieves an efficient allocation of the item, or the second-price sealed-bid auction is an efficient auction. Note that the first-price sealed-bid auction may not be an efficient auction since the bidders in this auction tend to bid below their values and the highest bid might not be the highest value.

Example 6.1 We continue with Example 5.1, which involved three mobile users and one service provider. The mobile users' bids are equal to their private values of the bandwidth unit, $b_i = v_i, i = 1, \ldots, 3$. Recall that the private values of the bandwidth unit to users 1, 2, and 3 are $v_1 = \$1.5$, $v_2 = \$1.2$, and $v_3 = \$1.0$, respectively. The auction is implemented in a single round, as shown in Figure 6.1. First, the mobile users submit their bids to the service provider. Upon receiving the bids, the service provider determines the winner and the price that the winner pays. Note that here the highest bid is $b_1 = \$1.5$, and the second-highest bid is $b_2 = \$1.2$. Determining the winner is done in the same way as in the first-price sealed-bid auction, in which the user with the highest bid is the winner. In particular, user 1 is the winner of the bandwidth unit since it has the highest bid, $b_1 = \$1.5$. Based on the second-price rule, the price that user 1 pays the service provider is equal to the second highest bid, $p = b_2 = \$1.2$. Thus, the payoff that user 1 receives is $\pi_1 = b_1 - p = \$1.5 - 1.2 = \$0.3 > 0$. As such, user 1 receives a positive payoff, $\$0.3$. Clearly, this provides the users with an incentive to participate in the auction. Moreover, we observe that the price that user 1 pays, $p = b_2 = 1.2$, is independent from the user's bid, $b_1 = 1.5$. In other words, the bid that user 1 submits does not impact the price that the user pays. Thus, to have high winning probability, the user tends to bid its true value of the bandwidth unit (i.e., truthful bidding). In the following section, we explain that the truthful bidding is the dominant strategy of the users when participating in the second-price sealed-bid auction.

6.1.2 Dominant Strategy and Nash Equilibrium

Similar to the English, Dutch, and first-price sealed-bid auctions, finding the Nash equilibrium is an important challenge in the second-price sealed-bid auction. In this section, we show that the Nash equilibrium in the second-price sealed-bid auction can be determined through finding a dominant strategy of the users in the auction. Indeed, a dominant strategy for a user in an auction is the bidding strategy that yields the highest payoff for the user regardless of what the other users choose. If all users have their dominant strategies, then these strategies constitute a *dominant strategy equilibrium* or Nash equilibrium of the auction. As such, the key issue is to determine the dominant strategy in the second-price sealed-bid auction. The following shows that truthful bidding is a dominant strategy in the auction.

We consider again the bandwidth resource trading market in Section 5.2.2. Recall that the market has N mobile users (i.e., bidders), which compete for a single bandwidth unit from the service provider (i.e., the seller). Each user has its own value v_i of the bandwidth unit.

Let b_i denote the bid of user i for the bandwidth unit. Based on the second-price rule of the second-price sealed-bid auction, the payoff of the user is given by

$$\pi_i = \begin{cases} v_i - \max_{j \neq i} b_j, & \text{if } b_i > \max_{j \neq i} b_j \\ 0, & \text{if } b_i < \max_{j \neq i} b_j \end{cases} \quad (6.1)$$

The expression in (6.1) means that user i wins in the auction if its bid is greater than the maximum bid of the other users. The user pays the service provider the second-highest bid and receives a payoff of $v_i - \max_{j \neq i} b_j$. Conversely, the user loses in the auction and receives a zero payoff if its bid is less than the maximum bid of the other users. We can ignore the case $b_i = \max_{j \neq i} b_j$ since this case occurs with a low probability.

We now prove that truthful bidding, $b_i = v_i$, is a dominant strategy of the user. For this purpose, we consider two other strategies that the user can select. The first is the overbidding strategy, meaning that the user bids $b_i > v_i$. For this strategy, b_i is called an *overbid*. The second is the underbidding strategy, meaning that the user bids $b_i < v_i$. For this strategy, b_i is called an *underbid*. To prove that truthful bidding is a dominant strategy, we need to show that the payoff of the user when selecting the truthful bidding is always higher than or at least equal to that of the user when selecting either the overbidding or underbidding strategy. As such, we can say that the truthful bidding dominates the overbidding and underbidding strategies, or the overbidding and underbidding strategies are dominated by truthful bidding.

First, we compare the payoffs of the user when it selects the overbidding with an overbid $b_i > v_i$ and the truthful bidding with a truthful bid $b_i = v_i$. There are three corresponding cases:

- If $\max_{j \neq i} b_j < v_i$, then user i wins the bandwidth unit when it selects the overbid as well as the truthful bid. The payoffs of the user for the two strategies are the same and are given by $\pi_i = v_i - \max_{j \neq i} b_j$. As such, increasing the bid does not change the payoff of the user.

- If $b_i < \max_{j \neq i} b_j$, then $v_i < \max_{j \neq i} b_j$, and user i does not win the auction with either the overbid or the truthful bid. The payoffs of the user for the two strategies are the same: both are zero. In other words, increasing the bid does not change the payoff of the user.
- If $v_i < \max_{j \neq i} b_j < b_i$, then only the overbidding strategy would win the bandwidth unit, and the payoff of the user for this strategy is $\pi_i = v_i - \max_{j \neq i} b_j$. It can be seen that if the user selects the overbidding strategy, its payoff is negative since the user paid more than its value for the bandwidth unit. If the user selects the truthful bidding, its payoff is zero since the user loses in the auction. As such, the payoff of the user when selecting the truthful bidding strategy is higher than that of the user when selecting the overbidding strategy.

Second, we compare the payoffs of the user when it selects the underbidding strategy, $b_i < v_i$, and the truthful bidding strategy, $b_i = v_i$. We also have three potential cases:

- If $v_i < \max_{j \neq i} b_j$, then $b_i < \max_{j \neq i} b_j$ for both strategies. The user loses in the auction with the overbid as well as the truthful bid. Thus, both strategies yield a zero payoff.
- If $b_i > \max_{j \neq i} b_j$, then user i wins the bandwidth unit with both the underbid and the truthful bid. The payoffs are the same and are given by $\pi_i = v_i - \max_{j \neq i} b_j$.
- If $b_i < \max_{j \neq i} b_j < v_i$, then the user wins the bandwidth unit if it selects the truthful bidding strategy. The payoff for this strategy is $\pi_i = v_i - \max_{j \neq i} b_j$. This payoff is positive since the user paid a price less than its value of the bandwidth unit. If the user selects the underbidding strategy, the user loses in the auction and receives a zero payoff. As such, the payoff of the user when selecting truthful bidding is higher than that of the user when selecting underbidding.

The aforementioned argument shows that truthful bidding dominates the other strategies of underbidding and overbidding. Thus, we have the following proposition.

PROPOSITION 6.1 *In a second-price sealed-bid auction, for the user with value v, bidding its true value, $b_i = v$, is a dominant strategy.*

Proposition 6.1 can be interpreted as stating that the payoff of the user when it selects truthful bidding is never worse than that of the user when it selects other strategies. Thus, truthful bidding is a dominant strategy in the second-price sealed-bid auction. The dominant strategy is guaranteed regardless of what the other users select, and truthful bidding is called a *weakly dominant strategy*.

DEFINITION 6.2 *For the user i with value v, strategy $b_i = v$ is weakly dominant if the strategy weakly dominates all other strategies of user i.*

Definition 6.2 means that the weakly dominant strategy of the user is the best response regardless of what other users select. Therefore, if all users select a weakly dominant strategy, the dominant strategies constitute a dominant strategy equilibrium, which is also the Nash equilibrium of the auction.

DEFINITION 6.3 *In a second-price sealed-bid auction with N users, if each user i selects a weakly dominant strategy b_i, then $\boldsymbol{b} = (b_1, b_2, \ldots, b_N)$ is the Nash equilibrium of the auction.*

It is worthwhile to compare the dominant strategy in the second-price sealed-bid auction and the equilibrium strategy in the English auction. We first recall the use of the English auction in the resource trading market, as presented in Section 4.1.2. In this auction, the service provider initially sets a low price of the bandwidth unit and then increases the price in the next rounds. The equilibrium strategy of the user is to determine the price, which is also the bid of the user, at which the user will drop out from the auction. The user drops out if the price reaches its expectation of the value of the bandwidth unit. The user with the highest value for the bandwidth unit wins the auction and pays the service provider the second-highest bid (see (4.6)). To be consistent with the second-price sealed-bid auction, the values of the bandwidth unit to the users are assumed to be independent of all other users' values, and the expectation of the value for the bandwidth unit can be replaced by their own values. As such, the user drops out if the price reaches its value of the bandwidth unit, meaning that the bid of the user is equal to its value of the bandwidth unit. In other words, the equilibrium strategy of the user in the English auction is its truthful bidding. Thus, we can say that the equilibrium strategy in the English auction is also the dominant strategy in the second-price sealed-bid auction. Moreover, it can be seen that the winner in the English auction pays the service provider the second-highest bid. Therefore, the English auction and the second-price sealed-bid auction have the same payment and hence the same revenue. In fact, the winner of the English auction often pays a price slightly greater than the second-highest bid. As shown in Example 4.1, user 1, which has the highest value for the bandwidth unit, wins the auction and pays the service provider a price of $1.205. The price is slightly greater than $1.2, which is the second-highest bid. However, when the service provider sets a small price step, such as $0.001, then we can say that the price, that user 1 pays (e.g., $1.201) is likely equal to the second-highest bid (i.e., $1.2). In summary, when the values of the bandwidth unit to the users are independent of each other, then the second-price sealed-bid auction and the English auction have the same truthful bidding strategy and the same revenue. The English auction is thus sometimes referred to as an *open second-price auction*.

In general, the second-price sealed-bid auction is simple to implement, and it exhibits desirable properties such as efficiency and truthfulness. However, while this auction is more studied in economic literature, it is uncommon in practice. For computer networks, the second-price sealed-bid auction can be used for cloud resource allocation in edge cloud computing [222] and friendly jamming power allocation to improve physical layer security in cognitive radio networks [151].

6.1.3 Second-Price Sealed-Bid Reverse Auction

In Section 5.3, we presented the first-price sealed-bid reverse auction, which is the combination of the first-price sealed-bid auction and the reverse auction. Similarly, when the reverse auction is combined with the second-price sealed-bid auction, we have the second-price sealed-bid reverse auction. In general, the second-price sealed-bid reverse auction and the first-price sealed-bid reverse auction have similar means of winner determination but different payments. We consider again the resource trading market in Section 5.3 to explore how the second-price sealed-bid reverse auction works.

The market consists of multiple service providers (i.e., the sellers) and one user (i.e., the buyer). Each service provider has a bandwidth unit for trading. The service providers compete to attract the user's demand by submitting asks a_i to the user. The asks are the prices that the service providers are willing to be paid by the user. In particular, the asks can be determined to maximize the profit of the service providers. After receiving the asks, the user selects the service provider with the lowest ask as the winner as follows:

$$\hat{i} = \arg \min_i a_i \tag{6.2}$$

The user purchases the bandwidth from the winning service provider $\hat{i}$. The user pays the service provider a price that is determined as follows:

$$p = \min_{i \neq \hat{i}} a_i \tag{6.3}$$

The expression in (6.3) means that the price that the service provider receives is equal to the second-lowest ask. Since the price that the service provider receives is higher than the ask that it submits, the service provider has more incentive to participate in the auction. Also, the service provider tends to submit its true value so as to increase its winning probability and receive a higher payoff. Here, the true value is the minimum price that the service provider accepts to trade its bandwidth. Therefore, similar to the second-price sealed-bid auction, truthful bidding is also a dominant strategy in the second-price sealed-bid reverse auction.

6.1.4 Development of Second-Price Sealed-Bid Auction for Computer Networks

This section discusses (i) the application of the second-price sealed-bid reverse auction for sensing task allocation in WSNs for IoT, (ii) the application of the second-price sealed-bid reverse auction for task scheduling in edge computing, and (iii) the application of the second-price sealed-bid auction for physical layer security in wireless networks.

Task Allocation in IoT

In Section 5.4.2, we discussed how to use the first-price sealed-bid reverse auction for the sensing task allocation in WSNs for IoT. For convenience, we briefly describe the auction approach here. In the auction approach, multiple sensors act as sellers, and the server is the data buyer. Upon receiving sensing data collection requests, such as for data monitoring of a specific area, from its customers, the server broadcasts the task message (i.e., task description) to all the sensors in the area. After receiving the task message, the sensors that are interested in the task calculate their asks (i.e., task prices) for completing the task according to (5.12). The sensors then submit their truthful asks (i.e., real costs) to the server, and the server selects the sensor with the lowest ask as the winner for executing the sensing task.

It can be observed that the ask given in (5.12) is actually the cost of the sensor for executing the sensing task. Naturally, the sensor does not submit its real cost to the server since the sensor receives a zero payoff even if it wins the auction. This may reduce the efficiency of the task allocation. To address this issue, the second-price sealed-bid reverse auction can be used in which submitting truthful asks is a dominant strategy of the sensors. The following discussion presents the use of the second-price sealed-bid reverse auction for sensing task allocation in WNSs for IoT, as proposed in [223].

The model used in the proposed scheme is similar to that in [213], which is shown in Figure 5.3. Specifically, the model consists of one server acting as the data buyer and N sensors that provide sensing data. The algorithm of the proposed scheme is essentially similar to that in [213] (see Section 5.4.2). The main difference is that the sensors submit their optimal asks rather than their real costs, as in [213]. Therefore, this section merely presents how the sensors determine the optimal asks.

The optimal asks are determined based on the payoff functions of the sensors. Since the payoff functions include the real cost for executing the sensing task, we rewrite the cost formulation given in (5.12) as follows:

$$C_i = \left(C_i^{com} + a_i^B\right)\left(1 + e^{\frac{\epsilon(t^D - t)}{-(t_i^R - t)}}\right) \tag{6.4}$$

where C_i^{com} is the bandwidth cost of sending the sensing data to the server, t^D is the task deadline, and t_i^R is the release time that the sensor can start executing the sensing task. a_i^B is the base price, which is given by

$$a_i^B = \frac{E_{task}\alpha}{1 - e^{-E_i/\beta}} \tag{6.5}$$

where E_{task} is the task size (i.e., the expected energy for completing the sensing task), E_i is the remaining energy of sensor i, and α and β are scaling parameters.

The objective of the cost formulation given in (6.4) is to meet the deadline of the task, improve the reliability, and achieve energy balance among the sensors. Apart from the cost, we consider the following parameters.

- Reserve price p^B: This is the budget of the server. It is the maximum price that the server can pay the sensor for executing the sensing task. Reserve price p^B is assumed to be publicly known.
- Ask a_i: This is the price offered by sensor i.
- Probability distribution $Q(a_i, p^B, C_i)$: This is the probability distribution of sensor i's belief about the server's preference of the winning ask. The server rationally prefers a lower ask, because a lower ask has a higher probability of winning the auction. Thus the function can be expressed in a form of truncated and decreasing geometric distribution. As such, the perceived probability decreases monotonically with the ask as follows:

$$Q(a_i, p^B, C_i) = \begin{cases} \frac{(1-\gamma)\gamma^{(a_i - C_i)}}{1 - \gamma^{(p^B + 1)}}, & \text{if } C_i \leq a_i \leq p^B \\ 0, & \text{otherwise} \end{cases} \tag{6.6}$$

where $0 < \gamma < 1$ is the distribution parameter.

$Q(a_i, p^B, C_i)$ is considered to be the winning probability of sensor i. Therefore, the payoff function of the sensor is determined as follows:

$$u_i = (a_i - C_i)Q(a_i, p^B, C_i) = \begin{cases} (a_i - C_i)\frac{(1-\gamma)\gamma^{(a_i-C_i)}}{1-\gamma^{(p^B+1)}}, & \text{if } C_i \leq a_i \leq p^B \\ 0, & \text{otherwise} \end{cases} \quad (6.7)$$

By using the second-order derivative, u_i is simply proved to be a concave function of a_i if $C_i \leq a_i \leq p^B$. Then the sensor can determine its optimal ask a_i^* in the range as follows:

$$a_i^* = \arg\max_{a_i} u_i \quad (6.8)$$

As such, the optimal asks $\mathbf{a}^* = (a_1^*, a_2^*, \ldots, a_N^*)$ constitute a unique Nash equilibrium of the auction. Note that the unique Nash equilibrium may not be a continuous function of the ask since the sensor's payoff function is discontinuous, or non-smooth. In particular, the unique Nash equilibrium can be all zero at any ask outside of the range $[C_i, p^B]$. We thus have the following theorem.

THEOREM 6.4 *If asks are selected from the range $[C_i, p^B]$, there exists a unique Nash equilibrium. Otherwise, no Nash equilibrium exists. C_i and p^B, respectively, are the lower and upper bounds of the asks.*

Assume that the ask is selected from the range $[C_i, p^B]$. Then the payoff function u_i is continuous and differentiable in the range. By taking the first-order derivative of u_i with respect to a_i, the optimal ask is given by

$$a_i^* = \frac{-1}{\ln \gamma} + C_i \quad (6.9)$$

Note that if $0 < \gamma < 1$, then the term $\ln \gamma$ is negative and $\epsilon = \frac{-1}{\ln \gamma}$ is a positive value. Thus, the optimal ask of the sensor is higher than its real cost with an amount of $\epsilon = \frac{-1}{\ln \gamma}$. However, as discussed in [223], the value of $\epsilon = \frac{-1}{\ln \gamma}$ is actually small compared with C_i. For example, while the value of the cost C_i is 40, the value of ϵ is 0.62 at $\gamma = 0.2$. Thus, we have the following theorem.

THEOREM 6.5 *[223] The optimal ask a_i^* of the sensor is a value near the ask's lower bound, the real cost C_i.*

After determining the optimal asks a_i^*, the sensors submit them to the server for the winner determination and the payment. The server determines the winner as follows:

$$\hat{i} = \arg\min_i a_i^* \quad (6.10)$$

The price paid to the winner is given by

$$p = \min_{i \neq \hat{i}} a_i^* \quad (6.11)$$

In general, since the ask is close to the real cost, the proposed auction achieves the truthfulness. Moreover, the winning sensor receives a price that is higher than its ask, its real cost. Thus, the payoff of the sensor is positive, and we say that the proposed scheme guarantees the individual rationality.

LEMMA 6.6 *[223] The proposed scheme guarantees individual rationality and truthfulness – in other words, incentive compatibility.*

It can be seen from (6.9) that apart from the real cost, the optimal ask is a function of the distribution parameter γ. This parameter is locally set by the sensor, and thus the sensors may have different γ. In general, if γ is large, the optimal ask is high. As a result, its payoff is high if the sensor wins the auction. However, the winning probability of the sensor is lower. Since the sensor does not know the optimal asks of the other sensors, it is challenging for the sensor to set γ. One simple solution is that the sensor can adaptively adjust γ after each round of the reverse auction depending on whether the sensor wins or loses in the previous rounds. In particular, if the sensor loses in the previous round of the auction, then the ask of the sensor may be too high. In this case, the sensor should reduce γ so that it has a higher winning probability in the next round. On the contrary, if the sensor wins the auction in a certain number of rounds, meaning that the sensor still has a high winning probability, it can increase γ to receive a higher payoff. In particular, let $\Delta \gamma$ denote the step for adjusting γ and n_{win} denote the number of recent rounds that the sensor wins. Then, the sensor can update the value of its γ in the next round as follows:

$$\gamma^{(k+1)} = \begin{cases} \gamma^{(k)} + \Delta\gamma, & \text{if } n_{win} \geq \eta \\ \gamma^{(k)} - \Delta\gamma, & \text{otherwise} \end{cases}$$

where η is the predefined constant. η may be different among the sensors. In particular, the sensor that defines a large value of η implies that the sensor increases γ only after winning a number of rounds of the auction. Such a sensor is said to be a "careful" sensor. Determining the optimal value of η generally is challenging and can be considered in future work.

Task Scheduling in Edge Computing

Edge computing is an important emerging technology that may be able to solve a variety of issues affecting cloud data centers, such as peak usage, high operational costs, bandwidth bottlenecks, and service interruption. Volunteer computing systems are an edge computing model in which a large number of distributed computers contribute their computing resources for processing users' tasks. However, due to the limited resources available from the volunteered computers, the efficiency of resource allocation schemes in the systems needs to be guaranteed. The second-price sealed-bid auction, which achieves both efficiency and truthfulness, can be used as a potential solution. This section presents the use of the second-price sealed-bid reverse auction for task scheduling in the volunteer computing systems. Such an approach is proposed in [224].

The model consists of users (i.e., customers) and multiple volunteered computers. The users request the volunteered computers for executing the users' computing tasks. The volunteered computers can be connected with each other over a peer-to-peer overlay network on the Internet. In such a peer-to-peer network, any volunteered computer

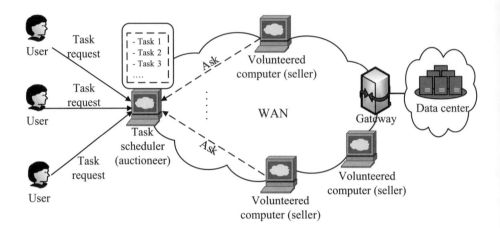

Figure 6.2 Computing task scheduling in volunteer computing systems using the second-price sealed-bid reverse auction [224].

can act as a task scheduler or a resource contributor. As illustrated in Figure 6.2, the task scheduler receives and schedules the users' tasks, and the resource contributors execute the tasks. Within this auction model, the task scheduler is the auctioneer, and the resource contributors, or just contributors for short, are the sellers. We assume that each customer requests one computing task. The following presents the task scheduling process based on the second-price sealed-bid reverse auction.

- First, the customers send computing task request messages to the task scheduler. The request message of task t_j includes the following information.
  - The budget $B(t_j)$: This is the maximum price that customer j can pay for executing its task t_j.
  - The demand vector $\mathbf{D}(t_j)$: This vector contains demands for resource types, such as CPU and network bandwidth. Assuming that the customer requires M resource types for executing the task, we have $\mathbf{D}(t_j) = (D^1(t_j), \ldots, D^M(t_j))$, where $D^m(t_j)$ is the demand of resource type m for task t_j.
  - The priority price $p^{\text{prior}}(t_j)$: This is the price that the customer needs to pay such that the queuing time of executing task t_j is shorter, meaning that the priority for executing the task is higher.
- Upon receiving the task request messages, the task scheduler puts them in a queue and sorts the tasks in descending order based on the priority prices. The tasks are then sequentially selected for being executed according to their priority prices $p^{\text{prior}}(t_j)$.
- For task t_j selected from the queue, the task scheduler seeks contributors whose resources satisfy the demand vector $\mathbf{D}(t_j)$. In particular, the task scheduler adopts the proactive index diffusion. The CAN algorithm proposed in [225] is used

to search the contributors. In this algorithm, the task scheduler sends a query message including the demand vector $\mathbf{D}(t_j)$ for the task to its neighbors. Each neighbor then broadcasts the query message in the network to find more available contributors. This finding continues until the number of hops is greater than a time-to-live threshold.

- Contributor i, which receives $\mathbf{D}(t_j)$, verifies whether the contributor's resources satisfy the demand vector. Let $\mathbf{c}_i(t_j) = (c_i^1(t_j), \ldots, c_i^M(t_j))$, where $c_i^m(t_j)$ is the capacity of available resource type m (e.g., the bandwidth). Then the contributor checks the following condition:

$$c_i^m(t_j) \geq d_i^m(t_j), m = 1, \ldots, M \tag{6.12}$$

- The contributor that satisfies condition (6.12) will reply to the task scheduler with an *availability message*. The availability message consists of an identifier, e.g., IP address, of the contributor and resource prices. The resource prices are expressed by $\mathbf{p}_i^{\text{unit}}(t_j) = (p_i^1(t_j), \ldots, p_i^M(t_j))$, where $p_i^m(t_j)$ is the price per unit of mth resource, e.g., the bandwidth.

- Multiple contributors can satisfy condition (6.12), and thus the task scheduler may receive multiple availability messages. The task scheduler calculates the price that the customer pays the contributor if this contributor is selected as follows:

$$a_i(t_j) = p^{\text{prior}}(t_j) + \mathbf{p}_i^{\text{unit}}(t_j)\mathbf{D}^T(t_j) \tag{6.13}$$

where $\mathbf{X}^T$ denotes the transpose of matrix $\mathbf{X}$.

- $a_i(t_j)$ is considered to be the ask, or asking price, of contributor i for executing task t_j. The task scheduler only selects contributors with asks that satisfy the budget condition, $a_i(t_j) \leq B(t_j)$.

- The task scheduler sorts the selected contributors in an ascending order of their asks. The task scheduler selects the contributor with the lowest ask as the winner for executing the task as follows:

$$\hat{i} = \arg\min_i a_i(t_j) \tag{6.14}$$

- The winning contributor receives a price from the customer for executing the task. The price is the second-highest ask, which is determined as follows:

$$p = \min_{i \neq \hat{i}} a_i(t_j) \tag{6.15}$$

Using the second-price rule incentivizes the contributors to truthfully reveal their true expectations on the task prices.

- Finally, the task is assigned to the winning contributor for being executed.

It can be observed that the proposed algorithm can guarantee the win-win solution for both the customers and the contributors. In particular, from the customers' side, their tasks are executed based on their expected resource demands, while they pay prices lower than their budgets. From the contributors' side, the prices that they receive from the customers for executing the tasks are higher than their expectations due to the second-price rule. The simulation results in [224] show that the proposed scheme outperforms the task schedule based on random-selection strategy in terms of social welfare, referring to the total utility of the customers and contributors. However, the task scheduler that acts as the auctioneer does not receive any payoff from the auction. This may discourage the task scheduler from participating in the auction.

Physical Layer Security in MANET

MANET is decentralized, which leaves it vulnerable to network attacks. One of the most common attacks is eavesdropping, in which attackers (i.e., eavesdroppers or wiretappers) attempt to acquire important and private information of legitimate users. To protect the wireless communication against eavesdropping attacks, traditional techniques, such as cryptographic techniques [226], can be used. However, these traditional techniques often require centralized authorities as well as encryption/decryption algorithms that are not suitable or even may not work in MANET due to resource constraints. As presented in Section 2.3, physical layer security is proposed as an alternative solution that exploits physical characteristics of the wireless channels, such as channel gains and noise. In particular, a legitimate source employs friendly jamming power from friendly jammers to maximize the transmission rate, also called the *secrecy capacity* or *secrecy rate*, of reliable information transmitted from the source to an intended destination at which the eavesdropper is unable to decode the information. This problem can be regarded as a power allocation issue, which has been efficiently solved by auction theory. In this section, we present the use of the second-price sealed-bid auction as a means to profice physical layer security in MANET and achieve both efficiency and truthfulness. The proposed scheme is described in [227].

The model consists of N friendly jammers, one source and its destination, and one eavesdropper, as shown in Figure 6.3. The source is willing to use power from the friendly jammers to maximize its secrecy rate. Also, the friendly jammers want to use the bandwidth of the source for their own data communication. Thus, the source as the seller sells its bandwidth to the friendly jammers. The friendly jammer as a buyer (i.e., a bidder) pays the source a price that is the friendly jammer's power. The bandwidth is assumed to be assigned to only one friendly jammer; thus the friendly jammers need to compete with each other. The friendly jammers compete on the bandwidth by submitting their bids to the source. Here, the bids are the power that the friendly jammers can offer the source. The source then selects the friendly jammer that maximizes the source's secrecy rate as the winner of the auction. The winning friendly jammer receives the bandwidth from the source and provides the source with the power that produces the second-highest secrecy rate. More specifically, the auction proceeds as follows.

The source divides its available bandwidth into two fractions. The bandwidth fraction of $1 - \alpha$ is assigned to the winning friendly jammer, and the remaining bandwidth of α is used for the source's communication. The source selects the friendly jammer based on how the friendly jammer can improve the source's secrecy rate, and the friendly jammer determines its bid, or power, so as to maximize its utility. Thus, we first define the secrecy rate of the source and the utility of the friendly jammers. Note that the secrecy

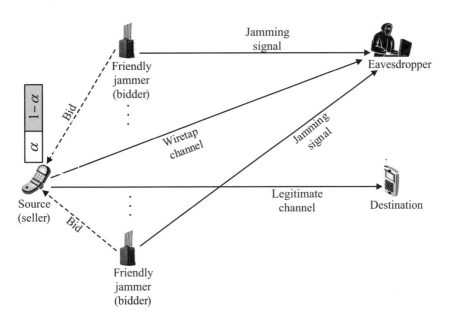

Figure 6.3 Physical layer security in MANET using second-price sealed-bid auction [227].

rate of the source is regarded as the secrecy rate of the channel, communication, or link between the source and its intended destination.

For convenience, the source is labeled by "S," its intended destination is labeled "D," friendly jammer i is labeled "J_i", and the eavesdropper is labeled "E". As discussed in Section 2.3, the achievable secrecy rate R_S of the source when it uses power P_{J_i} from friendly jammer J_i is defined as follows:

$$R_S(P_{J_i}) = \alpha \left[\log_2 \left(1 + \frac{h_{SD} P_S/\alpha}{\sigma^2 + h_{J_i D} P_{J_i}}\right) - \log_2\left(1 + \frac{h_{SE} P_S/\alpha}{\sigma^2 + h_{J_i E} P_{J_i}}\right) \right]^+ \quad (6.16)$$

where $[a]^+ = \max(0, a)$. The parameters and variables in (6.16) are defined as follows:

- h_{SD} is the power channel gain between the source and its intended destination.
- h_{SE} is the power channel gain between the source and the eavesdropper.
- $h_{J_i D}$ is the power channel gain between friendly jammer i and the eavesdropper.
- P_S and P_{J_i} are the transmit power of the source and friendly jammer i, respectively. In particular, P_S of the source is normalized with α to satisfy the average transmit power constraint.
- σ^2 is the independent additive white Gaussian noise variance. The links in the network physically are at the same location, so σ^2 is assumed to be the same for all links in the network.

Note that the source employs the friendly jammer only if the power provided by this jammer improves the current secrecy rate of the source. Thus, the source sets the lowest acceptable rate as $R_S^0 = R_S(P_{J_i} = 0)$, which can be considered to be a *reserve price* of the auction.

To determine the utility of friendly jammer i, we assume that the friendly jammer wins the auction. Then, it receives $1 - \alpha$ of access bandwidth from the source and pays the source the price, which is its power. Thus, the utility of the friendly jammer is defined as the difference between its achievable rate and the power:

$$U_{J_i}(P_{J_i}) = (1 - \alpha) \log\left(1 + \frac{h_{J_i} P_{J_i}}{\sigma^2}\right) - c P_{J_i} \tag{6.17}$$

where c is the cost per unit transmit power, and h_{J_i} is the power channel gain between friendly jammer i and its intended destination, the destination of the friendly jammer. By taking the second-order derivative of $U_{J_i}(P_{J_i})$ with respect to P_{J_i}, we can prove that $U_{J_i}(P_{J_i})$ is a concave function of P_{J_i}. Therefore, when P_{J_i} is large, the value of $U_{J_i}(P_{J_i})$ is zero and then negative. We can determine the power range in which $U_{J_i}(P_{J_i})$ is positive as follows. $U_{J_i}(P_{J_i})$ is positive within the range $P_{J_i} \in (0, P_{J_i}^{limit})$, where $P_{J_i}^{limit}$ is a solution of $U_{J_i}(P_{J_i}) = 0$ and given by

$$P_{J_i}^{limit} = \frac{\sigma^2}{h_{J_i}}(-\frac{1}{k}\mathcal{W}_{-1}(-ke^k) - 1) \tag{6.18}$$

where $k = \frac{c\sigma^2}{h_{J_i}} \frac{1}{1-\alpha} \ln 2$, and $\mathcal{W}_{-1}(\cdot)$ is the lth branch of the multi-valued Lambert W function [228].

Assume that the friendly jammers determine the power, denoted as $P_{J_i}^{bid}$, as their dominant bidding strategies. The friendly jammers submit the power as their bids to the source. Upon receiving the bids $P_{J_i}^{bid}$ of the friendly jammers, the source selects the friendly jammer that maximizes the source's secrecy rate as follows:

$$\hat{i} = \arg \max_i R_S(P_{J_i}^{bid}) \tag{6.19}$$

Note that $R_S(P_{J_i}^{bid})$ needs to be greater than the source's lowest acceptable rate, $R_S(P_{J_i}^{bid}) > R_S^0$. Otherwise, no jammer is chosen. According to the Vickrey auction's payment policy, the winner pays the seller a price equal to the second-highest bid. Correspondingly, the winning friendly jammer pays the source the power that produces the second-highest rate. The second-highest rate is determined by

$$R_S^2 = \max\left(\max_{i \neq \hat{i}} R_S(P_{J_i}^{bid}), R_S^0\right) \tag{6.20}$$

When the source receives rate $R_S = R_S^2$, we say that the payment exactly follows the Vickrey auction. The following shows that the source can receive a higher rate, $R_S \geq R_S^2$. The main reason is that even if the winning friendly jammer pays a higher price, its utility may not decrease. We have the following lemma.

LEMMA 6.7 *[229] $R_S(P_{J_i})$ is a concave function of P_{J_i} and achieves the maximum at*

$$[227] P_{J_i}^{max} = \frac{\sigma^2(h_{SE} - h_{SD})}{h_{SD} h_{J_i E} - h_{SE} h_{J_i D}} \tag{6.21}$$

$$+ \sqrt{\frac{\sigma^2 h_{SE} h_{SD}(h_{J_i E} - h_{J_i D})}{h_{J_i E} h_{J_i D}(h_{SD} h_{J_i E} - h_{SE} h_{J_i D})}} \sqrt{\frac{\sigma^2(h_{J_i E} - h_{J_i D})}{(h_{SD} h_{J_i E} - h_{SE} h_{J_i D})}} + \frac{P_S}{\alpha}$$

6.1 Second-Price Sealed-Bid Auction

Proving Lemma 6.7, the concavity of $R_S(P_{J_i})$ and the expression of $P_{J_i}^{max}$, is done using the second- and first-order derivatives as presented in [229]. Note that the utility functions of the friendly jammers are also concave functions of P_{J_i}. Thus, if we consider the secrecy rate of the source as its utility, the utility functions of the source and the friendly jammers are concave in P_{J_i} and have slopes of an equal signum. In particular, the utilities of the source and the friendly jammers may increase or decrease together; that is, their utilities do not conflict with each other. This allows the friendly jammer to pay a more flexible price than in the original Vickrey auction. For example, the friendly jammer can pay the source the power that produces a secrecy rate $R_S > R_S^2$ as long as the utility of the friendly jammer does not decrease. We have the following definition.

DEFINITION 6.8 *The friendly jammer can provide the source any power that produces the secrecy rate of the source at least the second-highest rate, $R_S \geq R_S^2$, where R_S^2 is given in (6.20).*

The modification enables both the source and the friendly jammer to achieve higher utilities compared with the original Vickrey auction. Now we discuss how the friendly jammer selects its dominant bidding strategy, $P_{J_i}^{bid}$.

In general, the friendly jammer can determine $P_{J_i}^{bid}$ by finding a value of P_{J_i} that maximizes its utility, $U_{J_i}(P_{J_i})$. However, the obtained power $P_{J_i}^{bid}$ (i) may be out of the permissible power range of the friendly jammer or (ii) may not make the secrecy rate of the source be at a maximum. Thus, some important values of P_{J_i} are introduced as follows:

- $P_{J_i}^B$: This is the power budget of friendly jammer J_i.
- $P_{J_i}^{lim}$: This is the power at which the utility $U_{J_i}(P_{J_i})$ of the friendly jammer is zero, $U_{J_i}(P_{J_i}^{lim}) = 0$. Note that $U_{J_i}(P_{J_i})$ is negative at $P_{J_i} > P_{J_i}^{lim}$ due to the concavity of $U_{J_i}(P_{J_i})$.
- $P_{J_i}^{max}$: This is the power at which the secrecy rate of the source achieves the maximum. $P_{J_i}^{max}$ is determined according to (6.21).

The dominant bidding strategy of the friendly jammer is then determined by considering the following cases:

- If $P_{J_i}^{max} \leq \min(P_{J_i}^{lim}, P_{J_i}^B)$, then $P_{J_i}^{max}$ is a dominant bidding strategy of the friendly jammer: $P_{J_i}^{bid} = P_{J_i}^{max}$. The reason is that (i) the utility of the friendly jammer is not negative as $P_{J_i}^{max} < P_{J_i}^{lim}$, (ii) the power that the friendly jammer provides the source is not out of the permissible power range since $P_{J_i}^{max} < P_{J_i}^B$, and (iii) the winning probability of the friendly jammer is high since $P_{J_i}^{max}$ produces the maximum secrecy rate of the source.
- If $P_{J_i}^{max} > \min(P_{J_i}^{lim}, P_{J_i}^B)$, since selecting $P_{J_i}^{max}$ yields a negative utility for the friendly jammer or is out of permissible power range, then the dominant bidding strategy of the friendly jammer is to select $\min(P_{J_i}^{lim}, P_{J_i}^B)$.

Based on the aforementioned argument, we have the following theorem.

THEOREM 6.9 *[224] The dominant bidding strategy of friendly jammer J_i is*

$$P_{J_i}^{bid} = \min(P_{J_i}^{max}, P_{J_i}^{lim}, P_{J_i}^{B}) \qquad (6.22)$$

where $P_{J_i}^{max}$ and $P_{J_i}^{lim}$ are given in (6.21) and (6.18), respectively.

Theorem 6.9 means that the friendly jammer has a high chance of winning and achieves a non-negative utility by selecting $P_{J_i}^{bid}$ given in (6.22). Note that $P_{J_i}^{bid}$ is the power that the friendly jammer submits to the source for winner determination; it is not the power that the friendly jammer provides the source if the friendly jammer wins. As stated in Definition 6.8, the friendly jammer is allowed to provide the source with any power that produces the secrecy rate of the source and is at least the second-highest rate. Let $\hat{i}$ denote the winning friendly jammer. Then, the power $P_{J_{\hat{i}}}^*$ that the friendly jammer provides the source is the solution of the following problem:

$$\arg\max_{P_{J_{\hat{i}}}} \quad U_{J_{\hat{i}}}(P_{J_{\hat{i}}}) \qquad (6.23)$$

$$\text{s.t.} \quad R_S(P_{J_{\hat{i}}}) \geq R_S^2$$

where $U_{J_{\hat{i}}}(P_{J_{\hat{i}}})$ is given in (6.17), and R_S^2 is given in (6.20). As presented in [227], the solution of (6.23) is given by

$$P_{J_{\hat{i}}}^* = \begin{cases} P_M, & \text{if } P_{J_{\hat{i}}}^{opt} \geq P_M \text{ and } R_S^2 > 0 \\ P_m, & \text{if } P_{J_{\hat{i}}}^{opt} \leq P_m \text{ and } R_S^2 > 0 \\ P_{J_{\hat{i}}}^{opt}, & \text{if } P_m \leq P_{J_{\hat{i}}}^{opt} \leq P_M \text{ and } R_S^2 > 0 \\ P_{J_{\hat{i}}}^{opt}, & \text{if } R_S^2 = 0 \end{cases}$$

where P_m and P_M, for $P_m \leq P_M$, are the roots of the quadratic equation $R_S(P_{J_{\hat{i}}}) = R_S^2$, and $P_{J_{\hat{i}}}^{opt}$ is the solution of the following problem:

$$\arg\max_{P_{J_{\hat{i}}}} \quad U_{J_{\hat{i}}}(P_{J_{\hat{i}}}) \qquad (6.24)$$

$$\text{s.t.} \quad 0 \leq P_{J_{\hat{i}}} \leq P_{J_{\hat{i}}}^{B}$$

where $U_{J_{\hat{i}}}(P_{J_{\hat{i}}})$ is given in (6.17). In particular, $P_{J_{\hat{i}}}^{opt}$ is given by

$$P_{J_{\hat{i}}}^{opt} = \left[\frac{1-\alpha}{c \ln 2} - \frac{\sigma^2}{h_J} \right]_0^{P_{J_{\hat{i}}}^{B}} \qquad (6.25)$$

where $[a]_{a_{min}}^{a_{max}} = \min(\max(a_{min}, a), a_{max})$.

As observed from (6.25), the power $P_{J_{\hat{i}}}^{opt}$ that the friendly jammer provides to the source is proportional to the bandwidth fraction $1-\alpha$ that the friendly jammer receives. Thus, if the source assigns a small bandwidth fraction $1-\alpha$ to the friendly jammer, then only a small amount of power or even no power, $P_{J_{\hat{i}}}^{opt} = 0$, is provided to the source. This aims to prevent the source from reserving an unfairly large amount of bandwidth α.

The simulation results of the proposed scheme are given in [227]. As shown, the proposed scheme outperforms the physical layer security scheme based on the Stackelberg game [230] when the number of jammers is between 1 and 5. In particular, the difference between them becomes much larger when the number of friendly jammers is higher. The reason is that the proposed scheme uses an auction, which makes the competition among the friendly jammers higher. Thus, as the number of friendly jammers increases, there is a slight or no difference between the maximum secrecy rate and the second-highest one. In other words, the winning friendly jammer may need to provide the source with nearly the maximum secrecy rate. In fact, it will be interesting to consider how the secrecy rate of the source is improved if more friendly jammers are selected to provide power. This can be performed using the generalization of the second-price sealed-bid auction for multiple items, the VCG auction. The theory of the VCG auction and its applications in computer networks are presented in the next sections.

6.2 Vickrey–Clarke–Groves Auction

In this section, we (i) define the **Vickrey–Clarke–Groves (VCG)** auction, (ii) formally describe the VCG auction, (iii) prove the dominant strategy of the VCG auction, and (iv) present the virtues of the VCG auction.

6.2.1 Definition

The VCG auction is a generalization of the Vickrey auction with multiple items. Indeed, in the case of a single item, the VCG auction is exactly the same as the Vickrey auction discussed in Section 6.1. Thus, the VCG auction process is essentially similar to the Vickrey auction. Specifically, the model typically consists of multiple bidders and one seller. This seller can act as an auctioneer. The bidders submit their bids to the seller. The bids are the prices that the bidders are willing to pay the seller. The seller determines the winner and the price that the winner needs to pay the seller. However, the VCG auction often has more items for trading, so there can be multiple winners. Also, each winner can receive more than one item. Moreover, in the VCG auction, the winners and the prices are determined so as to maximize the social value of the system. Thus, the winner and price determination in the VCG auction may have much higher computational complexity than those in the Vickrey auction. Further details of the winner and price determination are given in the next section. The general idea is presented in the following discussion.

To determine the winner, after receiving the bids, the seller makes all potential combinations of bids. Among the potential combinations of bids, the seller selects the best bid that satisfies the following conditions:

- The number of items requested by the corresponding bidders does not exceed the total number of items that the seller provides.

- The total bid, comprising the total price or total value, of the best combination is the highest.
- Each item is assigned to at most one bidder.

The winner determination in the VCG auction is formulated as a knapsack problem. This problem is a combinatorial optimization that determines the number of bids for a given limit number of items so that the total bid is as large as possible. The total bid is also the total value since the VCG auction exhibits truthfulness: the users' bids are their true values of the items. The combination of the selected bids is called the *best combination* of the auction, and those bidders that have the selected bids are the winners of the auction.

The prices that the winners pay the seller are determined in a socially optimal manner. In general, each winner is charged a price equal to the loss of the social value, also known as the social welfare, due to the winner's getting the item. In general, the loss of the social value is measured by how much social cost/opportunity cost the existence of the winner harms/hurts the other bidders. This actually is the winner's externality. Note that the social value is often associated with a specific combination of bids, and it is typically defined as the total bid of the combination. For example, given bids b_1 and b_2, then the social value of bid combination (b_1, b_2) is $b_1 + b_2$.

In general, the VCG auction has three major advantages.

- The auction is applicable for trading multiple items.
- This auction has one attractive property, incentive compatibility. With this property, bidders bid their true values of the items and do not need to know how much other bidders are willing to bid.
- The VCG auction guarantees the maximization of social welfare.

Despite these advantages, the VCG auction has some shortcomings, such as low revenue for the seller and vulnerability to collusion among the bidders. Specifically, the winner determination in this auction is a combinatorial optimization problem. Thus, the winner determination problem can be exceedingly complex to solve, as the number of bidders is large. However, in computer networks, this challenge can be avoided – for example, by using linear programming models and applying weighted bipartite matching algorithms, which can be optimally solved in polynomial time. Thus, the VCG auction is still used as an effective solution to a number of issues. Section 6.2.6 reviews the applications of the VCG auction for task allocation [231], optimal incentive-driven design [98], resource sharing [232], and wireless caching [233]. The following section provides a formal description of the VCG auction.

6.2.2 Description

We consider a spectrum trading market as follows:

- The market consists of N mobile users as the bidders and one service provider as the seller. Let $\mathcal{N}$ denote the set of the mobile users, $\mathcal{N} = \{1, \ldots, N\}$.
- The service provider trades a set $\mathcal{M}$ of M bandwidth units, $\mathcal{M} = \{t_1, \ldots, t_M\}$.

- Let $v_i(t_j)$ denote the value of bandwidth unit t_j to user i, and let $b_i(t_j)$ denote user i's bid for bandwidth unit t_j. In a simple market setting, we assume that each user wants to buy one bandwidth unit, and the value of every bandwidth unit in set $\mathcal{M}$ is identical to the user; that is, the user does not differentiate between the bandwidth units. Thus, we can express $v_i(t_j)$ as v_i and $b_i(t_j)$ as b_i.

After receiving the bids of the users, the service provider determines the winners of the bandwidth units and the prices that the winners need to pay. For the winner determination, the service provider makes all potential combinations of bids. The service provider then selects the combination that has the highest total bid as the best combination. The users that have the selected bids are the winners of the auction. For example, in the case with $N = 3$ users and $M = 2$ bandwidth units, the potential bid combinations are (b_1, b_2), (b_1, b_3), and (b_2, b_3). The service provider calculates the total bid for each combination: $b_1 + b_2$ for (b_1, b_2), $b_1 + b_3$ for (b_1, b_3), and $b_2 + b_3$ for (b_2, b_3). The combination with the highest total bid is selected as the best combination for the two bandwidth units.

In fact, for the market setting in which each user bids for one bandwidth unit, the service provider can simply determine the best combination as follows. It first sorts the users in a descending order of their bids, and then selects the first M users as the winners of the bandwidth units. The best combination includes the bids of the winners. We assume that user i is one of the winners, and we need to determine the price that the user pays the service provider for receiving bandwidth unit.

As defined in Section 6.2.1, the price actually is the externality imposed by user i on the other users. To determine the externality, we need to determine two metrics [234]. Note that, for convenience, we consider the auction including all N users to be the *original* auction and the auction with none of the users to be the *modified* auction.

- The first metric is the social value, or total bid, of the best combination when user i does not participate in the auction. To calculate the first metric, we remove user i from the original auction and then determine the best combination of bids of this modified auction. Then, the first metric, the social value, is the sum of bids in the best combination. Let $V_{\mathcal{N}\setminus\{-i\}}$ denote the social value of the best combination of the modified auction.
- The second metric is the social value, or total bid, without user i's bid of the best combination of the original auction. As such, to calculate the second metric, we determine the best combination of bids of the original auction. Then, the sum of bids excluding user i's bid in the best combination is the second metric, denoted as $V_{\mathcal{N}}^{-b_i}$.

Finally, the price that user i pays the service provider is given by

$$p_i = V_{\mathcal{N}\setminus\{-i\}} - V_{\mathcal{N}}^{-b_i} \qquad (6.26)$$

The payoff that user i receives is given by

$$\pi_i = v_i - p_i = v_i - (V_{\mathcal{N}\setminus\{-i\}} - V_{\mathcal{N}}^{-b_i}) \qquad (6.27)$$

Again, the difference between $V_{\mathcal{N}\setminus\{-i\}}$ and $V_{\mathcal{N}}^{-v_i}$ given in (6.26) can be interpreted as the externality imposed by user i on other users or as the opportunity cost of the bandwidth won by user i. Note that user i pays p_i only if it is among the winners of the auction. If the user loses the auction, it does not need to pay and receives a zero payoff: $\pi_i = 0$.

To clarify this process, we consider the case in which $N = 3$ users (i.e., 1, 2, and 3) and $M = 2$ bandwidth units (i.e., t_1 and t_2). The users submit their bids (i.e., b_1, b_2, and b_3) to the service provider. Upon receiving the bids, the service provider determines the winners and the payments as follows:

- The service provider makes all potential combinations of bids for the two bandwidth units. The potential combinations are (b_1, b_2), (b_2, b_3), and (b_1, b_3).
- The service provider calculates the total bid in each potential combination, and thus we have $b_1 + b_2$, $b_2 + b_3$, and $b_1 + b_3$.
- The service provider selects the combination that has the highest total bid as the best combination of the original auction. The users with the selected bids are the winners of the auction.
- We assume that user 1 with bid b_1 is one of the two winners, meaning that $b_1 > \min(b_2, b_3)$. According to (6.26), the price that user 1 pays the service provider is determined based on the two metrics $V_{\mathcal{N}\setminus\{-1\}}$ and $V_{\mathcal{N}}^{-b_1}$.
- To determine the first metric $V_{\mathcal{N}\setminus\{-1\}}$, we remove user 1 from the original auction. The modified auction now consists of users 2 and 3. There are two items, and thus the best combination of the modified auction is (b_2, b_3). Then, $V_{\mathcal{N}\setminus\{-1\}}$ is the sum of bids in the best combination: $V_{\mathcal{N}\setminus\{-1\}} = b_2 + b_3$.
- For the term $V_{\mathcal{N}}^{-b_1}$, we consider the original auction, and the best combination of this auction is $(b_1, \max(b_2, b_3))$. $V_{\mathcal{N}}^{-b_1}$ is the sum of bids in the best combination excluding user 1's bid, $V_{\mathcal{N}}^{-b_1} = \max(b_2, b_3)$.
- The price that user 1 pays is given by

$$p_1 = b_2 + b_3 - \max(b_2, b_3) \tag{6.28}$$

- The payoff that user 1 receives is $\pi_1 = v_1 - p_1$, which can be generally expressed by

$$\pi_1 = \begin{cases} v_1 - (b_2 + b_3 - \max(b_2, b_3)), & \text{if } b_1 > \min(b_2, b_3) \\ 0, & \text{if } b_1 < \min(b_2, b_3) \end{cases} \tag{6.29}$$

The case $b_1 = \min(b_2, b_3)$ can be ignored since it occurs with a low probability.

In particular, if $b_3 < b_2$, the price that user 1 pays is given by

$$p_1 = b_3 \tag{6.30}$$

Correspondingly, the payoff of user 1 is given by

$$\pi_1 = \begin{cases} v_1 - b_3, & \text{if } b_1 > b_3 \\ 0, & \text{if } b_1 < b_3 \end{cases} \tag{6.31}$$

It can be seen from (6.31) that user 1 pays the service provider the price equal to the bid of the loser of the auction. For the case with multiple losers, the price that the winner pays is the highest bid among bids of the losers. Thus, if there is only a single bandwidth unit for trading, the payment rule of the VCG auction is exactly the second-price rule of the Vickrey auction with a single item, as discussed in Section 6.1.

6.2.3 Dominant Strategy

This section discusses the dominant bidding strategy of users in the VCG auction. We prove that the dominant bidding strategy of the users is to bid their true value, $b_i = v_i$, when participating in the auction. The proof can be found in the literature, such as [235] and [236]. To make the argument easier to follow, we prove the dominant bidding strategy in the simple market setting as discussed in Section 6.2.2, which includes $N = 3$ users and $M = 2$ bandwidth units. Without loss of generality, we prove that the dominant strategy of user 1 is to bid its true value, $b_1 = v_1$. Similar to the Vickrey auction, we consider two other strategies that user 1 can select: an overbid and an underbid. Then, we compare the payoff of the user when it selects the two strategies with its payoff when it selects the truthful bid.

First, we compare the payoffs of the user when it engages in overbidding with an overbid $b_1 > v_1$ and in truthful bidding with a truthful bid $b_1 = v_1$. There are three corresponding cases.

- If $\min(b_2, b_3) < v_1$, then user 1 is one of the two winners when it selects the overbid and the truthful bid. The payoffs of the user for the two strategies are the same and are given by $\pi_1 = v_1 - (b_2 + b_3 - \max(b_2, b_3))$ (see (6.29)). In this case, bidding an overbid does not change the user's payoff.
- If $b_1 < \min(b_2, b_3)$, then $v_1 < \min(b_2, b_3)$. Thus, user 1 does not win the auction with the overbid as well as the truthful bid. The payoffs of the user for the two strategies are the same: zero. As such, overbidding does not change the user's payoff.
- If $v_1 < \min(b_2, b_3) < b_1$, then user 1 is one of the two winners when selecting the overbid. The payoff of the user for this strategy is $\pi_1 = v_1 - (b_2 + b_3 - \max(b_2, b_3))$. It is easy to check that this payoff is negative since the user paid more than its value of the bandwidth unit For example, if $b_2 > b_3$, then $\pi_1 = v_1 - b_3 < 0$ since $v_1 < b_3$. If the user engages in truthful bidding, its payoff is zero since the user loses in the auction. As such, the payoff of the user when selecting truthful bidding is higher than when selecting overbidding.

A similar argument shows that the payoff of the user when selecting the underbidding strategy, $b_1 < v_1$, is less than or equal to that when selecting the truthful bidding strategy, $b_i = v_i$.

In summary, the payoff of the user that engages in truthful bidding is always higher than or at least equal to that of the user that selects either the overbidding or underbidding strategies.

6.2.4 Examples

This section provides examples of using the VCG auction in the bandwidth trading markets.

Example 6.2 We consider again the bandwidth trading market with three mobile users and two bandwidth units. In the market, we assume that

- User 1 wants to buy one bandwidth unit and submits bid $v_1 = b_1 = \$1.5$ to the service provider for the bandwidth unit.
- User 2 wants to buy one bandwidth unit and submits bid $v_2 = b_2 = \$1.2$ to the service provider.
- User 3 wants to buy one bandwidth unit and submits bid $v_3 = b_3 = \$1.0$ to the service provider.

Upon receiving the bids, the service provider makes potential combinations of bids. The potential combinations of bids are (v_1, v_2), (v_1, v_3), and (v_2, v_3). The service provider calculates the total bid for each combination as follows:

- The total bid of (v_1, v_2) is $\$1.5 + \$1.2 = \$2.7$.
- The total bid of (v_1, v_3) is $\$1.5 + \$1.0 = \$2.5$.
- The total bid of (v_2, v_3) is $\$1.2 + \$1.0 = \$2.2$.

Since (v_1, v_2) has the highest total bid, it is selected as the best combination of the original auction. Correspondingly, users 1 and 2 are the winners of the auction.

Now, we determine the prices that the winners, users 1 and 2, pay the service provider. For each user, we need to calculate two metrics given in (6.26).

- For user 1: We remove user 1 from the original auction, and the modified auction has two users, 2 and 3, and two bandwidth units. The best combination of the modified auction is (v_2, v_3), which has a total bid of $\$1.2 + \$1.0 = \$2.2$. The best combination of the original auction is (v_1, v_2), and the second metric is the total bid excluding user 1's bid. Thus, the second metric is $\$0 + \$1.2 = \$1.2$. The price that user 1 pays is $\$2.2 - \$1.2 = \$1.0$. User 1 receives a payoff of $\$1.5 - \$1.0 = \$0.5 > 0$.
- For user 2: Similarly, we remove user 2 from the auction, and the best combination of the modified auction is (v_1, v_3), which has a total bid of $\$1.5 + \$1.0 = \$2.5$. The total bid excluding user 2's bid is the best combination of the original auction: $\$1.5 + \$0 = \$1.5$. Thus, the price that user 2 pays is $\$2.5 - \$1.5 = \$1.0$. User 2 receives a payoff of $\$1.2 - \$1.0 = \$0.2 > 0$.

The revenue of the service provider is the sum of the prices paid by the users 1 and 2: $\$1.0 + \$1.0 = \$2.0$.

We can see that the value of $\$1.0$ that users 1 and 2 pay the service provider is actually bid b_3 of user 3. Thus, the revenue of the service provider is low if user 3 submits a low bid.

In practice, the users can buy more than one bandwidth unit. The following example considers such a scenario.

Example 6.3 Consider again the market in which user 3 wants to buy more than one bandwidth unit. Specifically, user 3 wants to buy two bandwidth units. User 3 submits a bid of $v_3 = \$1.8$ to the service provider for the two bandwidth units. In this case, there are two potential combinations of bids, (v_1, v_2) and (v_3). The total bid of combination (v_1, v_2) is $\$1.5 + \$1.2 = \$2.7$ and the total bid of (v_3) is $\$1.8$. Thus, (v_1, v_2) is selected as the best combination since it has the highest total bid. Correspondingly, users 1 and 2 are the winners of the original auction. The prices that users 1 and 2 pay are determined as follows:

- For user 1: We remove user 1 from the auction, and the modified auction includes users 2 and 3. In the auction, the potential combination is (v_3). Thus, user 3 is the winner, and (b_3) is the best combination of the modified auction. The total bid of the best combination of the modified auction is $\$1.8$. The total bid excluding user 1's bid of the best combination of the original auction is $\$0 + \$1.2 = \$1.2$. Thus, the price that user 1 pays is $\$1.8 - \$1.2 = \$0.6$. User 1 receives a payoff of $\$1.5 - \$0.6 = \$0.9 > 0$.
- For user 2: We remove user 2 from the auction. In the modified auction including users 1 and 3, user 3 is the winner. The best combination is (v_3), which has a total bid of $\$1.8$. The total bid excluding user 2's bid is the best combination of the original auction: $\$1.5 + \$0 = \$1.5$. Thus, the price that user 2 pays the service provider is $\$1.8 - \$1.5 = \$0.3$. User 2 receives a payoff of $\$1.2 - \$0.3 = \$0.9 > 0$.

The revenue of the service provider is $\$0.6 + \$0.3 = \$0.9$.

6.2.5 Virtues

The VCG auction has the following important properties.

- Incentive compatibility or truthfulness: As proved in Section 6.2.3, truthful bidding is the dominant strategy of the users in the VCG auction. Accordingly, the dominant strategy of each user is to report its true value for the bandwidth unit without knowing how much other bidders bid. As such, the bandwidth units are guaranteed to be allocated to the users with the highest values. As a result, the allocation approaches that use the VCG auction are efficient.
- Social welfare: This is guaranteed since given a set of the bandwidth units, the allocation, or winner determination, of the VCG auction selects users that have the highest total bid, or highest total value.
- Individual rationality: This means that the payoff of the users is not negative when participating in the auction. As discussed in Section 6.2.3, truthful bidding ensures that the payoffs of the users are not negative. This property provides the users with an incentive to participate in the auction.

Given these properties, the VCG auction has been used to solve several issues in computer networks, as described in the next section.

6.2.6 Development of VCG Auction for Computer Networks

This section presents applications of the VCG auction for (i) task assignment in mobile device clouds, (ii) resource sharing in massive MIMO, (iii) collaborative caching in video streaming systems, and (iv) spectrum allocation in cognitive radio networks.

Task Assignment in Mobile Device Clouds

Mobile device clouds are one of the edge computing models in which a set of nearby mobile devices is used for executing computing tasks. Mobile device clouds can be considered to be a client-assisted cloud system. Such a system aims to reduce the network traffic and resource burden at servers at volunteer computing systems, cloudlets, and remote data centers. As a result, mobile device clouds can support real-time applications. However, executing computing tasks incurs extra costs, such as battery outages and bandwidth consumption. Thus it is essential to design an incentive mechanism that compensates for the costs to the task requesters (i.e., the customers or buyers) and provides rewards to the mobile devices (i.e., the sellers). The VCG auction can be used to achieve individual rationality, meaning a non-negative payoff for the mobile devices, and truthfulness, meaning efficient task allocation and low cost for the task requesters.

This section discusses how to design the incentive mechanism based on the VCG auction for task allocation in mobile device clouds. This mechanism is proposed in [231]. Since the considered model consists of multiple sellers and one buyer, the auction is actually a VCG reverse auction. The VCG reverse auction is similar to the original VCG auction, except that the roles of the buyers and the sellers are reversed. Also, the VCG reverse auction is similar to the Vickrey reverse auction (see Section 6.1.3) for multiple items.

System model: The system model is shown in Figure 6.4.

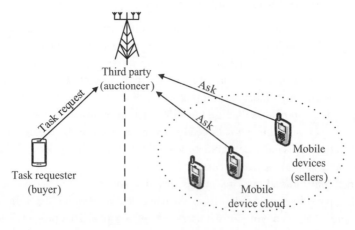

Figure 6.4 Mobile device cloud.

6.2 Vickrey–Clarke–Groves Auction

- There is a set of N mobile devices, the sellers, in the system that are willing to participate in the task execution. Let $\mathcal{N} = \{1,\ldots,N\}$ denote the set of the devices.
- There is one task requester, the buyer. The task requester has a set $\mathcal{M}$ of M indivisible tasks, $\mathcal{M} = \{t_1,\ldots,t_M\}$, to be allocated to mobile devices in the mobile device cloud.
- A trusted third party acts as the auctioneer. The third party determines the winning devices for the tasks and the corresponding prices paid to the winners.
- The tasks are assumed to be homogeneous, meaning that the tasks require the same amount of resources, such as CPU, battery, and bandwidth. Let w denote the amount of resources required for a task, where w is the same for every task.
- The mobile device has limited resources. Let W_i denote the maximum amount of resource that mobile device i can provide for executing the tasks.
- The mobile device submits a set of its asks for the tasks. The set is denoted as $\mathcal{A}_i = \{a_{i,1},\ldots,a_{i,M}\}$, where $a_{i,j}$ represents the ask submitted by mobile device i for task j. The ask refers to the cost for executing task t_j. The mobile device is selfish, so it may misreport its real cost for executing the task to improve its payoff. Let $c_{i,j}$ denote the real cost of device i for executing task t_j. $a_{i,j}$ may not be the same as $c_{i,j}$.
- For convenience, let $\mathcal{A} = \bigcup_{i=1}^{N} \mathcal{A}_i$.
- Let $x_{i,j}$ define a variant associated with device i. $x_{i,j} = 1$ if device i wins the auction for executing task t_j, and $x_{i,j} = 0$ if the device loses the auction for executing task t_j.

Problem formulation: The winner determination problem is to determine the value, 0 or 1, of $x_{i,j}$ so as to minimize the sum of the submitted asks for executing tasks at mobile devices. Note that the total amount of resources required of each device for executing the tasks is not greater than its available resource. Moreover, each task is assigned to at most one device, since the task is indivisible. Thus, the winner determination problem is formulated as follows:

$$\min \sum_{i=1}^{M} \sum_{j=1}^{N} x_{i,j} a_{i,j} \quad (6.32)$$

$$\text{s.t.} \sum_{j=1}^{M} x_{i,j} w \leq W_i, \forall i \in \mathcal{N} \quad (6.33)$$

$$\sum_{i=1}^{N} x_{i,j} = 1, \forall t_j \in \mathcal{M} \quad (6.34)$$

Assume that $x_{i,j} = 1$. We say that ask $a_{i,j}$ is selected and device i wins the auction for executing task t_j. Based on the selected asks, the auctioneer determines the price $p_{i,j}$ paid to device i for executing task t_j. Then, the payoff of device i for executing task t_j is defined as follows:

$$\pi_{i,j} = p_{i,j} - x_{i,j} c_{i,j} \qquad (6.35)$$

In general, the proposed scheme needs to be designed to achieve the following properties: individual rationality, truthfulness, and computational efficiency. The VCG auction can guarantee truthfulness and individual rationality. However, it is hard to achieve computational efficiency since the winner determination problem of the VCG auction is a combinatorial optimization. To address this challenge, the winner determination problem is converted into finding the minimum weight matching in an auxiliary bipartite graph, which can be optimally solved in polynomial time. Then, the payment rule of the VCG auction is used to guarantee the individual rationality and the truthfulness. Further details are presented in the following.

Winner determination: To use the bipartite graph algorithm, we need to show how to construct the bipartite graph from the auction model. Recall that the bipartite graph is typically denoted as $G = (\mathcal{H}, \mathcal{L}, \mathcal{E})$, where $\mathcal{H}$ and $\mathcal{L}$ are two non-overlapping sets that contain the vertices of the graph, and $\mathcal{E}$ is the set of all edges between the vertices of the two sets. Assume that each edge is assigned with a cost, or weight. Then running a minimum weight matching algorithm produces a *matching* set, which consists of edges in $\mathcal{E}$ that satisfies two conditions. The first condition is that the edges have the total lowest cost, and the second condition is that the edges have not shared endpoints – that is, shared vertices.

For the context considered here, the bipartite graph is shown in Figure 6.5 and constructed as follows:

- The maximum amount of resources that device i can provide is W_i, and the amount of resources required for every task is w. Thus, the maximum number of tasks that i can execute is $\lfloor \frac{W_i}{w} \rfloor$. Let $M^i = \min\left\{\lfloor \frac{W_i}{w} \rfloor, M\right\}$. The constraint in (6.33) can be rewritten as follows:

$$\sum_{j=1}^{M} x_{i,j} \leq M^i \qquad (6.36)$$

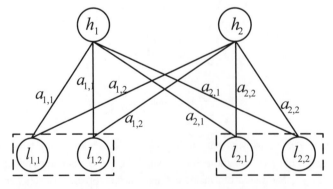

Figure 6.5 Illustration of a bipartite graph in which $N = 2$, $M = 2$, $M^1 = M^2 = 2$ [231].

6.2 Vickrey–Clarke–Groves Auction

- For $\mathcal{H}$: $\mathcal{H}$ is a set of vertices h_j. Each vertex h_j represents task t_j. There are M tasks, and thus $\mathcal{H} = \{h_1, \ldots, h_M\}$.
- For $\mathcal{L}$: $\mathcal{L}$ is a set of vertices $l_{i,m}, 1 \leq m \leq M^i$, where M^i is the maximum number of tasks that device i can execute. Thus, $\mathcal{L} = \{l_{1,1}, \ldots, l_{1,M^1}, \ldots, l_{N,1}, \ldots, l_{N,M^N}\}$.
- For $\mathcal{E}$: $\mathcal{E}$ is a set of edges between vertices h_j and $l_{i,m}$. Thus, we have $\mathcal{E} = \{(h_j, l_{i,m}) | h_j \in \mathcal{H}, l_{i,m} \in \mathcal{L}\}$.
- For each edge $(h_j, l_{i,m})$: We assign a weight equal to ask $a_{i,j}$ to the edge.

It is proved in [231] that finding a minimum weight bipartite matching in graph $G = (\mathcal{H}, \mathcal{L}, \mathcal{E})$ is equivalent to finding the solution of the winner determination problem given in (6.32). Running existing minimum weight matching algorithms [237] in graph $G = (\mathcal{H}, \mathcal{L}, \mathcal{E})$ has polynomial-time computational complexity, equivalent to computational efficiency. After running the algorithms, we have a matching set that consists of edges with the total lowest weight. Assume that $(h_j, l_{i,j})$ is one of the edges in the matching set. Then $x_{i,j} = 1$, and we say that ask $a_{i,j}$ of device i is selected for task t_j. As such, the achieved matching set is the solution of the winner determination problem in (6.32), which guarantees the minimum overall cost of the task allocation.

Price determination: We need to determine the price that the task requester pays device i for executing task t_j. The price determination is based on the payment strategy of the VCG auction given in (6.26), but bids are replaced by asks. Accordingly, we calculate two metrics:

- The first metric is denoted as $C_{\mathcal{A} \setminus \{a_{i,j}\}}$. We remove ask $a_{i,j}$ from the original auction and find the solution, the ask selection, in the modified auction. Then $C_{\mathcal{A} \setminus \{a_{i,j}\}}$ is the sum of these selected asks.
- The second metric, denoted by $C_\mathcal{A}$, is the sum of the selected asks in the original auction. Thus, this second metric is the sum of selected asks excluding ask $a_{i,j}$: $C_\mathcal{A} - a_{i,j}$.

Then, the payment that device d_i receives for executing task t_j is determined as

$$p_{i,j} = C_{\mathcal{A} \setminus \{a_{i,j}\}} - (C_\mathcal{A} - a_{i,j}) \tag{6.37}$$

The payoff that device d_i receives for executing task t_j is given by

$$\begin{aligned}\pi_{i,j} &= p_{i,j} - c_{i,j} \\ &= C_{\mathcal{A} \setminus \{a_{i,j}\}} - (C_\mathcal{A} - a_{i,j}) - c_{i,j}\end{aligned} \tag{6.38}$$

The payment rule given in (6.37) essentially follows the payment strategy of the VCG auction. Intuitively, the device has an incentive to truthfully submit its bid, referring to its real cost for executing the task, since the payment that the device receives does not depend on the submitted ask. The proof of the truthfulness of the payment rule is presented in [231]. We have the following lemma.

LEMMA 6.10 *[231] The proposed scheme based on the VCG auction guarantees the truthfulness of the asks.*

Lemma 6.10 implies that the device achieves the best payoff only if it submits a real cost, $a_{i,j} = c_{i,j}$. Therefore, the payoff in (6.38) of device i for executing task t_j can be expressed by

$$\pi_{i,j} = C_{\mathcal{A}\setminus\{a_{i,j}\}} - (C_{\mathcal{A}} - a_{i,j}) - c_{i,j} \qquad (6.39)$$
$$= C_{\mathcal{A}\setminus\{a_{i,j}\}} - C_{\mathcal{A}}$$

Note that $C_{\mathcal{A}}$ is the sum of selected asks in the original auction, and $C_{\mathcal{A}\setminus\{a_{i,j}\}}$ is the sum of selected asks in the modified auction, referring to the original auction without the participant $a_{i,j}$. Therefore, $C_{\mathcal{A}}$ is "more optimal" than $C_{\mathcal{A}\setminus\{a_{i,j}\}}$, in that $C_{\mathcal{A}}$ must be less than or equal to $C_{\mathcal{A}\setminus\{a_{i,j}\}}$: $C_{\mathcal{A}} \leq C_{\mathcal{A}\setminus\{a_{i,j}\}}$. As a result, the payoff $\pi_{i,j}$ of the device is non-negative, and we have the following lemma.

LEMMA 6.11 *[231] The proposed scheme based on the VCG auction achieves individual rationality.*

Apart from truthfulness and individual rationality, the proposed scheme achieves computational efficiency. Indeed, the winner determination problem is solved with an existing minimum weight matching algorithm. Given the number of tasks M and the number of devices N, the complexity of this algorithm is at most $(N^3 M^3)$. Moreover, the complexity of determining the payments for all the selected bids is $(N^3 M^4)$. The proposed scheme can be thus implemented in polynomial time, which is computationally efficient.

To validate the performance of the proposed scheme, the simulation is implemented in [231]. For individual rationality, the simulation results show that the price paid to the device is no less than the bid that the device submits. In regard to truthfulness, the device achieves the maximum payoff when it submits its real cost for executing the task. The payoff is lower, as the difference between the submitted bid and the real cost is larger. In terms of computational efficiency, the proposed scheme has a higher overall execution cost but a lower running time than does the optimal task allocation scheme, the exhaustive searching algorithm. For the future work, the dynamic nature, or movement, of the devices can be considered.

Resource Sharing in Massive Multiple-Input Multiple-Output Systems

This section presents the application of the VCG auction for resource allocation in massive Multiple-Input Multiple-Output (MIMO) systems. As presented in Section 2.4, massive MIMO is a key technology of the 5G wireless network, in which each base station is equipped with a number of antennas. The antennas can be shared among a number of users through slicing technologies, an approach called network virtualization. In particular, the slicing technologies partition/slice/virtualize physical resources such as antennas, bandwidth, and power, of the base station into *spatial streams*. The spatial streams are then allocated to users based on their demands. However, given a high density of users, how to efficiently allocate the spatial streams to the users is a critical issue. To achieve efficiency as well as desirable economic properties, the VCG auction can be adopted for resource allocation in massive MIMO systems as proposed in [232].

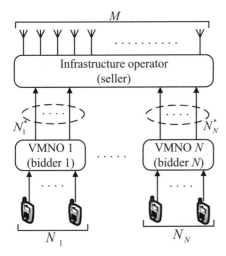

Figure 6.6 Spatial streaming sharing in massive MIMO, where VMNO stands for virtual mobile network operator [232].

The system model is shown in Figure 6.6 and includes the following entities:

- There are one infrastructure provider, the seller or the auctioneer, and N mobile virtual network operators, bidders or buyers.
- The infrastructure provider owns and operates a massive MIMO system, which can be a massive MIMO base station. The massive MIMO system has M antennas and $\tilde{M}$ spatial streams for trading. Correspondingly, the maximum number of users that can be served is $\tilde{M}$.
- Each mobile virtual network operator i can serve N_i users. Thus, the total number of users of N mobile virtual network operators is $\sum_{i=1}^{N} N_i$.
- Assume that the infrastructure provider can estimate channel gains of the users, for example, based on pilot signals.

The general idea of the proposed scheme is as follows. First, each mobile virtual network operator determines an optimal bid so as to maximize its payoff. The bid includes the number of spatial streams (i.e., the number of served users) and the price per spatial stream. The mobile virtual network operators then submit their bids to the infrastructure provider. Based on the bids, the infrastructure provider determines the winners and the corresponding prices that the winners pay according to the VCG auction. Further details are presented here:

- The infrastructure provider sends information, including the channel gain estimates of users and the maximum number of antennas M, to the mobile virtual network operators.
- Based on the information, each mobile virtual network operator i calculates the data rate $R_{i,j}$ that user j can achieve. In general, the data rate $R_{i,j}$ increases with the number of antennas and decreases with the number of served users.

- The mobile virtual network operator then determines an optimal bid, which comprises its bidding strategy, so as to maximize its own payoff. The bid consists of the optimal number of spatial streams and the value of each spatial stream to the operator. Let $N_i^* \leq N_i$ denote the optimal number of spatial streams, and let v_i denote the value of one spatial stream to mobile virtual network operator i. The mobile virtual network operator determines the values of the two variables, N_i^* and v_i as follows:

  - First, the mobile virtual network operator constructs its payoff. The payoff is a function of the revenue that it receives for serving its users and the price that it pays the infrastructure operator for receiving the spatial streams. In particular, the payoff is given by

  $$\pi_i = r_i(N_i) - p_i N_i \tag{6.40}$$

  where p_i is the price that the mobile virtual network operator needs to pay the infrastructure operator for receiving one spatial stream, and $r_i(N_i)$ is the revenue that the mobile virtual network operator receives from serving its N_i users. In general, r_i is proportional to the data rates $R_{i,j}$ that the users can achieve. One potential formulation of r_i is $r_i = p^0 \sum_{j=1}^{N_i} R_{i,j}$ [232], where p^0 is the price per unit of data rate that the mobile virtual network operator receives from its users. In fact, the mobile virtual network operator does not know p_i until the infrastructure provider determines the winners and prices. However, the mobile virtual network operator can estimate a value $\hat{p}_i$ of p_i. For example, $\hat{p}_i$ can be estimated from the last auction. The payoff of the mobile virtual network operator can be thus expressed by

  $$\pi_i = p^0 \sum_{j=1}^{N_i} R_{i,j} - \hat{p}_i N_i \tag{6.41}$$

  - Then, the optimal number of spatial streams N_i^* is determined as follows:

  $$N_i^* = \arg\max_{N_i} \pi_i \tag{6.42}$$

  $$= \arg\max_{N_i} \left(p^0 \sum_{j=1}^{N_i} R_{i,j} - \hat{p}_i N_i \right)$$

  Equation (6.42) can be simply solved by calculating values π_i for different values of N_i and then selecting N_i^* to maximize π_i.

  - Let $\pi_i^* = p^0 \sum_{j=1}^{N_i^*} R_{i,j}$ be the payoff at $N_i = N_i^*$. Then π_i^* is the maximum payoff that the mobile virtual network operator expects when deciding to buy the N_i^* spatial stream. Thus, the value v_i per spatial stream of the mobile virtual network operator can be determined by

  $$v_i = \frac{p^0 \sum_{j=1}^{N_i^*} R_{i,j}}{N_i^*} \tag{6.43}$$

- After calculating N_i^* and v_i, the virtual mobile network operators submit the combination bids (N_i^*, v_i) to the infrastructure provider. The infrastructure provider adopts the VCG auction to determine the winners and the prices as follows:

    - The winner determination is implemented in multiple rounds. In particular, in the first round, the virtual mobile network operator that has the highest value v_i and $N_i^* < \tilde{M}$ is selected as the winner. N_i^* spatial streams are then allocated to this winner. The winner selection process is repeated for the remaining virtual mobile network operators until the total number of spatial streams allocated to the winners is equal to or greater than the maximum number of spatial streams of the infrastructure provider: $\sum_i N_i^* \geq \tilde{M}$. Let $\mathcal{N}$ be the set of the winners.
    - The price that winner i pays the infrastructure provider for receiving N_i^* spatial streams is determined according to the payment strategy of the VCG auction. Let $\mathcal{N}'$ denote the set of winners of the auction when virtual mobile network operator i does not participate in the auction. The price for N_i^* spatial streams is given by

$$\sum_{j \in \mathcal{N}'} N_j^* v_j - \left(\sum_{j \in \mathcal{N}} N_j^* v_j - N_i^* v_i \right) \qquad (6.44)$$

The simulation results provided in [232] show that the proposed scheme outperforms the equally divided spectrum allocation in terms of the average data rate of each virtual mobile network operator as well as the total system data rate. However, compared with the optimal assignment scheme, the proposed scheme has a slight performance loss. The reason may relate to the estimation of $\hat{p}_i$, which may not be accurate when it is taken from the last auction. The inaccurate estimation of $\hat{p}_i$ results in an inaccurate determination of the optimal number of spatial streams N_i^*, which affects the performance. To address this issue, learning algorithms such as Q-learning can be used to get a more accurate value of $\hat{p}_i$.

Collaborative Caching in Wireless Video Streaming Systems
To facilitate video content delivery and to improve the user experience in wireless video streaming systems, wireless service providers can deploy cache servers at base stations close to users for storing video content. However, given the constraints of wireless bandwidth and the mobility of the users, it is challenging to fully accommodate the dynamic and unpredictable demands of users. Therefore, it is critical for the service providers to collaborate with each other. For example, a service provider can use its bandwidth and storage resources to assist other service providers to stream video content required by their users. However, the service providers are selfish, so collaborative mechanisms need to be designed so as to guarantee non-negative payoffs for the service providers, truthfulness, and social welfare maximization. The VCG auction can achieve the social welfare maximization goal while guaranteeing individual rationality and truthfulness. Thus, this auction can be used to design the collaborative mechanism, as proposed in [233].

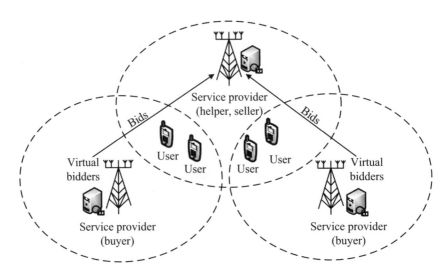

Figure 6.7 Collaborative caching using VCG auction.

The system model shown in Figure 6.7 consists of entities working as follows:

- There are one service provider, the seller or *helper*, which assists N other service providers, the buyers, to serve the video content requests of their users. Let $\mathcal{N} = \{1, \ldots, N\}$ be the set of the service providers.
- Let $\mathcal{K} = \{1, \ldots, K\}$ be a set of available video channels, or video programs, that the helper can serve.
- The helper needs to use its bandwidth and storage resources to deliver the video channels. Typically, the bandwidth resources are scarcer and have much higher cost than do the storage resources. Thus, only bandwidth resources are auctioned.
- Let W denote the available bandwidth capacity of the helper, and let W^0 denote a bandwidth unit. Then, the number of homogeneous bandwidth units that the helper can use to serve other service providers is $M = \lfloor \frac{W}{W^0} \rfloor$.
- For a given service provider in set $\mathcal{N}$, its users can request different video programs as well as different numbers of bandwidth units. Users that belong to the same service provider and request the same video program $k \in \mathcal{K}$ can be collectively represented by a *virtual bidder*. Let b_i^k denote the *virtual bidder*, which represents all users of service provider i requesting m_i^k bandwidth units on video program k. As such, the service provider may have multiple virtual bidders, and each virtual bidder may have multiple bids corresponding to the different numbers of bandwidth units of the users.
- The bids that virtual bidder b_i^k submits to the helper can be expressed by $v_i^k = \{(0,0), (1, v_i^k(1)), \ldots, (m_i^k, v_i^k(m_i^k))\}$, where $v_i^k(m_i^k)$ represents the value of the virtual bidder when receiving m_i^k bandwidth units.

After receiving the bids from the virtual bidders of the service providers, the helper determines the values $\hat{m}_i^k, \forall i \in \mathcal{N}, \forall k \in \mathcal{K}$ that maximize the social welfare as follows:

6.2 Vickrey–Clarke–Groves Auction

$$\max \sum_{k \in \mathcal{K}} \sum_{i \in \mathcal{N}} v_i^k(m_i^k) \qquad (6.45)$$

$$\text{s.t.} \sum_{k \in \mathcal{K}} \sum_{i \in \mathcal{N}} \hat{m}_i^k \leq M$$

For the problem in (6.45), m_i^k and $\hat{m}_i^k$ are integer values, and thus the problem involves integer programming, which is generally NP-hard. Solving such a problem typically incurs high computational complexity. However, as shown in [238], the problem in (6.45) can be solved in polynomial time if (i) the helper knows the value function $v_i^k(m_i^k)$ of the virtual bidders and (ii) the function is concave.

Fortunately, in video streaming systems, the helper can determine the value function $v_i^k(m_i^k)$ based on *user stream quality*. The user stream quality for video channel k is proportional to the bandwidth allocated to the channel and inversely proportional to the number of users concurrently watching the channel. Let n_i^k be the number of users of virtual bidder i concurrently watching video channel k.

Assume that prior to the collaborative caching, the amount of bandwidth that service provider i allocates to its users for video channel k is W_i^k. Then, the user streaming quality can be defined as follows [233]:

$$q_i^k = \gamma \left(\frac{W_i^k}{R_k n_i^k} \right)^\alpha \qquad (6.46)$$

where R_k is the streaming rate of video channel k, γ is an adjustable scaling parameter, and $0 < \alpha < 1$ indicates how fast the streaming quality can be improved as the allocated bandwidth increases.

Now, consider the collaborative caching solution and assume that the helper allocates m_i^k bandwidth units to the users for watching video channel k. Then, based on the streaming quality formulation given in (6.46), value function $v_i^k(m_i^k)$ is defined as follows:

$$v_i^k(m_i^k) = \gamma (m_i^k)^{1-\alpha} \left(\left(\frac{W_i^k + e_i W^0 m_i^k}{R_k} \right)^\alpha - \left(\frac{W_i^k}{R_k} \right)^\alpha \right) \qquad (6.47)$$

where $0 < e_i \leq 1$ represents the degradation of the value of the virtual bidder due to the differentiation (e.g., the delay) of areas between service provider i and the helper. Equation (6.47) indicates how the streaming quality of virtual bidder b_i^k improves as it receives m_i^k bandwidth units from the helper.

We take the first and second derivatives of $v_i^k(m_i^k)$ with respect to m_i^k. As shown in [233], the first derivative $\frac{\partial v_i^k}{\partial m_i^k} > 0$ means that the value function monotonically increases as the number of allocated bandwidth units increases. The second derivative $\frac{\partial^2 v_i^k}{\partial^2 m_i^k} < 0$ means that the value function is concave and possesses the *downward-sloping property* [238]. Therefore, the optimization problem given in (6.45) can be solved in polynomial time, as shown in Algorithm 1. This algorithm is essentially a greedy algorithm that finds the value of m_i^k so as to maximize the marginal value gain of virtual bidders. The algorithm has a computational complexity of $\mathcal{O}(NKM)$.

Algorithm 1 Winner determination for VCG auction–based collaborative caching [233].

$m = \lfloor \frac{W}{W^0} \rfloor$
if $m \geq \sum_{k \in \mathcal{K}} \sum_{i \in \mathcal{N}} m_i^k$ **then**
$\quad \hat{m}_i^k = m_i^k, \forall i \in \mathcal{N}, \forall k \in \mathcal{K}$
else
$\quad \hat{m}_i^k = 0, \forall i \in \mathcal{N}, \forall k \in \mathcal{K}$
$\quad$**While** $m > 0$ **do**
$\quad\quad \{i, k\} = \arg \max_{i \in \mathcal{N}, k \in \mathcal{K}} \left(v_i^k(\hat{m}_i^k + 1) - v_i^k(\hat{m}_i^k) \right)$
$\quad\quad \hat{m}_i^k = \hat{m}_i^k + 1$
$\quad\quad m = m - 1$
$\quad$**end While**
end if

Assume that virtual bidder b_i^k wins $\hat{m}_i^k$ bandwidth units in the auction. The price that the virtual bidder pays the helper is determined according to the payment policy of the VCG auction as follows:

$$p_i^k = \max_{v_i^k(\hat{m}_i^k)=0} \sum_{k \in \mathcal{K}} \sum_{i \in \mathcal{N}} v_i^k(m_i^k) - \left(\sum_{k \in \mathcal{K}} \sum_{i \in \mathcal{N}} v_i^k(\hat{m}_i^k) - v_i^k(\hat{m}_i^k) \right) \quad (6.48)$$

On the right side of (6.48), the first term is the total value, the social value, of the winners in the auction in which virtual bidder b_i^k does not submit bid $v_i^k(\hat{m}_i^k)$, i.e., $v_i^k(\hat{m}_i^k) = 0$. The second term is the total value of the winners in the auction excluding bid $v_i^k(\hat{m}_i^k)$.

The payoff that the virtual bidder receives is $\pi_i^k = v_i^k(\hat{m}_i^k) - p_i^k$ and is further expressed by

$$\pi_i^k = \sum_{k \in \mathcal{K}} \sum_{i \in \mathcal{N}} v_i^k(\hat{m}_i^k) - \max_{v_i^k(\hat{m}_i^k)=0} \sum_{k \in \mathcal{K}} \sum_{i \in \mathcal{N}} v_i^k(m_i^k) \quad (6.49)$$

Note that in (6.49), $\sum_{k \in \mathcal{K}} \sum_{i \in \mathcal{N}} v_i^k(\hat{m}_i^k)$ is the total value of the optimal allocation, and $\max_{v_i^k(\hat{m}_i^k)=0} \sum_{k \in \mathcal{K}} \sum_{i \in \mathcal{N}} v_i^k(m_i^k)$ is the total value of an allocation when $v_i^k(\hat{m}_i^k) = 0$. Thus, the total value of the latter is not greater than that of the former. This can be understood as implying that adding one more bid, $v_i^k(\hat{m}_i^k) \geq 0$, into the auction will never decrease the social welfare. Therefore, $\pi_i^k \geq 0$, meaning that the payoff of virtual bidder b_i^k is non-negative, and the proposed scheme supports individual rationality. Moreover, as proved in [233], the payoff π_i^k when the virtual bidder bid its true value $v_i^k(\hat{m}_i^k)$ is not lower than the payoff $\pi_i^{\prime k}$ when the virtual bidder bids a false value $v_i^{\prime k}(\hat{m}_i^k)$: $\pi_i^k \geq \pi_i^{\prime k}$. In other words, the proposed scheme exhibits truthfulness.

To evaluate the performance of the proposed collaborative caching scheme, its simulation is implemented in [233]. The performance is evaluated in terms of overall streaming quality and *cache hit rate*. In particular, the overall streaming quality refers to the percentage of online users that experience a smooth playback, and the cache hit rate is a probability that the requested content can be found in the cache. The simulation results show that the cache hit rate of the proposed scheme is three times higher than that of the independent cache, in which the service provider allocates only its own resources to local users without collaborative caching. Moreover, the proposed scheme improves the overall streaming quality up to 40% compared with the independent caching scheme. However, when the available bandwidth capacity W of the helper is low, the overall streaming quality decreases since requests of other service providers may not be satisfied.

Spectrum Allocation in Cognitive Radio Networks
In cognitive radio, licensed (primary) users lease underutilized channels to unlicensed (secondary) users to improve the primary users' revenue and resource utilization. To achieve the full potential of cognitive radio, it is essential to design incentive mechanisms for the primary users. This section presents an incentive mechanism based on an optimal auction for the spectrum sharing in a cognitive radio network. The scheme is proposed in [239].

In general, the optimal auction uses the VCG auction for the winner and price determination. However, the VCG auction may result in low revenue for the primary users. Thus, the concept of *virtual values* from Myerson's optimal mechanism [194] is introduced. Accordingly, bids of the secondary users are first transformed into virtual bids, and then the VCG auction is used to determine the winners and the corresponding prices based on the virtual bids. The use of virtual bids aims to maximize the expected revenue of the primary users, while the VCG auction is employed to achieve desirable economic properties, such as truthfulness. Note that to enhance the spectrum efficiency, secondary users in different cells can use the same channels under the interference constraints.

System model: The system model is shown in Figure 6.8.

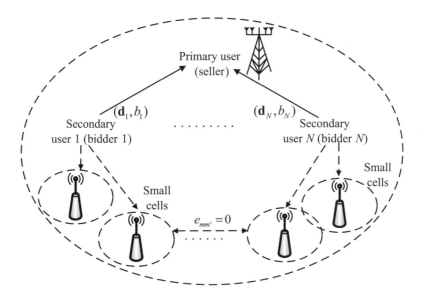

Figure 6.8 Spectrum allocation in cognitive radio networks using VCG auction.

- There is one primary user, the seller, which has K channels covering a certain region. The region is divided into M cells.
- When adjacent cells use the same channel, they may interfere with each other. Given a channel, a graph $F = (\mathcal{M}, \mathcal{E})$ is used to represent the region and the interference relations among the cells. Here, $\mathcal{M}$ is a set of M cells, and $\mathcal{E}$ is a set of interference relations among the cells. In particular, $\mathcal{E} = \{e_{m,m'}\}_{M \times M}$, where $e_{m,m'} = 0$ indicates that there is no interference between cells m and m' when they use the same channel, and $e_{m,m'} = 1$ indicates that there is interference between the two cells.
- There are N secondary users, the buyers or bidders, in the auction. Secondary user i submits a combination bid $(\mathbf{d}_i, b_i)$, where $\mathbf{d}_i$ is the channel demand vector and b_i is the bidding price for the demand. In particular, $\mathbf{d}_i = (d_i^1, \ldots, d_i^M)$, where d_i^m represents the number of channels that secondary user i requests in cell m. Here, the request is strict, meaning that the winners receive either all requested channels or nothing.
- Let v_i denote the value of spectrum resource $\mathbf{d}_i$ to secondary user i. Note that v_i is private information that is known only to secondary user i; it is not known to other secondary users. However, by observing historical values, the primary user (i.e., the seller) is assumed to have knowledge of the distribution function of values, denoted by $F_i(v_i)$.
- The primary user needs to determine the winners and the corresponding prices. Let $\mathbf{x} = (x_1, \ldots, x_N)$ be a winner determination vector, and let $\mathbf{p} = (p_1, \ldots, p_N)$ be the price vector.

Virtual value: Given the knowledge of the value distribution $F_i(v_i)$, the primary user can adopt Myerson's optimal mechanism [194] to construct virtual values $\phi_i(v_i)$. The use of the virtual values aims to maximize the primary user's expected revenue. The virtual values $\phi_i(v_i)$ of the secondary users are defined as follows:

DEFINITION 6.12 [239] *Secondary user i with value v_i has a virtual value given by*

$$\phi_i(v_i) = v_i - \frac{1 - F_i(v_i)}{f_i(v_i)} \tag{6.50}$$

where $f_i(v) = \frac{dF_i(v)}{dv}$ is the density function of v.

Assume that the distribution $F_i(v_i)$ satisfies the monotone hazard rate assumption, meaning that $\frac{f_i(v)}{1-F_i(v_i)}$ is a monotone non-decreasing function of v_i. Then, virtual value $\phi_i(v_i)$ is a monotone non-decreasing function of v_i. This means that the secondary user with a high value has a high virtual value, and the secondary user with a low value has a low virtual value. Also, the method for determining the winners using $\phi_i(v_i)$ is equivalent to the approach using v_i.

VCG auction: Now, the VCG auction is applied to determine the winners and the prices as follows:

6.2 Vickrey–Clarke–Groves Auction

- Given bids $((\mathbf{d}_1, b_1), \ldots, (\mathbf{d}_N, b_N))$ from the secondary users, the primary user computes virtual bids $((\mathbf{d}'_1, b'_1), \ldots, (\mathbf{d}'_N, b'_N))$, where $\mathbf{d}'_i = \mathbf{d}_i$ and $b'_i = \phi(b_i)$, and where $\phi(b_i)$ is determined according to (6.50).
- Based on the virtual bids $\phi(b_i)$, the primary user determines the winners $\mathbf{x}'$ that maximize the total virtual bid of all the secondary users under the strict allocation and interference constraints. The problem is expressed as follows:

$$\max_{\mathbf{x}'} \sum_{i=1}^{N} \phi_i(b_i) x'_i \quad (6.51)$$

$$\text{s.t.} \sum_{k=1}^{K} a_i^{m,k} = x'_i d_i^m, m = 1, \ldots, M, i = 1, \ldots, N \quad (6.52)$$

$$\sum_{i=1}^{N} a_i^{m,k} \leq 1, m = 1, \ldots, M, k = 1, \ldots, K \quad (6.53)$$

$$\sum_{i=1}^{N} a_i^{m,k} \cdot \sum_{i=1}^{N} a_i^{m',k} = 0, \forall e_{m,m'} = 1, k = 1, \ldots, K \quad (6.54)$$

where

- $a_i^{m,k}$ is feasible allocation associated with secondary user i. $a_i^{m,k} = 1$ means that channel k at cell m is allocated to the user, and $a_i^{m,k} = 0$ means that channel k at cell m is not allocated to the user.
- The constraint in (6.52) ensures strict allocation for the secondary users. This means that the requests of the secondary users cannot be partially satisfied.
- The constraint in (6.53) means that a given channel in a given cell can be allocated to at most one secondary user.
- The constraint in (6.54) refers to the interference constraint. More specifically, a given channel cannot be allocated to two cells m and m' if they cause interference with each other, such that the interference relation among the cells $e_{m,m'} = 1$.

We consider a simple case in which there is a single channel, $K = 1$, and each secondary user bids on only a single cell. Then, the problem given in (6.51) actually determines an independent set of secondary users, the secondary users using the channels without interference, so as to maximize the total virtual bid. If we consider the virtual bid of each secondary user as its weight, then the problem aims to determine an independent set of secondary users to maximize the total weight. The problem of finding such a set is called a *weighted independent set problem*. The weighted independent set problem is NP-hard. The intuitive approach for solving the problem is the approximate algorithm.

- Assume that $\mathbf{x}'$ is the solution of the problem in (6.51), which is the winner determination vector based on virtual bids $\phi_i(b_i)$. As stated earlier, $\mathbf{x}'$ is also the winner determination vector $\mathbf{x}$: $\mathbf{x} = \mathbf{x}'$.

- The virtual price that winner i pays the primary user is determined according to the pricing rule of the VCG auction:

$$p'_i = \max_{\mathbf{x}_{-i}} \sum_{j \neq i} x_j \phi_j(b_j) - (\max_{\mathbf{x}} \sum_j x_j \phi_j(b_j) - x_i \phi_i(b_i)) \qquad (6.55)$$

where $\mathbf{x}_{-i}$ represents the vector of winners excluding secondary user i.

- Based on Myerson's optimal mechanism [194], the actual price that winner i pays the primary user is

$$p_i = \phi_i^{-1}(p'_i) \qquad (6.56)$$

Given this price, the payoff of winner i is $\pi_i = v_i - p_i$. It is shown in [194] that the payoff is positive: $\pi_i > 0$.

- Note that if secondary user i loses in the auction (i.e., $x_i = 0$), then $p_i = 0$ and $\pi_i = 0$.

Properties of the proposed scheme: The proposed scheme has the following properties:

- The proposed scheme guarantees a non-negative payoff for the secondary users, $\pi_i \geq 0$, and thus it demonstrates individual rationality.
- The proposed scheme guarantees truthfulness, meaning that each secondary user maximizes its payoff only by submitting its truthful value, $b_i = v_i$. This can be proved by showing that the payoff of the secondary user when it submits $b_i = v_i$ is always equal to or higher than that of the user when it submits $b_i \neq v_i$. The proof can be found in [239].
- For the expected revenue, it is shown in [194] that for any truthful auction scheme, the expected payment from secondary user i satisfies the following condition:

$$\mathbb{E}[p_i(b_i)] = \mathbb{E}[\phi_i(b_i)x_i(b_i)] \qquad (6.57)$$

This means that maximizing the total virtual bid is equivalent to maximizing the expected revenue of the primary user. Thus, we have the following theorem:

THEOREM 6.13 *[239] The proposed scheme maximizes the expected revenue of the primary user.*

To evaluate the performance of the proposed scheme, a simulation is implemented in [240]. The number of cells is 25, and the number of secondary users is $N = 5$. Each secondary user requests at most one channel in any cell. The distribution of the secondary user's value is $F(v) = v$ in the range of $(0, 1]$. The baseline scheme is a greedy algorithm based on bids b_i instead of virtual bids $\phi_i(b_i)$. As shown in the simulation results, the proposed scheme outperforms the baseline scheme in terms of revenue. In particular, as the number of channels increases, the revenue obtained from the proposed scheme remains stable while that obtained from the baseline scheme significantly decreases and eventually reaches zero. The reason is that in the baseline scheme, the winners pay the primary user based on the losers' bids. When the number of channels increases, the number of losers decreases or there may even be no losers.

Thus, the payments from the winners decrease. The proposed scheme is stable due to the reservation effect of the virtual value functions. The proposed scheme thus provides the primary user with an incentive to share its spectrum and makes dynamic spectrum access more practical. Future work may include an extension of the auction model in which there are multiple primary users (i.e., multiple sellers) and may consider other spectrum request formats.

6.3 Summary

This chapter contains two main parts. The first part introduces the second-price sealed-bid auction and its application in computer networks. In particular, we provide the definition of the second-price sealed-bid auction and discuss the dominant strategy as well as the Nash equilibrium in the auction. Then, we compare the dominant strategy in the second-price sealed-bid auction and the equilibrium strategy in the English auction to show that the two auctions have the same truthful bidding strategy and the same revenue. At the end of the first part, we present applications of the second-price sealed-bid auction for addressing important issues in computer networks, including task allocation in IoT, task scheduling in edge computing, and physical layer security in MANET. The second part of the chapter introduces a generalization of the second-price sealed-bid auction with multiple items, the VCG auction. We formally describe the VCG auction through an example from computer networks' perspective. Considering a specific case, we prove the dominant strategy in the VCG auction. Then, we provide examples to show how the VCG auction works.

7 Combinatorial Auction

7.1 Introduction

The first- and second-price sealed bid auctions discussed in the previous chapters can be considered to be traditional auctions in which homogeneous types of items, such as bandwidth units, are auctioned. This means that any bandwidth unit cannot be distinguished from other ones; thus, the value of a particular bidder, or user, is identical for every bandwidth unit. However, in practice, bidders often require heterogeneous types of items rather than homogeneous types of items. For example, in wireless networks, users may prefer a package of frequency bands and antennas rather than only bandwidth units, for example, to reduce the resource cost. In cloud networking, the payoff of the users is higher when they receive a package of computing resources and frequency bands together. In moving-target tracking applications, to increase the tracking accuracy, data requesters prefer to receiving measurement data from multiple sensors rather than from a single sensor. A combinatorial auction or package auction is an efficient solution to meet the requirements. In the combinatorial auction, the bidders are allowed to bid on combinations/packages of heterogeneous types of items. Upon receiving bids, the seller solves the winner determination problem to obtain an optimal allocation of different types of items to the bidders under constraints, such as a supply constraint. After solving the winner determination problem, the seller determines the prices that the winners pay for receiving the packages of items. This price determination depends on pricing rules that the seller selects. For example, the pricing rules can be based on the first-price sealed-bid auction or the VCG auction.

In general, compared with the traditional auctions, the combinatorial auction has the following major advantages:

- It maximizes the payoff for the bidders since they can fully express their preferences.
- It is suitable for allocating substitutable and complementary items.
- It improves resource efficiency.

However, the combinatorial auction has one big challenge: solving the winner determination problem. This problem is generally NP-hard, and a polynomial-time algorithm for finding the optimal allocation does not exist. The Lagrangian relaxation approach is commonly used to find approximate solutions. Moreover, iterative combinatorial auctions such as ascending proxy auction can be an effective solution.

The following sections present a detailed description of the combinatorial auction and its applications for computer networks. First, we provide definitions of substitutable and complementary items. Second, we introduce the types of bidding language used in combinatorial auctions and the winner determination problem. Third, we introduce iterative combinatorial auctions, which are proposed to address the challenges of the winner determination problem. Finally, we review applications of the combinatorial auction for computer networks. In particular, we discuss the adoption of the combinatorial auction for spectrum allocation in cognitive radio [241], virtualization of 5G massive MIMO [242], mobile data offloading in HetNets [129], and resource allocation in device-to-device communications [243].

7.2 Substitutable and Complementary Items

In a combinatorial auction, the items can be substitutes or complements. They are defined from bidders' perspective, in terms of how the bidders' winning the first item affects marginal values of the other items. For simplification, we assume that there are two items in the auction. Each item has its own individual value. We have the following definitions:

- The two items are substitutes if the value of a package/combination of the two items to bidders is lower than the sum of the individual values.
- The two items are complements if the value of a package/combination of the two items to bidders is greater than the sum of the individual values.

To further understand substitutable item and complementary item, we consider the following example.

Example 7.1 We consider a sensing application in IoT including one service provider, the seller, and one data requester, the buyer. The data requester is willing to buy measurements of air pollution within an area of interest. As shown in Figure 7.1, the area is divided into two subareas, and the service provider deploys three sensors to provide measurements of the two subareas. Sensors 1 and 2 are both deployed in subarea 1, and sensor 3 is deployed in subarea 2. The service provider offers the three measurements in a combinatorial auction. This auction allows the data requester to bid for each of individual measurements or for a combination of the measurements. In general, more measurements improve the accuracy of the air pollution estimate, so the data requester is willing to bid for a combination of the measurements rather than an individual one. Table 7.1 shows values of individual measurements and combinations of these measurements to the data requester.

- Measurements of sensor 1 and sensor 2 are for the same area, subarea 1. For the data requester, the sum of the individual values is greater than the value of a combination of the two measurements: $v_1 + v_2 = \$2 > \$1.5 = v_{(1,2)}$. Thus, the measurements of sensor 1 and sensor 2 are substitutes. This means that when the

Table 7.1 Substitutable and complementary items in IoT applications.

Sensor index	Measurement	Value (v)	Item types
1	Subarea 1	$1	Individual
2	Subarea 1	$1	Individual
3	Subarea 2	$2	Individual
(1, 2)	(Subarea 1, subarea 1)	$1.5	Substitutes
(1, 3)	(Subarea 1, subarea 2)	$4	Complements
(2, 3)	(Subarea 1, subarea 2)	$4	Complements

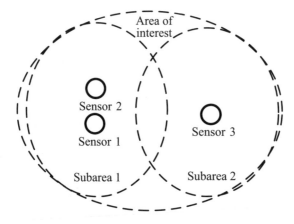

Figure 7.1 Sensor deployment in IoT.

data requester wins one measurement, such as the measurement of sensor 1, the marginal value of the measurement of sensor 2 is lower because the measurement of sensor 1 can substitute for the other one.

- However, measurements of sensor 1 and sensor 3 are considered to be complements. The reason is that the value of both measurements to the data requester is greater than the sum of the individual values: $v_1 + v_3 = \$3 < \$4 = v_{(1,3)}$. Similarly, the measurements of sensor 2 and sensor 3 are complements since $v_2 + v_3 = \$3 < \$4 = v_{(2,3)}$.

We consider another example to show that items in the combinatorial auction can be either substitutes or complements. A mobile network provider may be willing to pay $1,000 for a radio spectrum license for zone A and $1,500 for the same license in zone B. The licenses are substitutes if the mobile network provider is not willing to pay an amount greater than or equal to $1,000 + $1,500 = $2,500 for both of the licenses. However, in some cases, such as when the mobile network provider wants to deploy and provide network services in both zones, it is willing to pay an amount greater than $2,500 to obtain both licenses. The licenses would then be complements.

7.3 Single-Round Combinatorial Auction

7.3.1 Bidding Language

The combinatorial auction allows bidders to bid for combinations/subsets of items, and the number of possible bids in the auction is usually huge. For example, if there are M bandwidth units for trading in the auction, there are potentially $2^M - 1$ subsets of bandwidth units. Assume that a user (bidder) wants to bid for all the subsets. The user needs $2^M - 1$ real numbers to represent the bids for $2^M - 1$ subsets. When M is large (e.g., $M = 50$), the numbers of bids can be very large or even infinite. This significantly increases the communication and computation overhead in the network and may make the winner determination impractical. Thus, it is necessary to introduce a *bidding language* that allows the user to succinctly encode or express common bids [244]. A bidding language is a mapping from a mathematical space of possible bids into a set of finite strings of characters. The characters can be ASCII characters. The following discussion introduces bidding languages commonly used in combinatorial auctions in computer networks.

Atomic Bid

An atomic bid is a single-minded bid that expresses how much the bidder is willing to pay for a subset of items. The atomic bid is typically expressed by a tuple $(< \mathcal{M}, b >)$, where $\mathcal{M}$ is the subset of items that the bidder is willing to buy, and b is the price that the user is willing to pay for the subset. For convenience, the atomic bid is also expressed by $b(\mathcal{M})$ in some places in this book. For example, a service provider has a set $\mathcal{M}$ of one bandwidth unit W_0, one power unit P, and one antenna A for trading. We can denote the set as $\mathcal{M} = \{W_0, P, A\}$. Then, each user can submit an atomic bid in a tuple of $(< \{W_0, P\}, b_1 >)$, where $\{W_0, P\}$ is the subset of the bandwidth unit and the power unit that the user wants to buy, and b_1 is the price that the user is willing to pay for the subset. Also, we can express the atomic bid of this user as $b_1(\{W_0, P\})$.

Now, let v_1 be the value of the subset $\{W_0, P\}$ to the user, and assume that $b_1 = v_1$. Then the atomic bid can be expressed by $(< \{W_0, P\}, v_1 >)$. It can be seen that submitting a single atomic bid does not allow the user to express all of its values for subsets of resources. For example, suppose user 1 wants to bid for a subset including one antenna A with a value v'_1, $(< \{A\}, v'_1 >)$, apart from subset $\{W_0, P\}$. By using the atomic bidding language, the user is not allowed to submit two atomic bids $(< \{W_0, P\}, v_1 >)$ and $(< \{A\}, v'_1 >)$ simultaneously. Thus, it may be more efficient to combine the atomic bids. OR and XOR bids allow bidders to combine different atomic bids and thereby express a larger list of preferences (i.e., values).

OR Bid

The OR bid language, also known as an additive-OR bid, allows the bidder to submit an arbitrary number of atomic bids, consisting of a collection of tuples $(< \mathcal{M}_i, b_{\mathcal{M}_i} >)$, where $\mathcal{M}_i$ is a subset of the items, and $b_{\mathcal{M}_i}$ is the price that the bidder is willing to pay for the subset. Note that subsets $\mathcal{M}_i$ submitted by the bidder do not overlap with each other: $\forall i \neq j, \mathcal{M}_i \cap \mathcal{M}_j = \emptyset$. Thus, it can be said that the subsets are not mutually exclusive. Formally, the OR bid can be expressed by $(< \mathcal{M}_1, b_{\mathcal{M}_1} >)\text{OR} \cdots \text{OR}(< \mathcal{M}_K, b_{\mathcal{M}_K} >)$. By submitting the OR bid, the bidder shows that it is willing to receive any number of disjoint atomic bids and is willing to pay the sum of prices specified in the atomic bids. In particular, the bidder is willing to receive the maximum number of $M_1 + \cdots + M_K$ items and is willing to pay a price of $b_{\mathcal{M}_1} + \cdots + b_{\mathcal{M}_K}$.

As a specific example, we consider again the previously described resource market in which the user submits an OR bid of $(< \{W_0, P\}, \$2 >)\text{OR}(< \{A\}, \$1 >)$ to the service provider. This means that the user has a value of \$2 for the combination of the bandwidth unit and the power unit and has a value of \$1 for the antenna. The OR bid also implies that the user has a value of $\$2 + \$1 = \$3$ for all of them, $\{W_0, P, A\}$. Referring to the definitions of the complementary and substitutable items in Section 7.2, it can be seen that the OR bid represents atomic bids that include only complementary items. We have the following proposition.

PROPOSITION 7.1 *[244] An OR bid can represent all atomic bids that do not have any substitutable items; that is, for all $i \neq j$ and $\mathcal{M}_i \cap \mathcal{M}_j = \emptyset$, $v(\mathcal{M}_i \cup \mathcal{M}_j) \geq v(\mathcal{M}_i) + v(\mathcal{M}_j)$.*

XOR Bids

Similar to the OR bid, an XOR bid, also known as an exclusive-OR bid, allows bidders to submit multiple atomic bids. However, by submitting the XOR bid, the bidder indicates that it is willing to obtain at most one of the atomic bids. The price that the bidder is willing to pay is specified in the winning atomic bid. Formally, the XOR bid can be expressed by $(< \mathcal{M}_1, b_{\mathcal{M}_1} >)\text{XOR} \cdots \text{XOR}(< \mathcal{M}_K, b_{\mathcal{M}_K} >)$. The user may receive any of $|\mathcal{M}_1|, |\mathcal{M}_2|, \ldots, |\mathcal{M}_K|$ items, but is not willing to receive all of them. We say that the atomic bids in the XOR bid are mutually exclusive. In other words, the XOR bid is suitable for markets that trade substitutable items. Compared with the OR bid, use of the XOR bid does not face an *exposure problem*. To understand the exposure problem, we consider the following situation. A service provider, the bidder, is willing to win all spectrum licenses to deploy its network services in multiple areas. In other words, the spectrum licenses are complementary items. The service provider has an incentive to submit its bids higher than its values for the licenses. This may result in a risk of incurring losses to the service provider even if it wins the licenses. To avoid the losses, the service provider attempts to lower its bids. This situation is known as the exposure problem. Since the OR bid is used for the complementary items, the exposure problem often occurs with the OR bid rather than the XOR bid.

Apart from the OR and XOR bidding languages, there are other combinations of OR and XOR bids, such as OR-of-XORs bids and XOR-of-ORs bids. In general, these bidding languages allow bidders to represent a larger list of their preferences. However, in particular for the computer networks, mobile users are non-processional traders, and it may be difficult for them to design such complicated bidding languages. Thus, we do not present these bidding languages in detail in this book. Further details of the languages are found in [244].

7.3.2 Winner Determination Problem

In this section, we introduce the definition of the winner determination problem in the combinatorial auction and then provide an example that explains how to solve the problem.

We consider a cellular network with N mobile users, $I = (1, \ldots, N)$, which share a set $\mathcal{M}$ of M channels provided by a service provider. The users can bid for individual channels or packages/subsets of the channels. Assume that user i wants to buy a package S of channels. The value of the package to user i is expressed by $v_i(S)$, and the bid of the user is $b_i(S)$, the price that the user is willing to pay. Note that $b_i(S)$ may not be the same as $v_i(S)$. The users submit the bids to the service provider. Upon receiving the bids from all the users, the service provider determines the winning bids so as to maximize its revenue. More specifically, the service provider determines a feasible combination of bids that guarantees that the sum of the bids is the highest and that each channel is allocated at most to one user. This allocation problem is called a *winner determination problem* in a combinatorial auction. Formally, the winner determination problem is expressed by [202]

$$\max_{x_i(S)} \sum_{i \in I} \sum_{S \subseteq \mathcal{M}} b_i(S) x_i(S) \tag{7.1}$$

$$\text{s.t.} \sum_{S \supseteq \{j\}} \sum_{i \in I} x_i(S) \leq 1, \forall \{j\} \subseteq \mathcal{M} \tag{7.2}$$

$$\sum_{S \subseteq \mathcal{M}} x_i(S) \leq 1, \forall i \in I \tag{7.3}$$

$$x_i(S) \in \{0, 1\}, \forall S \subseteq \mathcal{M}, \forall i \in I \tag{7.4}$$

where

- $x_i(S)$ is a binary variable such that $x_i(S) = 1$ represents that user i wins subset S and $x_i(S) = 0$ represents that the user does not win the subset.
- The objective in (7.1) refers to revenue maximization. If revenue maximization is achieved, then the allocation is an efficient allocation.
- The constraint in (7.2) ensures that each channel is allocated to at most one user. When this constraint is satisfied, we say that the allocation is a feasible allocation.
- The constraint in (7.3) means that each user obtains at most one winning bid, meaning that the bids here are mutually exclusive, using the XOR bidding language.

The solution to problem (7.1)–(7.3) may depend on the specific context. We consider the following example.

Example 7.2 Consider a spectrum trading market that consists of one service provider and three mobile users. The service provider has three bandwidth units for trading, and the bandwidth units are assumed to be homogeneous. Let W_1, W_2, and W_3 denote bandwidth units 1, 2, and 3, respectively. There are totally $2^3 - 1 = 7$ possible subsets of the three bandwidth units: $\{W_1\}, \{W_2\}, \{W_3\}, \{W_1, W_2\}, \{W_1, W_3\}, \{W_2, W_3\}$, and

Table 7.2 Winner determination in combinatorial auction [202].

S	$b_1(S)$	$b_2(S)$	$b_3(S)$
$\{W_1\}$	$2	$1.5	**$2**
$\{W_2\}$	$1	**$2**	$1
$\{W_3\}$	$1	$1	**$2**
$\{W_1, W_2\}$	$2	$2	$2
$\{W_1, W_3\}$	$2.5	$1.5	$2.75
$\{W_2, W_3\}$	$1.5	$1.5	$1.5
$\{W_1, W_2, W_3\}$	$3	$4	$3

$\{W_1, W_2, W_3\}$. Each mobile user submits its atomic bids for the bandwidth units. The atomic bid includes a subset that the user wants to buy and the corresponding price that the user is willing to pay. We assume that each user bids for all seven of the possible subsets. Table 7.2 shows the information for the users' atomic bids. After receiving the atomic bids, the service provider determines the winning bids and the corresponding payments.

To determine the winning bids, the service provider can simply list the possible combinations of the atomic bids. Then, it selects the combination that produces the highest total price and satisfies the constraints in (7.2) for a feasible allocation, and in (7.3) for the XOR bid condition. We discuss some following combinations to understand how the service provider determines the winning bids.

- One possible combination is $(b_3(\{W_1, W_3\}), b_2(\{W_2\}))$, in which subset $\{W_1, W_3\}$ is allocated to user 3 and subset $\{W_2\}$ is allocated to user 2. The sum of prices for this combination is $b_3(\{W_1, W_3\}) + b_2(\{W_2\}) = \$2.75 + \$2 = \$4.75$. This combination is feasible, meaning that the constraint in (7.2) is satisfied, since the same bandwidth unit is not allocated to different users. Also, the combination meets the condition of XOR bid: the constraint in (7.3) is satisfied, since each user wins at most one bid. However, the combination is not an efficient allocation because it does not achieve the highest sum of prices. Therefore, this combination is not a solution to the winner determination problem.
- Another possible combination is $(b_3(\{W_1\}), b_2(\{W_2\}), b_2(\{W_3\}))$, in which subset $\{W_1\}$ is allocated to user 3, $\{W_2\}$ is allocated to user 2, and $\{W_3\}$ is allocated to user 3. The revenue of the service provider obtained from the combination is $b_3(\{W_1\}) + b_2(\{W_2\}) + b_3(\{W_3\}) = \$5$. This combination allows the seller to obtain the highest sum of prices. However, it does not satisfy the constraint in (7.3), the XOR bid condition, since user 3 wins two different bids. Therefore, this combination is not a solution to the winner determination problem.
- By checking all the possible combinations, it is seen that the winning bids should be $b_1(\{W_1\})$, $b_2(\{W_2\})$, and $b_3(\{W_3\})$ since this combination produces the highest sum of prices, $b_1(\{W_1\}) + b_2(\{W_2\}) + b_3(\{W_3\}) = \$5$, and satisfies both the constraints in (7.2) and (7.3).

- Users 1, 2, and 3 are considered to be the winners of the auction since each of them has one winning bid.

After solving the winner determination problem, the service provider calculates the prices that the winning users need to pay. The price determination depends on the service provider's objective. For example, if the service provider aims to maximize its revenue, the price rule of the first-price auction can be used. If the service provider wants to maximize the social welfare and achieve economic properties, such as truthfulness, the price policy of the VCG auction can be used.

From Example 7.2, it is seen that as the number of bandwidth units for trading increases, the number of possible combinations increases exponentially and the winner determination problem becomes more complicated. It is known that the winner determination problem in the single-round combinatorial auction is an NP-complete problem, and solving it requires significant computation time and resources. Some common methods entail using approximate algorithms. More details of these methods are discussed in Section 7.5. Another effective method is to change the single-round combinatorial auction to a series of iterative combinatorial auctions, an approach presented in the next section.

7.4 Iterative Combinatorial Auctions

In iterative combinatorial auctions, bidders are allowed to bid for individual items or packages of items in multiple rounds. At the end of each round, the seller provides information regarding the provisional allocation of the items and the prices reached in the round. The bidders thus have knowledge of their rivals' bids and may change their own bids in the next rounds; in other words, they benefit from a *price discovery*. The following discussion presents two common types of iterative combinatorial auctions: the ascending proxy auction and the clock-proxy auction.

7.4.1 Ascending Proxy Auction

In an ascending proxy auction, bidders do not directly submit their bids to the seller. Instead, the bidder submits its bids to a *proxy bidder* for packages of items. Here, the proxy bidder can be an electronic proxy agent or an auctioneer, and the bid for one package is the price that the bidder is willing to pay the seller for the package. Let $b_i(S)$ denote the bid for package S. The bidder may have multiple bids $b_i(S)$. Note that bid $b_i(S)$ may not be the value $v_i(S)$ of the package to the bidder.

In each round, the seller sets standing prices for all possible packages. Based on bids $b_i(S)$ from the bidder and the standing prices, the proxy bidder submits a unique bid on the bidder's behalf to the seller. This *proxy bid*, denoted as $b_i^{proxy}(S)$, is actually the standing price set by the seller for package S. By submitting $b_i^{proxy}(S)$, the proxy bidder expects that the bidder can achieve the highest payoff if the bidder wins the auction. The proxy bidder determines the proxy bids for other bidders in the same way. Upon receiving the proxy bids, the seller solves the winner determination problem – for example, by choosing the highest proxy bids as the winning bids. Since the number of proxy bids is often smaller than the number of the bidders' bids, the winner determination problem in the ascending proxy auction becomes much simpler than that in the single-round combinatorial auction. The seller announces the provisional winners and corresponding payments in each round. If the bidder loses in the round, the proxy bidder then determines again the proxy bid for the bidder in the next round so that the loser has a chance to win the auction. The process continues until no new proxy bids are submitted, and the auction ends. Since each bidder is satisfied with one of its bids, such that the substitute condition is satisfied, the bidder can use the XOR bidding language to encode its bids when it submits them to the proxy bidder.

Example 7.3 We consider a bandwidth trading market including a service provider and multiple mobile users (i.e., bidders). For simplification, we assume that the service provider has two bandwidth units, W_1 and W_2, for trading. Thus, the possible packages of the bandwidth units are $\{W_1\}$, $\{W_2\}$, and $\{W_1, W_2\}$. Table 7.3 illustrates a bidding process of user i in the ascending proxy auction.

- Assume that user i submits its bids for the packages to the proxy bidder as follows: $b_i(\{W_1\}) = \$4$, $b_i(\{W_2\}) = \$3$, and $b_i(\{W_1, W_2\}) = \$14$.
- In round 1, the service provider sets standing prices for the packages as $p^1(\{W_1\}) = \$1$, $p^1(\{W_2\}) = \$2$, and $p^1(\{W_1, W_2\}) = \$9$, respectively. Based on the bids $b_i(\mathcal{W})$ and the standing prices $p^1(\mathcal{W})$, the proxy bidder determines a proxy bid for the user. Refer to Table 7.3, the potential payoffs that the user can obtain with the bids are \$3 for package $\{W_1\}$, \$1 for $\{W_2\}$, and \$5 for $\{W_1, W_2\}$. Perhaps bidding for package $\{W_1, W_2\}$ produces the highest payoff for the user if it wins. Thus, the proxy bidder selects $b_i^{proxy}(\{W_1, W_2\}) = \$9$ as a proxy bid of user i and submits it to the service provider. Note that \$9 is also the standing

Table 7.3 Bidding process of user i in an ascending proxy auction [202].

Possible packages S	$\{W_1\}$	$\{W_2\}$	$\{W_1, W_2\}$
User's bid: $b_i(S)$	$4	$3	$14
Round 1			
Standing price: $p^1(S)$	$1	$2	$9
Potential payoff	$4 − $1 = $3	$3 − $2 = $1	$14 − $9 = $5
Proxy bid: $b_i^{proxy}(S)$			$9
Round 2			
Standing price: $p^2(S)$	$1	$2	$10
Potential payoff	$4 − $1 = $3	$3 − $2 = $1	$14 − $10 = $4
Proxy bid: $b_i^{proxy}(S)$			$10

price set by the service provider for the package. The proxy bidder does the same for the remaining bidders. After receiving the proxy bids, the service provider solves the winner determination problem. If user i is not among the provisional winners, the proxy bidder determines again the proxy bid for the user in round 2.
- In round 2, assume that the service provider keeps the standing prices for packages $\{W_1\}$ and $\{W_2\}$ but increases the standing price of package $\{W_1, W_2\}$ to $10. Then, the potential payoffs that the user can obtain with the bids are $3 for package $\{W_1\}$, $1 for $\{W_2\}$, and $4 for $\{W_1, W_2\}$. Bidding package $\{W_1, W_2\}$ still produces the highest payoff, so the proxy bidder selects $b_i^{proxy}(\{W_1, W_2\}) = \$10$ as a proxy bid in round 2 and submits it again to the service provider.
- This bidding process is repeated until no proxy bid is submitted, a point reached when the user wins one of the packages or the payoff of the user becomes negative.

As shown in Example 7.3, the proxy bid that the proxy bidder submits is always lower than the bid that the user submits/expects. Thus, the payoff of the user is always positive if the user wins the auction. Intuitively, the ascending proxy auction supports truthfulness and individual rationality. This auction is similar to the Vickrey auction for multiple items, also known as the VCG auction. However, unlike the VCG auction, the ascending proxy auction avoids a low revenue for the service provider (i.e., the seller) by introducing the standing prices. As a result, the ascending proxy auction eliminates shill bidding, a situation in which the service provider introduces spurious (fake) bids. The fake bids increase the competition in the auction and increase the final price, which defrauds the users.

7.4.2 Clock-Proxy Auction

In the ascending proxy auction, each user/bidder submits its atomic bids to the proxy bidder. The atomic bid includes a package/set of resources and the price that the user is willing to pay for the package. As such, the user's problem is to decide which resources to buy, how much of each resource, and the corresponding prices. The problem may become much more complicated when the number of resources is large. Price discovery mechanisms such as ascending clock auctions can be adopted along with the ascending proxy auction to help the users more easily make their decisions.

The authors in [245] develop a *clock-proxy auction* that includes two phases: the clock phase and the proxy phase. In particular, the clock phase is implemented by using an ascending clock auction with multiple rounds.

- First, the service provider sets the initial prices for resources.
- Given these prices, each user indicates its demand, or package of resources that the user is willing to buy. The service provider considers the package and the total prices of resources specified in the package as an atomic bid of the user in the round.
- Upon receiving the demands for its resources, the service provider increases the prices of those resources with excess demands.

Table 7.4 Clock phase process for two bandwidth units, $\overline{Q}_W = 2$.

t	p_W^t	$Q_{1,W}^t$	$Q_{2,W}^t$	Q_W^t	$\overline{Q}_W$
0	$0.1	2	2	4	2
1	$0.2	2	1	3	2
2	$0.3	2	0	2	2

Table 7.5 Clock phase process for two antennas, $\overline{Q}_A = 2$.

t	p_A^t	$Q_{1,A}^t$	$Q_{2,A}^t$	Q_A^t	$\overline{Q}_A$
0	$0.1	2	2	4	2
1	$0.2	1	2	3	2
2	$0.3	0	2	2	2

- Given the new prices, the users again indicate their demands.
- This process is repeated until there are no excess demands for the resources.
- When the clock phase finishes, the proxy phase starts. It is implemented by using the ascending proxy auction as presented in Section 7.4.1.
- After the two phases finish, the service provider collects all atomic bids received in the clock phase and all proxy bids received in the proxy phase to determine the winning bids.

In this type of auction, the users indicate only their demands for the resources, without determining and submitting the prices of those resources. Therefore, the users can make their decisions more easily as they need less information.

To further understand the clock-proxy auction process, we next consider a resource trading market that includes one service provider and two mobile users, users 1 and 2. The service provider has two types of resources, type W bandwidth and type A antenna. Assume that the number of bandwidth units for trading is 2, $\overline{Q}_W = 2$, and the number of antennas is 2, $\overline{Q}_A = 2$. The process of the clock phase for the two types of resources is shown in Tables 7.4 and 7.5 and described here:

- In round $t = 0$, the service provider sets the initial prices of one bandwidth unit and one antenna: $p_W^0 = p_A^0 = \$0.1$. In this round, user 1 bids for two bandwidth units, $Q_{1,W}^0 = 2$, and two antennas, $Q_{1,A}^0 = 2$. Similar to user 1, user 2 bids for two bandwidth units and two antennas: $Q_{2,W}^0 = Q_{2,A}^0 = 2$. Thus, the total demand for bandwidth is $Q_W^0 = Q_{1,W}^0 + Q_{2,W}^0 = 4 > 2 = \overline{Q}_W$, and the total demand for antennas is $Q_A^0 = Q_{1,A}^0 + Q_{2,A}^0 = 4 > 2 = \overline{Q}_A$. As such, the demands for both bandwidth units and antennas are greater than the supply in this round.
- At the begin of round $t = 1$, the service provider increases the prices of the bandwidth unit and antenna: $p_W^1 = p_A^2 = \$0.2$. At this price, user 1 still keeps

its demand for bandwidth but reduces its demand for antennas: $Q^1_{1,W} = 2$ and $Q^1_{1,A} = 1$. User 2 reduces its demand for bandwidth but keeps its demand for antennas: $Q^1_{2,W} = 1$ and $Q^1_{2,A} = 2$. However, it can be seen that the demands for both bandwidth and antennas still exceed the supply.

- In round $t = 2$, the service provider increases the prices of both resources: $p^2_W = p^2_A = \$0.3$. The total demand for bandwidth is $Q^2_W = Q^2_{1,W} + Q^2_{2,W} = 2 + 0 = 2$, and the total demand for antennas is $Q^2_A = Q^2_{1,A} + Q^2_{2,A} = 0 + 2 = 2$. The demands, for both bandwidth and antennas are equal to the supply: $Q^2_W = 2 = \overline{Q}_W$ and $Q^2_A = 2 = \overline{Q}_A$. Thus, the service provider does not increase the prices of the resources, and the clock phase terminates.

Note that in any round, the total price that user i is willing to pay is the sum of the price per resource unit multiplied by the number of resources demanded for each type. For example, in round $t = 0$, the total price that user 1 is willing to pay for the bandwidth and antenna is $p^0_W Q^0_{1,W} + p^0_A Q^0_{1,A} = 0.1 \times 2 + 0.1 \times 2 = \$0.4$. We can interpret that as follows: in round $t = 0$, user 1 is willing to pay a price of $\$0.4$ for a package/set of two bandwidth units and two antennas which can be expressed as an atomic bid, $(< \{W, W, A, A\}, \$0.4 >)$. Similarly, in round $t = 1, 2$, the atomic bids of user 1 are $(< \{W, W, A\}, \$0.6 >)$ and $(< \{W, W\}, \$0.6 >)$, respectively.

When the clock phase finishes, the proxy phase starts. This phase uses the ascending proxy auction, discussed in Section 7.4.1. Specifically, each user submits its atomic bids to a proxy bidder. The packages specified in the atomic bids may be different from those that the user submits in the clock phase. If the package is among the group of packages in the clock phase, the price for this package must be greater or equal to the price set by the service provider for the package in the clock phase. For example, suppose user 1 wants to bid for $\{W, W\}$ in the proxy phase. Since $\{W, W\}$ is among the packages in the clock phase, the price for this package must be greater than or equal to $\$0.6$. After receiving the atomic bids from the user, in each round, the proxy bidder submits a proxy bid on the user's behalf to the service provider. The proxy bid is calculated based on the atomic bids and the standing price set by the service provider so as to guarantee the highest profit for the user. If the user loses in the current round, the proxy bidder determines again the proxy bid for the user in the next round. The process continues until no new proxy bids are submitted, at which point the proxy phase ends.

After the clock phase and the proxy phase finish, the service provider collects all atomic bids and proxy bids submitted by the users in the two phases. Note that for a particular user, the atomic bids and proxy bids are considered to be mutually exclusive (i.e., XOR bidding language). The service provider then determines the combination of feasible atomic bids that maximizes its revenue, thereby solving the winner determination problem. Since the bids are mutually exclusive, solving the winner determination problem can be implemented similarly to Example 7.2.

In general, the clock-proxy auction has some similarities with the English–Dutch auction, which was presented in Section 4.5. Both auctions consist of two phases, and the first phase is an ascending clock auction. By including the ascending clock auction, the users can obtain the preference information of their rivals (price discovery), which enables the users to easily make bidding decisions. However, the clock-proxy auction and the English–Dutch auction are quite different in their motivation for using the second phase. In the English–Dutch auction, the second phase is the Dutch auction, which uses the price strategy of the first-price sealed-bid auction to maximize the service provider's revenue. In the clock-proxy auction, the second phase adopts the proxy auction, which uses a price strategy similar to that of the Vickrey auction (see Section 7.4.1). Such a price strategy motivates users to submit their real values for resources, thereby demonstrating truthfulness. This enables the clock-proxy auction to achieve the efficiency of resource allocation.

7.5 Development of the Combinatorial Auction for Computer Networks

In this section, we describe how the combinatorial auction is used to address resource management issues in computer networks. Specifically, the issues examined include (i) spectrum allocation in cognitive radio, (ii) virtualization of 5G massive MIMO, (iii) mobile data offloading in 5G HetNets, and (iv) resource allocation in D2D communication underlying cellular networks.

7.5.1 Spectrum Allocation in Cognitive Radio

Cognitive radio allows secondary users to use the underutilized spectrum of the primary users. In Chapters 4 and 6, we discussed adoption of the English auction and the VCG auction for allocating spectrum resources to the secondary users in cognitive radio networks. The spectrum resources are bandwidth units. In practice, the spectrum should be modeled in a time–frequency division manner, meaning that the spectrum consists of time slots and frequency bands. As such, when requesting the spectrum resources, the secondary users can specify time slots and frequency bands that meet their dynamic demands. However, this raises the resource allocation issues since the time slots and frequency bands are heterogeneous resources. Combinatorial auctions to allocate homogeneous types of resources can be adopted to guarantee feasibility and efficiency of the resource allocation in the cognitive radio. Such an approach is proposed in [241] and described next.

System Model
The system model consists of one primary user and a set $\mathcal{N}$ of N secondary users, $\mathcal{N} = \{1, \ldots, N\}$.

- The primary user has an underutilized spectrum for trading. The entire spectrum is divided into M pairs of time slots t_j and frequency bands f_j, as shown in Figure 7.2. For simplification, each pair (t_j, f_j) is mapped to a value of s_j, $j \in \{1, \ldots, M\}$. In the general case, we have (t_j, f_k), which is mapped to s_{jk}. The set of mapping values s_j is denoted as $\mathcal{M} = \{s_1, \ldots, s_M\}$. The primary user informs the secondary users of this mapping before the auction starts.

7.5 Development of the Combinatorial Auction for Computer Networks

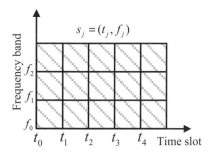

Figure 7.2 Illustration of frequency bands and time slots [241]. Here, $f_{j+1} = f_j + \Delta f$ and $t_{j+1} = t_j + \Delta t$.

- Each secondary user i submits its bid as $(< \mathcal{S}_i, b_i >)$ or $b_i(\mathcal{S}_i)$, where $\mathcal{S}_i \subseteq \mathcal{M}$ is a package of mapping values s_i indicated by pair (t_j, f_j), that secondary user i wants to buy, and b_i is the price that the secondary user is willing to pay the primary user for the package. This price may be not the same as the value v_i of the package to secondary user i.
- Upon receiving bids from all the secondary users, the primary user needs to determine the winning bids, thereby solving the winner determination problem. Also, it determines the price p_i that the secondary users with the winning bids need to pay. Note that the losing secondary users are charged a price of zero. Thus, the utility of secondary user i is defined as

$$u_i = \begin{cases} v_i - p_i, & \text{if secondary user } i \text{ wins} \\ 0, & \text{otherwise} \end{cases} \quad (7.5)$$

Winner Determination Problem Formulation

To determine the winning bids, the primary user formulates its winner determination problem. To achieve efficiency in resource allocation, an objective of the problem needs to be designed to maximize the values of the winning bids. Thus, the primary user formulates its winner determination problem as follows [241]:

$$\max_{x_i(\mathcal{S}_i)} \sum_{i \in \mathcal{N}} \sum_{\mathcal{S}_i \subseteq \mathcal{M}} b_i(\mathcal{S}_i) x_i(\mathcal{S}_i) \quad (7.6)$$

$$\text{s.t.} \sum_{\mathcal{S}_i \ni s_j} \sum_{i \in \mathcal{N}} x_i(\mathcal{S}) \leq 1, \forall s_j \in \mathcal{M} \quad (7.7)$$

$$\sum_{\mathcal{S}_i \subseteq \mathcal{M}} x_i(\mathcal{S}_i) \leq 1, \forall i \in \mathcal{N} \quad (7.8)$$

where

- $x_i(\mathcal{S}_i) \in \{0, 1\}$ is a decision variable: $x_i(\mathcal{S}_i) = 1$ represents that secondary user i wins package $\mathcal{S}_i$, and $x_i(\mathcal{S}_i) = 0$ otherwise.

- The objective in (7.6) refers to maximizing the sum of values of the selected (winning) bids.
- The constraint in (7.7) ensures that each mapping value s_j, pair (t_j, f_j), is allocated to at most one secondary user. If this constraint is satisfied, the allocation is a feasible allocation.
- The constraint in (7.8) means that each secondary user obtains at most one winning bid, meaning that the bids here are mutually exclusive (i.e., XOR bidding language).

Winner Determination

By using the maximum independent set problem [246], the problem in (7.6) is proved to be NP-hard. To solve the problem, we can use dynamic programming. Dynamic programming allows us to find an exact optimal solution, but it has high computational complexity. In particular, given N secondary users in the auction, the computational complexity of the dynamic programming method is $O(N^2)$ [241]. To match the dynamic environment of cognitive radio, the computational complexity needs to be reduced. Thus, approximate algorithms such as the polynomial-time greedy algorithm should be adopted to solve the winner determination problem. The proposed algorithm is described in Algorithm 2.

Algorithm 2 Greedy algorithm for the winner determination of combinatorial auction in cognitive radio [241].

1: Initialize: $\mathcal{L} = \emptyset$.
2: Collect bids ($< \mathcal{S}_i, b_i >$) from secondary users.
3: Compute normalized bids $b_i / \sqrt{|\mathcal{S}_i|}$.
4: Order submitted bids in a descending order of $b_i / \sqrt{|\mathcal{S}_i|}$ such that

$$\frac{b_1}{\sqrt{|\mathcal{S}_1|}} \geq \frac{b_2}{\sqrt{|\mathcal{S}_2|}} \geq \cdots \geq \frac{b_N}{\sqrt{|\mathcal{S}_N|}}. \tag{7.9}$$

5: For $i = 1 : N$, if $\mathcal{S}_i \cap (\bigcup_{k \in \mathcal{L}} \mathcal{S}_k) = \emptyset$, then $\mathcal{L} = \mathcal{L} \cup \{i\}$; else secondary user i's bid is denied.
6: Output: a set of winners $\mathcal{L}$.

Algorithm 2 consists of the following steps:

- First, the primary user collects bids from the secondary users.
- Then, the primary user calculates *normalized bids*, $b_i / \sqrt{|\mathcal{S}_i|}$, where $|\mathcal{S}_i|$ is the number of mapping values s_i specified in $\mathcal{S}_i$.
- The primary user orders the normalized bids in a descending order (see (7.9) for an example).
- In the ordered list, the algorithm sequentially checks every bid to see if it overlaps with all the previous winning bids. If it does not, the corresponding secondary user is added into the set $\mathcal{L}$ of winners. Otherwise, the secondary user is the loser, and its bid is denied.

7.5 Development of the Combinatorial Auction for Computer Networks

As observed in [241], the computational complexity of the ordering process, Step 4, is $O(N \log(N))$. In Step 5, the algorithm needs to go over only N bids to decide the final allocation, and the computational complexity of this step is $O(N \log(N))$. Therefore, we have the following theorem:

THEOREM 7.2 *[241] The proposed greedy algorithm has polynomial-time computational complexity.*

Pricing Mechanism

To improve the efficiency of the resource allocation, the prices for the winners need to be designed to achieve incentive compatibility, also known as truthfulness, which provides the secondary users with a motivation to reveal their true values for the resources. For this purpose, the pricing policy of the VCG auction can be used. However, it is well known that the VCG auction does not ensure incentive compatibility when the approximate algorithm, the greedy algorithm, is used to solve the winner determination problem. Alternatively, a pricing policy called *critical payment* can be adopted for the price determination. The critical payment approach is essentially similar to the VCG pricing policy. The critical payment is sometimes called *VCG-like pricing* [241].

- Assume that

    - Secondary user i with bid b_i is the winner in the auction.
    - Secondary user j with bid b_j is the loser.
    - Secondary user j is the winner if secondary user i does not participate in the auction.

- Then, the price that secondary user i pays the primary user is determined as follows:

    - Secondary user i pays zero if secondary user j does not exist.
    - If there exists secondary user j, then secondary user i pays $b_j \sqrt{|\mathcal{S}_i|}/\sqrt{|\mathcal{S}_j|}$.

Note that $\sqrt{|\mathcal{S}_i|}/\sqrt{|\mathcal{S}_j|}$ is used to normalize the price due to the use of the greedy algorithm. Thus, the price that secondary user i pays is actually the highest bid of the losers. This is similar to the price determined in the Vickrey or VCG auctions, which guarantees incentive compatibility.

Before showing the truthfulness of this payment scheme, we consider sufficient conditions that make a payment mechanism truthful. As proved in [241], we have the following lemma:

LEMMA 7.3 *[241] The combinatorial auction is truthful if and only if the following three conditions hold:*

- **Ex-post budget balance**: *The secondary users are all rational, so that their values for the resources are greater than or equal to the prices that they need to pay the primary user.*
- **Monotonicity**: *If secondary user i with normalized bid $b_i/\sqrt{|\mathcal{S}_i|}$ wins the auction, then it still wins with normalized bid $b_j/\sqrt{|\mathcal{S}_j|} \geq b_i/\sqrt{|\mathcal{S}_i|}$.*

- *Critical payment*: There exists a critical value of the bid that winning secondary user i needs to submit to win the auction. Then, the price that the winning secondary user pays does not depend on the bid that it submits.

For the proposed auction, we can verify the following points:

- The ex-post budget balance condition is always guaranteed since the secondary users are assumed to be rational bidders; that is, they bidders bid for the resources at prices lower than the values of the resources to the bidders.
- The monotonicity property holds since the secondary user can increase its order in the ranking, and its winning probability, by either increasing the price the secondary user pays or reducing the resource demand $|S_i|$ (see (7.9)).
- For the winning secondary user i, b_j is the minimum price that the secondary user needs to bid to win the requested resources. If secondary user i bids less than b_j, it loses the auction. Thus, b_j is called the *critical bid* of secondary user i, and the corresponding payment $b_j\sqrt{|S_i|}/\sqrt{|S_j|}$ is called the *critical payment* of secondary user i. We say that the payment scheme has a critical payment; that is, the critical payment condition is satisfied. Obviously, the price that winning secondary user i pays does not depend on the bid b_i that it submits, as the critical payment condition is satisfied.

In summary, the proposed payment scheme satisfies these conditions, and thus the proposed auction has incentive compatibility. The simulation results in [241] show that the proposed auction outperforms the baseline (i.e., the auction in which the winner determination is based on only the values of the secondary users) in terms of social welfare. However, the spectrum utilization ratio (i.e., the ratio of allocated resources to the total resources) of the proposed auction is lower than that of the baseline. This implies that to reach a better level of social welfare, the utilization ratio may be compromised. However, when the number of secondary users is large enough (e.g., 1000), the spectrum utilization ratio of the proposed auction is approximately 80%, which is acceptable in realistic systems. In future work, weights of resources, such as the time slot and frequency band, need to be taken into account in the packages of the secondary users to adapt to the secondary users' dynamic requests. Also, the combinations of OR bids and XOR bids can be considered, which allows the secondary users to represent a larger list of their preferences.

7.5.2 Virtualization of 5G Massive MIMO

Wireless virtualization in 5G allows an infrastructure provider to decompose network services/resources from its physical network infrastructure, which consists of massive MIMO base stations. The infrastructure provider can then provision the services/resources to mobile virtual network operators that do not own network resources. However, implementing the virtualization in 5G raises certain issues of resource allocation. The main issue is how to allocate multiple heterogeneous types of resources, such as channels, power, and antennas, to mobile virtual network

7.5 Development of the Combinatorial Auction for Computer Networks

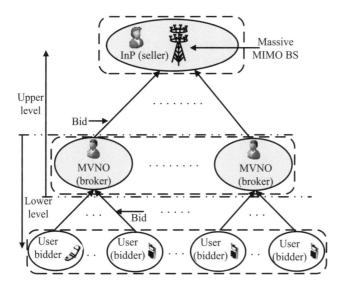

Figure 7.3 Combinatorial auction for multi-seller multi-buyer hierarchical model in massive MIMO systems [242]. InP and MVNO stand for infrastructure provider and mobile virtual network operator, respectively.

operators and both achieve efficiency and accommodate dynamic user demands. A combinatorial auction allows bidders to explicitly request bundles of heterogeneous types of resources. Therefore, the combinatorial auction can be adopted for the virtualization implementation in 5G to improve resource utilization and to satisfy the dynamic demands of the mobile virtual network operators. Such an approach is proposed in [242] and is described next.

System Model

The system model shown in Figure 7.3 consists of the following entities:

- One infrastructure provider owns one massive MIMO base station with A antennas. Also, this infrastructure provider has P power units and C channels for trading. Here, each channel has bandwidth W. The power units and channels are assumed to be homogeneous.
- There is a set $\mathcal{M}$ of M mobile virtual network operators, $\mathcal{M} = \{1, \ldots, M\}$. The mobile virtual network operators acting as brokers (i.e., middlemen) purchase the infrastructure services from the infrastructure provider to serve their own users.
- Each mobile virtual network operator m serves a set $\mathcal{N}_m$ of N_m users.

The resource allocation in the system model requires a two-level hierarchical framework. In particular, the infrastructure provider abstracts its physical resources into virtual resources, or isolated slices, for each mobile virtual network operator. The mobile virtual network operator then allocates resources within its slice to its subscribed users. To address the hierarchical resource allocation problem while satisfying the requirements of efficient resource allocation, a hierarchical combinatorial auction can be adopted. As presented in [242], a hierarchical auction consists of subauctions in two levels. The upper level uses a combinatorial auction in which the infrastructure provider is a seller and the mobile virtual network operators are the buyers/bidders. In the lower level, each mobile virtual network operator conducts a combinatorial auction in which the mobile virtual network operator is the seller and its users are the bidders. This means that there are M subauctions performed in the lower level. For convenience, the auction in the upper level is called an *upper-level auction* and that in the lower level is called a *lower-level auction*.

Bidding

Since the hierarchical auction consists of the upper-level auction and the lower-level auction, we need to consider how bidders in the two subsections bid.

Bids in lower-level auctions: Each mobile virtual network operator conducts a lower-level auction for allocating the resources to its users. Without loss of generality, we can consider the lower-level auction held by mobile virtual network operator m. In this auction, the mobile virtual network operator is the auctioneer, and its users are the bidders. In general, the resource demand of the user can be explicit, with the user specifying which type of resources it wants, or implicit, with the user simply indicating an intrinsic target rate. In the following, we consider only the users with explicit resource demands; the case involving users with implicit demands is further discussed in [242].

User n demands a resource package $\mathcal{S}_n$ from its mobile virtual network operator as follows:

$$\mathcal{S}_n = \{C_n, P_n, A_n\} \tag{7.10}$$

where C_n, P_n, and A_n, respectively, are the demanded number of channels, power units, and antennas of user n. Mobile virtual network operator m is removed from the terms/variables for simplicity, as indicated by index m.

Assume that a user is single-minded, meaning that it is only interested in $\mathcal{S}_n$. Then, the bid of the user is an atomic bid expressed as $(<\mathcal{S}_n, b_n>)$ (see Section 7.3.1), where b_n is the price that the user is willing to pay the mobile virtual network operator for the package. Note that the atomic bid can be expressed as $b_n(\mathcal{S}_n)$. Let $r_n(\mathcal{S}_n)$ denote the achievable rate when resource package $\mathcal{S}_n$ is allocated to the user. Then the value of the package to the user can be defined as $v_n = \gamma_n r_n(\mathcal{S}_n)$ [242], where γ_n is a constant.

Bids in upper-level auction: In the upper-level auction, the mobile virtual network operators are the bidders and the infrastructure provider is the seller. Different from the users, the mobile virtual network operators act as brokers, and they have no intrinsic demands. The mobile virtual network operators are thus not single-minded, and each mobile virtual network operator can submit multiple bids for possible packages of resources. The bid of mobile virtual network operator m is expressed as $\{\mathcal{S}_m, b_m(\mathcal{S}_m)\}$, where $\mathcal{S}_m = \{C_m, P_m, A_m\}$ is a tuple of the demanded number of channels C_m, the amount of transmit power P_m, and the demanded number of antennas A_m. P_m can be interpreted as the number of power units. We assume that the mobile virtual network operator accepts at most one bid, and thus the XOR bid is used to combine its bids.

Let p_m denote the total cost that the mobile virtual network operator pays the infrastructure provider for the resource package, and let p_n denote the price charged to user n in the lower-level auction. The value of resource package $\mathcal{S}_m$ to mobile virtual network operator m can be defined as follows:

$$v_m(\mathcal{S}_m) = \sum_n p_n - p_m \tag{7.11}$$

Winner Determination Problem Formulation

To determine winning bids in the hierarchical auction, we need to formulate the winner determination problems both for the infrastructure provider in the upper-level auction and for the mobile virtual network operators in the lower-level auctions.

Infrastructure provider: Mobile virtual network operator m submits bids $\{\mathcal{S}_m, b_m(\mathcal{S}_m)\}$. Upon receiving the bids from all the mobile virtual network operators, the infrastructure provider selects those bids that maximize the total value of the selected bids, such that efficient resource allocation is achieved. Thus, the winner determination problem for the infrastructure provider is formulated as follows [242]:

$$\max_{x_m(\mathcal{S}_m)} \sum_m \sum_{\mathcal{S}_m} b_m(\mathcal{S}_m) x_m(\mathcal{S}_m) \tag{7.12}$$

$$\text{s.t.} \sum_{\mathcal{S}_m} \sum_m C_m(\mathcal{S}_m) x_m(\mathcal{S}_m) \leq C \tag{7.13}$$

$$\sum_{\mathcal{S}_m} \sum_m A_m(\mathcal{S}_m) x_m(\mathcal{S}_m) \leq A \tag{7.14}$$

$$\sum_{\mathcal{S}_m} \sum_m P_m(\mathcal{S}_m) x_m(\mathcal{S}_m) \leq P \tag{7.15}$$

$$\sum_{\mathcal{S}_m} x_m(\mathcal{S}_m) \leq 1, \forall m \in \mathcal{M} \tag{7.16}$$

where

- $x_m(\mathcal{S}_m) \in \{0, 1\}$ is the decision variable; $x_m(\mathcal{S}_m) = 1$ represents that mobile virtual network operator m wins package $\mathcal{S}_m$ and $x_m(\mathcal{S}_m) = 0$ otherwise.
- C, P, and A are the maximum numbers of channels, power units, and antennas, respectively, that the infrastructure provider can provide. $C_m(\mathcal{S}_m)$, $P_m(\mathcal{S}_m)$, and $A_m(\mathcal{S}_m)$ are the numbers of channels, power units, and antennas specified in package $\mathcal{S}_m$.
- The constraints in (7.13), (7.14), and (7.15) guarantee that the total resource demanded in the winning packages does not exceed the available capacity.
- The constraint in (7.16) ensures that each mobile virtual network operator obtains at most one bid. This means that the bids are mutually exclusive, using XOR bidding language.

Mobile virtual network operators: In the lower-level auctions, each mobile virtual network operator is the auctioneer, and its users are the bidders. Upon receiving bids

$b_n(\mathcal{S}_n)$ from its users, the objective of the mobile virtual network operator is to maximize the sum value of wining bids. For simplification, it is assumed that [242]

- The channels are homogeneous for each user.
- For each channel, the maximum number of users that can be served simultaneously is N_C.
- The achievable rate of a user is independent of which other users are sharing the same channel.
- The number of antennas of the massive MIMO base station is large enough to serve any demand for the antennas of the users.

Given these assumptions, mobile virtual network operator m formulates its winner determination problem as follows:

$$\max_{x_n} \sum_{n} b_n(\mathcal{S}_n) x_n \tag{7.17}$$

$$\text{s.t.} \sum_{n} P_n(\mathcal{S}_n) x_n \leq \overline{P}_m \tag{7.18}$$

$$\sum_{n} C_n(\mathcal{S}_n) x_n \leq \overline{C}_m N_C \tag{7.19}$$

$$x_n \in \{0, 1\}, \forall n \tag{7.20}$$

where x_n is the decision variable indicating whether the resource demand of user n is satisfied, and ($P_n(\mathcal{S}_n)$ and $C_n(\mathcal{S}_n)$), respectively, are the numbers of power units and channels requested in package $\mathcal{S}_n$. $\overline{P}_m$ and $\overline{C}_m$, respectively, are the numbers of power units and channels that mobile virtual network operator m receives from the infrastructure provider in the upper-level auction.

Winner Determination

Similar to hierarchical games, the method of backward induction is used to solve the hierarchical auction. In particular, the winner determination problem in the lower-level auction is solved first, and then the winner determination problem in the upper-level auction is solved.

Winner determination in the lower-level auctions: The problem in (7.12) is an integer programming problem, which is NP-hard. Similar to [241], to solve the problem in polynomial-time, the greedy algorithm should be adopted to find an approximate optimal solution. The greedy algorithm is described in Algorithm 3 [242].

Algorithm 3 can be summarized as follows:

- Each user submits its bid to the mobile virtual network operator. The bid includes a package $\mathcal{S}_n$ of resources and a price b_n that the user is willing to pay the mobile virtual network operator for the package.
- Upon receiving the bids, the mobile virtual network operator calculates normalized bids, $b_n/\sqrt{|\mathcal{S}_n|}$, where $|\mathcal{S}_n|$ is the number of resources specified in $\mathcal{S}_n$. Since $\mathcal{S}_n$ consists of different types of resources, such as channel and power, $|\mathcal{S}_n|$ can be determined as a weighted sum of the number of resources with different types

---

Algorithm 3 Greedy algorithm to solve the winner determination problem for mobile virtual network operator m [242].

1: Input: Submitted bids $\{S_n, b_n\}$, N_m, $\overline{C}_m$, N_C, $\overline{P}_m$, ω_C, and ω_P.
2: Initialize: Set $C = 0$ and $P = 0$.
3: Compute $|S_n| = \omega_C C_n + \omega_P P_n$ for the submitted bids.
4: Compute normalized bids $b_n / \sqrt{|S_n|}$.
5: Order submitted bids in a descending order of $b_n / \sqrt{|S_n|}$:

$$\frac{b_1}{\sqrt{|S_1|}} \geq \frac{b_2}{\sqrt{|S_2|}} \geq \cdots \geq \frac{b_{N_m}}{\sqrt{|S_{N_m}|}} \tag{7.21}$$

6: For $n = 1 : N_m$, if $C + C_n \leq \overline{C}_m N_C$ and $P + P_n \leq \overline{P}_m$, then allocate C_n channels and P_n power units to user n.

---

of resources in package S_n. In particular, $|S_n| = \omega_C C_n + \omega_P P_n$, where ω_C and ω_P are the weights of the number of channels and the number of power units, respectively.

- The mobile virtual network operator then sorts the bids in a descending order of the normalized bids.
- The mobile virtual network operator selects a bid as the winning bid if (i) it has the highest normalized bid and (ii) the number of resources requested in the bid is smaller than the available resources.

Winner determination in the upper-level auction: Similar to the lower-level auction, the winner determination problem in the upper-level auction is an integer programming problem, which is NP-hard. Also, the objective of the problem is to maximize the total value of the selected bids. Thus, Algorithm 3 can also be used to solve the winner determination problem in the upper-level auction. Note that in the upper-level auction, the maximum numbers of channels, power units, and antennas that the infrastructure provider provisions are C, P, and A, respectively.

Price Determination

To achieve efficiency in resource allocation in the auction, a pricing policy for the winning bidders needs to guarantee incentive compatibility (i.e., truthfulness). Note that the two subauctions, the lower-level auction and the upper-level auction, have the same objective, the efficient resource allocation, and use the same winner determination algorithm, the greedy algorithm. Thus, the designed pricing policy can be applied to the bidders in both subauctions – the users in the lower-level auction and the mobile virtual network operators in the upper-level auction. Similar to [241], the VCG-like pricing can be applied. The proposed pricing policy is based on the definition of *blocking*.

DEFINITION 7.4 *[242] If bidder n with bid b_n is the winner and bidder j with bid b_j is a loser, then bidder n uniquely blocks bidder j if bidder j is the winner without the participation of bidder n in the auction.*

It can be seen from Definition 7.4 that bidder j actually is the critical bidder, as discussed in [241]. Correspondingly, bid b_j is the critical bid of the winning bidder n. Based on the definition, a *greedy price*, denoted by p_n^{greedy}, is determined as follows:

$$p_n^{\text{greedy}} = \frac{b_j}{\sqrt{|S_j|}}\sqrt{|S_n|} \qquad (7.22)$$

To avoid the sellers obtaining a low revenue, a base access price p^{base} can be adopted [242]. Thus, bidder n with winning bid b_n is charged with price p_n:

$$p_n = \max\{p^{\text{base}}, p_n^{\text{greedy}}\} \qquad (7.23)$$

As discussed in Section 7.5.1, p_n^{greedy} is actually the highest bid of the losers. In other words, it is the minimum price that bidder n needs to bid to win the requested package of resources. Thus p_n^{greedy} is the *critical value* of bidder n.

It is worth noting that the winner determination algorithm (Algorithm 3) is monotone. Specifically, if bidder n with normalized bid $b_n/\sqrt{|S_n|}$ wins the auction, then the bidder with normalized bid $b_m/\sqrt{|S_m|} > b_n/\sqrt{|S_n|}$ is also the winner of the auction. The bidder can increase its normalized bid, and in turn its order in the ranking, by either increasing the price that the bidder is willing to pay or reducing the number of resources, defined here as the number of channels and power units.

The proposed auction scheme has a monotone property in terms of allocation and has a critical value of the payment, thereby ensuring incentive compatibility (i.e., truthfulness) according to [247].

LEMMA 7.5 *[247] If the allocation algorithm in the auction is monotone, and each winning bidder pays the critical value, then the auction is incentive compatible.*

The simulation results in [242] show that the proposed auction scheme outperforms the fixed sharing resource scheme in terms of average social welfare (i.e., the sum value of all accepted bids), average resource utilization (i.e., the proportion of resources utilized), and the average user satisfaction (i.e., the ratio of users whose resource requests are satisfied). In future work, the fairness among users should be considered. In particular, fairness metrics can be taken into account in the users' value functions.

7.5.3 Mobile Data Offloading in 5G HetNets

In 5G HetNets, macro base stations consume the major part of the network energy, and thus the minimization of the consumed energy is needed. This goal can be achieved by mobile data offloading schemes. Mobile data offloading allows use of the resources of small cell base stations such as access points to offload the traffic of users from macro base stations. In particular, mobile data offloading (i) improves the QoS of the users, (ii) mitigates the traffic and bandwidth consumption over backhaul links, (iii) achieves higher energy efficiency, and (iv) significantly improves network capacity as well as coverage.

7.5 Development of the Combinatorial Auction for Computer Networks

However, macro base stations and access points often belong to different owners. Thus, mobile data offloading schemes need to be designed to (i) minimize the total cost from renting the access points, (ii) maximize the offloading of data from the macro base stations, and (iii) satisfy the resource constraints of the access points. These goals are considered to be the objective and constraints of an optimization problem that can be effectively solved by the combinatorial auction. The following discussion focuses on the application of combinatorial auction for mobile offloading in 5G HetNets, an approach proposed in [129]. Note that the model in [129] consists of a single buyer and multiple sellers, and the combinatorial auction should be a reverse combinatorial auction. The reverse combinatorial auction is essentially a combinatorial auction in which the roles of the seller and the buyer are reversed and asks are used instead of bids.

System Model
The system model is a HetNet that consists of the following entities:

- A mobile network operator owns macro base stations that serve a set $\mathcal{N}$ of its mobile users.
- There is a set $\mathcal{M}$ of M access points. Each access point is owned by one owner; thus, there are M owners.
- Let $\mathcal{N}_j \subseteq \mathcal{N}$ denote the set of mobile users that are in the radio range (i.e., the coverage) of access point $j \in \mathcal{M}$.
- Access point j has an unused capacity C_j that the owner of the access point is willing to lease. Let v_j denote the value of leasing the access point to the owner. v_j can be the maintenance and bandwidth cost of the access point, which is generally unknown to the mobile network operator.
- Each owner j as a seller submits an ask to the mobile network operator. The ask is expressed as $(< C_j, a_j >)$, where a_j is the price that owner j is willing to receive from leasing the capacity C_j of its access point to the mobile network operator.
- Let p_j denote the price that the mobile network operator pays owner j for using the capacity C_j. Then, the utility of the owner is defined as $u_j = p_j - v_j$ if access point j is selected, and 0 otherwise. u_j represents how much utility the owner gains from leasing the capacity of its access point.
- To assign the access points to the mobile users, the mobile network operator calculates a vector of channel utilizations associated with each mobile user. In particular, given the amount of traffic d_i of mobile user i, the vector is defined as $\mathbf{e}_i = (e_{i1}, \ldots, e_{iM})$, where e_{ij} represents the channel utilization of access point j when it is selected to offload the data traffic of mobile user i. e_{ij} is computed as $e_{ij} = d_i / r_{ij}$, where r_{ij} is the maximum achievable transmit rate of mobile user i when it is associated with access point j. In general, a small value of e_{ij} means that access point j should be selected since the access point allows the mobile user to achieve a high transmit rate.

Winner Determination Problem Formulation
Upon receiving the asks of the owners, the mobile network operator needs to select a subset of access points and a subset of its mobile users to be offloaded from the

macro base stations. The objective is to minimize the total cost from renting the access points, and to maximize the offloading of data connections from the macro base stations given the resource constraints of the access points. The mobile network operator thus formulates its winner determination problem as follows [129]:

$$\min_{x_j, y_{ij}} \left(\sum_{j \in \mathcal{M}} a_j x_j - \sum_{j \in \mathcal{M}} \sum_{i \in \mathcal{N}_j} \omega_0 y_{ij} \right) \quad (7.24)$$

$$\text{s.t. } y_{ij} \leq x_j, \qquad \forall j \in \mathcal{M}, \forall i \in \mathcal{N}_j \quad (7.25)$$

$$\sum_{j \in \mathcal{M}} y_{ij} \leq 1, \qquad \forall i \in \mathcal{N} \quad (7.26)$$

$$\sum_{i \in \mathcal{N}_j} y_{ij} e_{ij} \leq x_j, \qquad \forall j \in \mathcal{M} \quad (7.27)$$

$$\sum_{i \in \mathcal{N}_j} y_{ij} d_i \leq x_j C_j, \qquad \forall j \in \mathcal{M} \quad (7.28)$$

$$y_{ij} = 0, \qquad \forall j \in \mathcal{M}, \forall i \notin \mathcal{N}_j \quad (7.29)$$

$$x_j, y_{ij} \in \{0, 1\}, \qquad \forall j \in \mathcal{M}, \forall i \in \mathcal{N} \quad (7.30)$$

where

- $x_j, j \in \mathcal{M}$, is a decision variable; $x_j = 1$ represents that access point j is selected, and $x_j = 0$ otherwise.
- $y_{ij}, i \in \mathcal{N}, j \in \mathcal{M}$, is also a decision variable; $y_{ij} = 1$ indicates that mobile user i is assigned to access point j, and $y_{ij} = 0$ otherwise.
- The first term of the objective function in (7.24), $\sum_{i \in \mathcal{N}} a_j x_j$, represents the total cost that the mobile network operator pays the owners of access points. The second term, $\sum_{j \in \mathcal{M}} \sum_{i \in \mathcal{N}_j} \omega_0 y_{ij}$, is included to maximize the offloading of data connections from the macro base stations to the rented access points. ω_0 is the trade-off parameter between the two objectives.
- The constraint in (7.35) ensures that only those access points that win the auction are used to offload the traffic of mobile users.
- The constraint in (7.36) ensures that each mobile user is served by at most one access point.
- The constraints in (7.27) and (7.28) guarantee that the total traffic load served by each selected access point does not exceed the capacity of the access point.
- The constraint in (7.29) avoids the assignment of mobile users to an access point that does not cover the mobile users.

Greedy Algorithm for Mobile Data Offloading
The optimal reverse auction problem described in (7.24) is NP-hard. Similar to the aforementioned approaches [241] and [242], the greedy algorithm is adopted to solve the problem.

- Upon receiving the asks from the owners of the access points, the mobile network operator calculates the normalized asks associated with the access points. The normalized ask of an access point is the ratio of its submitted ask to the

Algorithm 4 Greedy algorithm for the winner determination of reverse combinatorial auction in HetNets [129].

Input: $\mathcal{M}, \mathcal{N}, a_j, C_j, d_i, e_{ij}$
Output: $x_j, y_{ij}, \hat{j}$
1: $\mathcal{L}_A \leftarrow sort(j \in \mathcal{M}, a_j / \sum_{i \in \mathcal{N}_j} e_{ij}, ascending)$;
2: $\mathcal{U} \leftarrow \mathcal{M}$;
3: **While** $(\mathcal{L}_A \neq \emptyset) \cap (\mathcal{U} \neq \emptyset)$ **do**
4:     $l \leftarrow j; j \leftarrow Next(\mathcal{L}_A); x_j \leftarrow 1$;
5:     **While** $\left(\sum_{i \in \mathcal{N}_j} y_{ij} e_{ij} \leq 1\right) \cap \left(\sum_{i \in \mathcal{N}_j} y_{ij} d_i \leq x_j C_j\right)$ **do**
6:         $\mathcal{L}_U \leftarrow sort(i \in \mathcal{N}_j, e_{ij}, ascending)$;
7:         $i \leftarrow next(\mathcal{L}_U)$;
8:         **if** $\sum_{k \in \mathcal{N}} y_{ik} = 0$ **then**
9:             $y_{ij} \leftarrow 1$;
10:            $\mathcal{U} = \mathcal{U} \setminus \{i\}$;
11:        **end if**
12:     **end While**
13: **end While**
14: **if** $\mathcal{L}_A = \emptyset$ **then**
15:     $\hat{j} \leftarrow l$;
16: **else**
17:     $\hat{j} \leftarrow next(\mathcal{L}_A)$;
18: **end if**

    channel utilization of the mobile users that the access point may serve, $a_j / \sum_{i \in \mathcal{N}_j} e_{ij}$. A small value of the normalized ask means that the cost for renting the access point is low, so that more mobile users can be served.
- The mobile network operator sorts the access points in an ascending order based on their normalized asks.
- The mobile network operator selects those access points with the lowest normalized asks as the winners to serve its mobile users.
- Given a selected access point $j \in \mathcal{M}$, the mobile network operator assigns mobile users to the access point. The assignment task needs to guarantee that (i) the number of mobile users to be offloaded is the maximum, meaning that mobile users with smallest e_{ij} should be selected, and (ii) the overall traffic demand of the mobile users does not exceed the capacity of the access point.
- After assigning the mobile users to the access points, the mobile network operators determines a critical access point $\hat{i} \in \mathcal{M}$. The critical access point is the first unselected access point or the last selected access point of the sorted list.

For each selected access point j, the corresponding owner j (i.e., the winner) receives a price p_j paid by the mobile network operator. p_j is determined as follows:

$$p_j = \frac{a_{\hat{j}}}{\sum_{i \in \mathcal{M}_{\hat{j}}} e_{i\hat{j}}} \sum_{i \in \mathcal{M}_{\hat{j}}} e_{i\hat{j}} \qquad (7.31)$$

It can be seen from (7.31) that the payment for the winner is determined based on the ask of the critical access point $\hat{i}$. As discussed in 7.5.1 and Sections 7.5.2, the critical payment is used to guarantee incentive compatibility (truthfulness) for the proposed auction.

The simulation results in [129] show that by accounting for channel utilization in the winner determination problem, the proposed auction can achieve an average bandwidth utilization of the selected access points higher than 75%. However, the overall cost paid by the mobile network operator in the proposed auction, which equates to the offloading cost, is higher than that obtained with the optimal exhaustive algorithm. In fact, the offloading cost of the proposed auction decreases as the number of mobile users increases. The reason is that the fixed bandwidth of each access point is distributed among more mobile users. Future work can consider general situations in which the presence of multiple mobile network operators makes the design of mobile offloading schemes more complicated.

7.5.4 Resource Allocation in D2D Communication Underlying Cellular Networks

In the previous examples, the combinatorial auction used for the resource allocation was a single-round auction. In this section, we show that an iterative combinatorial auction can be used for resource allocation to reduce the computation complexity of the single-round combinatorial auction. The approach is considered in the D2D communication underlying a cellular network as proposed in [243].

System Model and Problem Formulation

The network model shown in Figure 7.4 consists of one base station, multiple cellular users, and multiple D2D pairs.

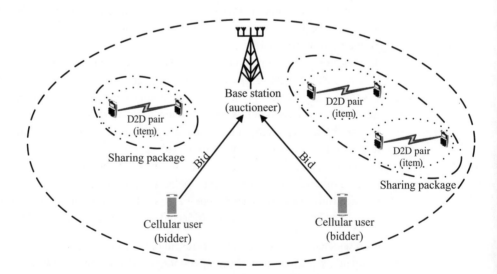

Figure 7.4 Combinatorial auction for resource sharing in D2D communication underlaying cellular networks [243].

7.5 Development of the Combinatorial Auction for Computer Networks

- The base station acts as a controller that has C radio channels.
- A total of N cellular users rely on the radio channels to communicate with the base station.
- The total number of D2D pairs is D. Let d denote the index of a D2D pair and $\mathcal{D}$ denote the set of all D2D pairs.
- To maximize the sum rate of the whole network, the base station allows the cellular users to share the radio channels with the D2D pairs.
- The D2D pairs can use the same channels if they are far enough away from each other. D2D pairs sharing a radio channel constitute a *sharing set* or *sharing package*. Assume that the D2D pairs constitute M such sharing packages. Let $\mathcal{S}_k$ denote sharing package k, $k \in \{1, \ldots, M\}$.

Since the cellular users share both the channels and the sharing packages with the D2D pairs, the D2D communications can cause interference to the cellular users' communication. This degrades the performance, quantified by the data rate, of the cellular users. To mitigate the performance degradation, the base station chooses which sharing package will be associated with which cellular user. However, this significantly increases the communication overhead in the network. Alternatively, each cellular user can locally select sharing packages as long as its performance gain is not negative when the sharing packages are allocated to the cellular user. The performance gain of cellular user i when selecting sharing package $\mathcal{S}_k$ is defined as follows:

$$v_i(\mathcal{S}_k) = \max\left(R_i(\mathcal{S}_k) - R_i^0, 0\right) \tag{7.32}$$

where R_i is the channel rate of cellular user i, and $R_i(\mathcal{S}_k)$ is the sum of the channel rate of cellular user i and those of the D2D pairs in package $\mathcal{S}_k^0$ when this package is allocated to the cellular user. The formulation details of $R_i(\mathcal{S}_k)$ and R_i^0 can be found in [243]. $v_i(\mathcal{S}_k)$ is defined as the value of package $\mathcal{S}_k$ to cellular user i.

Thus, the cellular users can be the bidders, the D2D pairs are the items, and the base station is the seller, or auctioneer. The users need to compete for the sharing packages to improve their utility. Recall that the sharing package consists of D2D pairs that share a channel. Since each cellular user uses one channel, the cellular user selects only one of the sharing packages. Thus, XOR bidding should be used; that is, the bidder, the cellular user, can submit multiple bids but wins at most one bid. Let $b_i(\mathcal{S}_k)$ denote the price that cellular user i is willing to pay for sharing package $\mathcal{S}_k$. The utility of the cellular user when receiving the sharing package is given by

$$u_i(\mathcal{S}_k) = v_i(\mathcal{S}_k) - b_i(\mathcal{S}_k) \tag{7.33}$$

Upon receiving bids from the cellular users, the base station needs to formulate the winner determination problem. The objective is to maximize the total price that the cellular users pay. Thus, the winner determination problem is defined as follows:

$$\max_{x_i(\mathcal{S}_k)} \sum_{i=1}^{N} \sum_{k=1}^{M} b_i(\mathcal{S}_k) x_i(\mathcal{S}_k) \tag{7.34}$$

$$\text{s.t.} \sum_{k=1}^{M} x_i(\mathcal{S}_k) \leq 1, \forall i \in \{1, \ldots, N\} \tag{7.35}$$

$$\sum_{d \in \mathcal{S}_k} \sum_{i=1}^{N} x_i(\mathcal{S}_k) \leq 1, \forall d \in \{1, \ldots, D\} \tag{7.36}$$

where

- $x_i(\mathcal{S}_k)$ is the decision variable; $x_i(\mathcal{S}_k) = 1$ represents that cellular user i wins set $\mathcal{S}_k$, and $x_i(\mathcal{S}_k) = 0$ otherwise.
- The constraint in (7.35) guarantees that each cellular user obtains at most one package. The constraint in (7.36) guarantees that each D2D pair, the item, is allocated to at most one cellular user.

Iterative Combinatorial Auction

This section shows that an iterative combinatorial auction can be used to achieve the objective in (7.34) while satisfying the constraints in (7.35) and (7.36). In general, the iterative combinatorial auction is implemented in multiple rounds as follows:

- In each round, the base station updates prices for each D2D pair according to the supply and demand rule.
- Cellular users that accept these prices bid for the sharing packages.
- If only one cellular user bids for the sharing package, the sharing package is allocated to the cellular user. The cellular user is removed from the auction since XOR bidding is used.
- If there is an excess demand for one D2D pair, the base station increases the price of any D2D pair and the cellular users bid again.
- The auction process is repeated until all D2D pairs are auctioned off or every cellular user receives its package.

In the proposed auction, the base station sets prices for each D2D pair in each round, and the cellular users submit their demands, identified as the sharing package $\mathcal{S}_k$, to the base station. Here, linear anonymous prices [248] are used, and the price of a sharing package is the sum of the prices of D2D pairs in the package. Thus, given the price $p^t(d)$ that the base station sets for D2D pair d in round t, the price of sharing package $\mathcal{S}_k$ is determined as follows:

$$p^t(\mathcal{S}_k) = \sum_{d \in \mathcal{S}_k} p^t(d) \tag{7.37}$$

In round t, if a cellular user bids for a sharing package $\mathcal{S}_k$, the base station understands that the price that the cellular user pays for package $\mathcal{S}_k$ is equal to $p^t(\mathcal{S}_k)$: $b_i(\mathcal{S}_k) = p^t(\mathcal{S}_k)$. In general, the price $p^t(d)$ is updated if more one cellular user bids for D2D pair d. The question is how the base station sets the prices for the D2D pairs at the beginning of the auction if the base station does not receive demands from the cellular users. The base station can set low prices for the D2D pairs at the beginning of the auction. However, a number of cellular users can bid for the D2D pairs and cause

7.5 Development of the Combinatorial Auction for Computer Networks

Algorithm 5 Iterative combinatorial auction algorithm for resource allocation in D2D communications [243].

1: In each round t, the base station updates prices for each item d as $p^t(d)$.
2: Cellular users bid for package S_k^t if $u_i(S_k^t) \geq 0$; that is, $v_i(S_k^t) \geq \sum_{d \in S_k^t} p^t(d)$.
3: The base station adds the bids into set $\mathcal{B}^t$. The base station checks each bid in the set.
4: If $S_k^t \cap S_m^t \neq \emptyset$ and $k \neq m$, then the base station increases the price of over-demanded item d as $p^{t+1}(d) = p^t(d) + \delta$. Otherwise, package S_k^t is allocated to cellular user i.
5: If $\exists d \in \mathcal{D}$ and $d \notin \mathcal{B}^t$, then the base station decreases the price of over-supplied item d in the next round as $p^{t+1}(d) = p^t(d) - \Delta$.
6: The auction algorithm repeats Steps 1 to 5 until all the D2D pairs are auctioned off or every cellular user wins its package.

---

interference within the cellular system. To avoid such a situation, the prices of the D2D pairs are initially set high.

The details of the proposed auction algorithm are shown in Algorithm 5. By using this algorithm, the proposed auction has the following important properties [243]:

- **Incentive compatibility**: This means that cellular user i should bid at the price equal to its value of the package: $v_i = \sum_{d \in S_k^t} p^t(d) = p^t(S_k^t)$. Indeed, if the cellular user bids for the package as $v_i < p^t(S_k)$, then even if the user wins the package, it receives a negative utility or payoff, $u_i^t(S_k^t) < 0$. If the cellular user does not bid for the package at $v_i = p^t(S_k^t)$ and waits until $v_i > p^t(S_k^t)$, then other cellular users may bid for the package and the cellular user loses an opportunity to win the package. Thus, the user cellular should bid the package at $v_i = p^t(S_k^t)$.
- **Convergence**: The proposed auction algorithm can converge in a finite number of rounds since (i) the prices of the items, comprising the D2D pairs, are updated according to the supply and demand rule and (ii) the number of packages of items is finite. The detailed proof can be found in [243].
- **Low complexity**: The proposed auction algorithm has lower complexity than the optimal (exhaustive) algorithm. As presented in [243], given the number of D2D pairs as D and the number of cellular users as N, the complexity of the exhaustive algorithm is $O(N^D)$, while that of the proposed auction algorithm is $O(N(2^D - 1) + t)$, where t is the total number of rounds.

The simulation results in [243] show that the proposed scheme can achieve the system sum rate close to the optimal algorithm. Also, the proposed scheme outperforms random allocation in terms of the system sum rate. Specifically, as the number of available radio channels increases, the system sum rate of the proposed scheme increases. The reason is that the number of cellular users and D2D pairs sharing the same channels decreases with an increase in the number of the available radio channels. This reduces the interference among them and improves the system sum rate. The simulation results also show that the proposed scheme can achieve up to 90% system efficiency, defined as the ratio of the system utility obtained by the proposed auction scheme to that obtained by the exhaustive algorithm. However, the proposed scheme still uses the base station as a central controller, which requires channel state information, feedback, and control signaling transmission. This increases signaling overhead in the network. Methods such as channel state information feedback compression and signal flooding can be adopted to reduce the overhead.

7.6 Summary

In this chapter, we introduce the combinatorial auction and its applications for computer networks. First, we provide definitions and examples of two items commonly used in the combinatorial auction, substitutable and complementary items. Second, we introduce types of bidding language, including atomic bids, OR bids, and XOR bids, that allow bidders to succinctly encode or express common bids in the combinatorial auction. Third, we present the basic definition of the winner determination problem in the combinatorial auction. Also, we provide an example to clarify the means used to solve the problem. Through the example, we show that one of the greatest challenges for combinatorial auctions is the high computational complexity required to solve the winner determination problem. To address this challenge, we present two iterative combinatorial auctions, the ascending proxy auction and the clock-proxy auction. Finally, we discuss how the combinatorial auction is used to address resource management issues in computer networks – namely, spectrum sharing in cognitive radio networks, virtualization of 5G massive MIMO, mobile data offloading in 5G HetNets, and resource allocation in D2D communication underlying cellular networks.

8 Double-Sided Auction

8.1 Introduction

In the previous chapters, we presented the forward and reverse auctions. The forward auction has only one seller and multiple buyers, and the reverse auction has multiple sellers and a single buyer. However, many realistic resource trading markets typically have multiple sellers (e.g., service providers) and multiple buyers (e.g., mobile users). In such a situation, we can use a *two-sided auction* or *double auction*. The double auction can be considered to be a combination of the forward auction and the reverse auction. The basic idea of the double auction is to match bids, or bidding prices, of the buyers and asks, or asking prices, of the sellers by assigning payments from the buyers to the sellers and allocating resources from the sellers to the buyers accordingly [42]. In this chapter, we further describe how to determine the winners and payments in the double auction. Note that there exist different double auctions, and their discussions are well presented in [202]. In this chapter, we particularly consider two double auctions that are commonly used in computer networks: the single-round double auction or sealed-bid double auction, and the multi-round double auction or continuous double auction. More specifically, we first provide definitions of the double auctions. Then, we give and analyze examples from computer networks' perspective to further understand the double auctions.

8.2 Single-Round Double Auction

In the single-round double auction model, buyers and sellers submit their bids and asks, respectively, to an auctioneer simultaneously in a single round (see Figure 8.1). Let b_i and a_j denote the bid (bidding price) and the ask (asking price) of buyer i and seller j, respectively. Here, b_i can be the maximum price that buyer i is willing to buy a resource unit, and a_j is the minimum price that seller j is willing to sell the resource unit. After receiving the bids and the asks, the auctioneer sorts the bids in descending order and the asks in ascending order. The auctioneer determines the largest index m at which the ask is still smaller than the bid. Based on the index, the auctioneer determines a clearing price, also known as a hammer price or transaction price, that clears the resource market. The sellers with the asks less than or equal to the clearing price and the buyers with bids higher than or equal to the clearing price are the winners of the

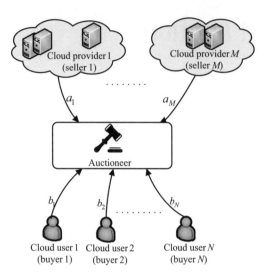

Figure 8.1 Double auction model.

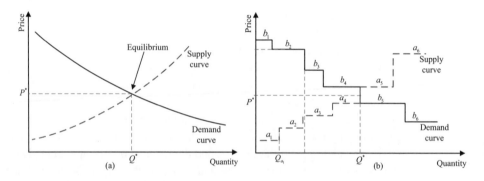

Figure 8.2 Comparison between (a) the supply and demand model from economics and (b) the single-round double auction [42].

single-round double auction. The auctioneer allocates resource units to the winning buyers. Also, it determines and assigns payments to the sellers.

In fact, the single-round double auction is similar to the supply and demand model [249] that is a part of the basic economic model. As shown in Figure 8.2(a), the supply and demand model includes demand and supply curves. The demand curve shows the relationship between the price of a resource unit and the resource quantity (i.e., the number of resource units) demanded by the buyers, and the supply curve presents the relationship between the price and the resource quantity supplied by the sellers. Thus, the supply and demand model represents how the demand and the supply of resources change as the prices of the resources change. In general, the sellers and the buyers tend to change the resource prices and the demand quantity to maximize their objectives of profit and utility, respectively. The supply and demand curves are used to determine a unique point of the market called *market equilibrium* or *supply* and demand equilibrium or economic equilibrium. The market equilibrium is the point at which the supply and demand curves intersect [16]. In Figure 8.2(a), the market equilibrium is (P^*, Q^*), where P^* and Q^* are the equilibrium price and equilibrium quantity, respectively. The market equilibrium implies that the quantity of resources demanded is equal to the quantity of resources supplied.

Similar to the standard supply and demand model, the single-round double auction can formulate the supply and demand curves from the asks and the bids of the sellers and the buyers, respectively. As shown in Figure 8.2(b), the x-axis represents the supplied resource units, and the y-axis represents the asks and the bids. For example, Seller 1 is willing to sell Q_{a_1} resource units at price a_1, and Buyer 2 is willing to buy Q_{b_2} resource units at price b_2, and so on. The intersection between the supply curve and the demand curve is the equilibrium point of the single-round double auction. The equilibrium point allows the auctioneer to determine the winners and the prices in the single-round double auction. To further understand how the auctioneer determines the winners and the prices in the single-round double auction, we consider the following example.

Example 8.1 We consider a cloud resource market that includes five cloud providers (the sellers), five cloud users (the buyers or the bidders), and one third party (the auctioneer). For simplification, we assume that each cloud provider has one cloud resource unit for selling, and each buyer is willing to buy one cloud resource unit. The cloud providers from 1 to 5 submit their asks as $a_1 = \$2$, $a_2 = \$3$, $a_3 = \$4$, $a_4 = \$7$, and $a_5 = \$9$ to the auctioneer, respectively. Simultaneously, the cloud users from 1 to 5 submit their bids as $b_1 = \$8$, $b_2 = \$6$, $b_3 = \$5$, $b_4 = \$4$, and $b_5 = \$3$ to the auctioneer, respectively. By using the single-round double auction, the auctioneer determines the winners as follows:

- The auctioneer sorts the buyers in descending order of their bids: $b_1 = \$8 > b_2 = \$6 > b_3 = \$5 > b_4 = \$4 > b_5 = \$3$.
- The auctioneer sorts the sellers in ascending order of their asks: $a_1 = \$2 < a_2 = \$3 < a_3 = \$4 < a_4 = \$7 < a_5 = \$9$.
- The auctioneer selects $m = 3$ as the largest index at which the ask, $a_3 = \$4$, is still smaller than the bid: $a_3 = \$5$.
- The winning bids are b_1, b_2, and b_3, and the winning asks are a_1, a_2, and a_3.
- Correspondingly, the first three sellers are the winning sellers, and the first three buyers are the winning buyers. As seen, with the single-round double auction, the number of winning buyers is equal to the number of winning sellers.

After determining the winners, the auctioneer assigns the resource units to the winning buyers. Also, it calculates the prices that the winning buyers need to pay and those that the winning sellers receive. The auctioneer can determine the prices based on uniform pricing and discriminatory pricing policies as presented in the next sections.

8.2.1 Uniform Pricing Policy

As the uniform pricing policy is used, all the buyers pay the same price P^* for the resource units and all the sellers receive the same price P^*. P^* can be any price within the interval between the ask a_m and the bid b_m, $a_m \leq P^* \leq b_m$, where m is the largest index at which the ask is still smaller than the bid. Consider Example 8.1 again. If $m = 3$, $a_m = \$4$, and $b_m = \$5$, then P^* is defined as $4 \leq P^* \leq 5$. For example, P^* can be $\$4.5$. As such, P^* is greater than the winning asks a_1, a_2, and a_3, and P^* is less than the winning bids b_1, b_2, and b_3. In other words, the cloud users pay prices lower than those they bid, and the cloud providers receive prices higher than those they submit. Therefore, both the cloud users and the cloud providers have non-negative utilities when they participate in the auction. We say that the double auction guarantees individual rationality, as defined in Section 6.2.5.

To specify how to set the price P^*, we can use the pricing policy of the k-double auction [250]. Let m be the largest index at which the ask is smaller than the bid. According to the k-double auction, the price P^* is calculated as follows:

$$P^* = kb_m + (1-k)a_m \qquad (8.1)$$

where k, $0 \leq k \leq 1$, is the factor. From Example 8.1, P^* is defined as

$$P^* = 4k + 5(1-k) \qquad (8.2)$$

With $0 \leq k \leq 1$, we have $\$4 \leq P^* \leq 5\$$. For $k = 0.1$, $P^* = \$4.9$; that means that the cloud users 1, 2, and 3 pay the same price $P^* = \$4.9$, and the cloud providers 1, 2, and 3 receive the same payment $P^* = \$4.9$. For $k = 0.5$, the price P^* is defined as $P^* = 0.5(b_m + a_m) = 0.5(4+5) = \$4.5$. In this case, the price P^* is the average of a_m and b_m, and thus the pricing policy is called the *average pricing mechanism*.

In general, by using the uniform pricing policy, the single-round double auction demonstrates the following properties:

- *Individual rationality:* The auction guarantees individual rationality since no participant, either buyer or seller, loses from participating in the auction. As discussed in Example 8.1, the utility of both the cloud users (buyers) and cloud providers (sellers) is non-negative when they participate in the auction.
- *Balanced budget:* The auction has a strong balanced budget or strong budget balance as the total payment from the buyers is transferred to the sellers. In particular, in Example 8.1, when the pricing policy of k-double auction with $k = 0.5$ is used, the total payment from the cloud users is $\$4.5 + \$4.5 + \$4.5 = \$13.5$, which is all transferred to the cloud providers.
- *Economic efficiency:* The auction exhibits this property since the resources are assigned to the buyers that value them the most. In Example 8.1, the cloud resource units are allocated to cloud users 1, 2, and 3, which have the highest bids.

However, the single-round double auction with the uniform pricing policy has two issues. First, it is not easy to set the factor k that trades off the utility of the buyers and the sellers. Second, this approach does not guarantee truthfulness, or incentive compatibility. The reason is that the price P^* is determined according to bid b_m of buyer m and ask a_m of seller m. To improve their utility, buyer m has an incentive to submit a bid lower than its value for the resource, and seller m has an incentive to submit an ask higher than its value for the resource. To address the issue, the concept of *critical payment* as discussed in Section 7.5.1 or McAfee's mechanism [251] can be adopted. For McAfee's mechanism, P^* is the average of the lowest bid of the losing buyers and the highest ask of the losing sellers: $P^* = 0.5(b_{m+1} + a_{m+1})$.

8.2.2 Discriminatory Pricing Policy

When the uniform pricing policy is used, the price P^* that the winning sellers receive is higher than the winning asks that they submit. Although the winning sellers have a positive utility, one intuitive question is why the winning buyers are charged a price that is lower than their bids (i.e., the prices that the buyers are willing to pay). Similarly, although P^* is lower than the winning bids that the winning buyers submit, the question is why the winning buyers must pay the winning sellers a price higher than those that the sellers submit/offer?

In discriminatory pricing, different prices can be set differently for the winning buyers and sellers. There are two discriminatory pricing policies [202]:

- *Pay-as-bid/pay-your-bid*: When using this policy, the auctioneer determines the prices for the winning buyers and the sellers based on the winning bids of the winning buyers. This means that the price that each winning buyer pays is its winning bid. Consider Example 8.1 again. The winning cloud user 1 pays the winning cloud provider 1 a price $b_1 = \$8$, the winning cloud user 2 pays the winning cloud provider 2 a price $b_2 = \$6$, and the winning cloud user 3 pays the winning cloud provider 3 a price $b_3 = \$5$. As such, the total payment that the winning cloud users pay is $b_1 + b_2 + b_3 = 8 + 6 + 5 = \$19$.
- *Pay-seller's ask*: When using this policy, the auctioneer determines the prices based on the winning asks of the winning sellers. This means that the price that each winning buyer pays is the winning ask of its winning seller. Continuing with Example 8.1, the winning cloud user 1 pays the winning cloud provider 1 a price $a_1 = \$2$, the winning cloud user 2 pays the winning cloud provider a price $a_2 = \$3$, and the winning cloud user 3 pays the winning cloud provider 3 a price $a_3 = \$4$. In this case, the total payment that the winning cloud users pay is $a_1 + a_2 + a_3 = 2 + 3 + 4 = \$9$.

The total payment when the pay-as-bid policy is used is higher than that when the pay-seller's-ask policy is used. The auctioneer can decide which policy is appropriate depending on the specific objectives of the auction design. For example, to maximize the revenue for the seller, the pay-as-bid policy can be used. However, this results in reducing the utility of the buyers, the cloud users, and they have no incentive to participate in the auction. In fact, the discriminatory pricing model can be determined by combining the pay-as-bid policy and the pay-seller's-ask policy as presented in [202].

In this case, the total payment that the winning buyers need to pay the winning sellers can be determined as $\sum_{i,j} (kb_i + (1-k)a_j)$, where $0 \leq k \leq 1$ is the factor; b_i and a_j are the winning bid of the winning buyer i and the winning ask of the winning seller j, respectively; and k represents the impact, or weight, of the buyers and sellers on the price determination. Returning to Example 8.1, the winning bids are $b_1 = \$8$, $b_2 = \$6$, and $b_3 = \$5$, and the winning asks are $a_1 = \$2$, $a_2 = \$3$, and $a_3 = \$4$. Given a factor $k = 0.7$, the total payment that the winning cloud users pay the winning cloud providers is $0.7(8+6+5) + (1-0.7)(2+3+4) = \$16$. As seen, $\$9 \leq \$16 \leq \$19$ guarantees that no cloud user pays more than the amount that the user bids, and no seller receives less than the amount that the seller offers.

8.3 Continuous Double Auction

In the single-round double auction, each winning buyer is assigned to a winning seller depending on the decision of the auctioneer. In practice, the winning buyer may not accept the assignment – for example, due to the high price that the buyer needs to pay. Therefore, the continuous double auction is adopted. The continuous double auction can be considered an "online" version of the single-round double action. It allows the auctioneer to decide the assignment without knowing what bids/asks will arrive in the future. This auction is thus applicable to future network environments that will be dynamic.

The continuous double auction can be performed in multiple periods, which helps the buyers and the sellers both learn the bidding strategy of their rivals and discover the real values of the resources. Each period r in the continuous double auction is called a *trading period* and has a fixed number of steps T. Each step $t \leq T$ consists of two substeps: the *bid–ask substep* is followed by a *trading substep*. The following terminology is used in the continuous double auction:

- Bid–ask substep: In this substep, buyers submit their bids, the prices that they are willing to pay, and the sellers submit their asks, the prices that they are willing to accept.
- Outstanding bid b_t^r: This is the maximum bid among the bids submitted in step t of period r.
- Outstanding ask a_t^r: This is the minimum ask among the asks submitted in step t of period r.
- The buyer with the outstanding bid and the seller with the outstanding ask are selected to participate in the trading substep.
- In the trading substep, we have the following cases:
    - If the buyer accepts the outstanding ask and the seller accepts the outstanding bid, a transaction between the buyer and the seller is carried out. This means that the buyer is assigned to the seller, and the transaction price P^* is the outstanding bid.

8.3 Continuous Double Auction

- Otherwise, the sellers and the buyers can continue to submit their asks and their bids in the bid–ask substep of the next step $t + 1$.
- The auction terminates when no bid or ask is submitted in the bid–ask step.

The following example explains how the continuous double auction works.

Example 8.2 We consider a cloud resource trading market that includes three cloud providers, four cloud users, and one auctioneer. Each cloud provider may have multiple cloud resource units for trading, and each cloud user may buy more than one cloud resource unit. The auction can be implemented in multiple periods, and each period consists of $T = 3$ steps. Each step has two substeps, the bid–ask substep and the trading substep. Note that the ask is the price per resource unit that the cloud provider is willing to accept, and the bid is the price per resource unit that the cloud user is willing to pay.

The auction starts with the first period, $r = 1$, and includes three steps as shown in Table 8.1.

- Step $t = 1$:
    - The cloud users and the cloud providers submit their bids and asks, respectively, to the auctioneer in the bid–ask substep.
    - The auctioneer selects the outstanding bid and the outstanding ask. Here, the outstanding bid is the bid $b_4 = 7$ submitted by cloud user 4, and the outstanding ask is the ask $a_1 = 12$ submitted by cloud provider 1.
    - Cloud user 4 and cloud provider 1 are selected to participate in the trading substep. However, cloud user 4 does not accept the ask $a_1 = 12$, and cloud provider 1 does not accept the bid $b_4 = 7$. Therefore, there is no transaction in this step.
- Step $t = 2$:
    - The cloud users and the cloud providers again submit their bids and asks to the auctioneer in the bid–ask substep. Note that the values of bids and the

Table 8.1 Continuous double auction process for cloud resource trading in the first period [202]. **A** and **NA** stand for "accepted" and "not accepted," respectively.

		First period $r = 1$							
t	Substep	b_1	b_2	b_3	b_4	a_1	a_2	a_3	P^*
1	Bid–ask	4	5	6	7	12	13	14	
	Trading				NA	NA			
2	Bid–ask	5	6	8	9	11	12	13	
	Trading				NA	NA			
3	Bid–ask	5	7	11	9	10	11	12	
	Trading			A		A			11

Table 8.2 Continuous double auction process for cloud resource trading in the second round [202]. **A** and **NA** stand for "accepted" and "not accepted," respectively.

		Second period $r = 2$							
t	Substep	b_1	b_2	b_3	b_4	a_1	a_2	a_3	P^*
1	Bid–ask	7	8	6	10	10	9	12	
	Trading				A		A		10
2	Bid–ask	7	9	6	5	10	8	12	
	Trading		A				A		9
3	Bid–ask	0	0	0	0	0	0	0	
	Trading								

asks may be different from those in the previous step, $t = 1$. For example, bid b_3 of cloud user 3 increases from 6 to 8, and ask a_1 of cloud provider 1 decreases from 12 to 11. This can be explained as follows: after the first step $t = 1$, the participants, which include both the cloud users and the cloud providers, learn about their rivals' bidding strategies as well as the values of the cloud resources.

- The auctioneer selects cloud user 4 with the outstanding bid and cloud provider 1 with the outstanding ask for the trading substep. However, there is still no transaction in this step.

• Step $t = 3$:

- In the bid–ask substep, the auctioneer selects the cloud user with the outstanding bid, $b_3 = 11$, and cloud provider 1 with the outstanding ask, $a_1 = 10$, to participate in the trading substep.
- In the trading substep, the outstanding bid is greater than the outstanding ask. Thus, cloud user 3 accepts the outstanding ask, and cloud provider 1 accepts the outstanding bid. Therefore, there is a transaction between them.
- Cloud user 3 buys cloud resource units from cloud provider 1 with the transaction price $P^* = b_3 = 11$.

The auction is repeated in the second period $r = 2$ as shown in Table 8.2. In this period, there are two transactions in step $t = 1$ and $t = 2$. At step $t = 3$, there is no bid and ask submitted from the participants; because the bids and the asks are zero, the auction terminates.

8.4 Development of Double Auction for Computer Networks

The double auction allows multiple buyers and multiple sellers to trade resources, and it has been widely applied in real markets. In this section, we discuss applications of the double auction to solve emerging issues in computer networks. In particular, we discuss how to apply the double auction for (i) sensing task allocation in participatory sensing [252], (ii) location privacy in participatory sensing [253], (iii) spectrum allocation in heterogeneous networks [254], and (iv) cloud resource allocation in cloud networking [255].

8.4.1 Sensing Task Allocation in Participatory Sensing

Mobile participatory sensing is a sensing paradigm of IoT in which portable smart devices, such as smartphones, are used to perform sensing tasks. However, performing the sensing tasks also incurs costs, such as power consumption, CPU and the bandwidth cost. Moreover, the smart devices may need to physically move to some specific locations to collect the sensing data required by the data requesters (i.e., customers). While the owners of smartphones, also known as mobile users, and the task requesters are rational, a major problem is how to encourage both of them to participate in the participatory sensing application. Both the Vickrey auction and the VCG auction, discussed in Chapter 6, can be used. However, these auctions only provide the incentive mechanism to either the mobile users or the data requesters. To address this problem, a double auction can be used. Such an approach can be found in [252], where the single-round double auction is adopted for sensing task allocation to achieve desirable economic properties. The proposed approach is described next.

System Model
The system model shown in Figure 8.3 includes the following entities:

- There is a cloud platform (i.e., a trustworthy platform) that acts as an auctioneer.
- There is a set $\mathcal{M}$ of M data requesters as buyers/bidders. Each data requester $j \in \mathcal{M}$ is assumed to have a sensing task t_j. Here, the sensing tasks can be air quality monitoring tasks that need to collect the air quality measurements in an area. Let $\mathcal{T} = \{t_1, \ldots, t_M\}$, where t_j is the sensing task of data requester j. Data requester j has the sensing time demand d_j to perform sensing task t_j. d_j can be interpreted as the number of sensing time units that the data requester demands from the cloud platform.

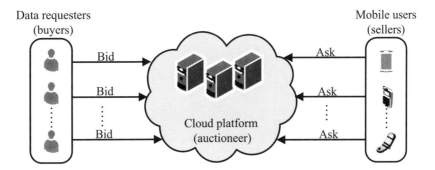

Figure 8.3 Double auction for mobile participatory sensing [252].

- There is a set $\mathcal{N}$ of N mobile users/sellers in the target area to perform the sensing tasks. Each mobile user $i \in \mathcal{N}$ has the available sensing time Q_i, comprising the maximum number of sensing time units that the mobile user can provide to perform the sensing tasks. Let $q_i \leq Q_i$ be the number of sensing time units that mobile user i provides and let C_i be the cost per sensing time unit of the mobile user.
- Note that sensing task t_j can be performed by more than one mobile user, such that the total sensing time unit of the mobile users satisfy the sensing time demand d_j of data requester j. Also, mobile user i can provide sensing time units to more than one data requester as long as $q_i \leq Q_i$.

To match each task j and mobile user i, the single-round double auction as presented in Section 8.2 is used. In this auction, the data requesters and the mobile users simultaneously submit their bids and asks, respectively, to the cloud platform. The bids and the asks are defined as follows:

- The bid of data requester j includes a bidding price b_j, representing the price per sensing time unit that the data requester is willing to pay, and the sensing time demand d_j.
- The ask of mobile user i includes an asking price a_i, representing the price per sensing time unit that the mobile user is willing to receive, and its available sensing time Q_i.

After receiving the bids and the asks, the cloud platform determines (i) the winning mobile users/sellers and the winning data requesters/buyers, (ii) the set of payments $\mathcal{P} = (p_1, \ldots, p_N)$ including prices paid to the winning mobile users, and (iii) the set of charges $\mathcal{Y} = (y_1, \ldots, y_M)$ including prices that the data requesters need to pay for receiving sensing time units. Given the payments and the charges, the utility of the mobile users and the data requesters can be defined as follows.

The utility of mobile user i is the difference between its payment and the total cost as follows:

$$u_i^s = \begin{cases} p_i - C_i q_i, & \text{if } q_i > 0 \\ 0, & \text{otherwise} \end{cases} \quad (8.3)$$

The utility of data requester j is defined as follows:

$$u_j^b = \begin{cases} v_j d_j - y_j, & \text{if } d_j \text{ is satisfied} \\ 0, & \text{otherwise} \end{cases} \quad (8.4)$$

Problem Formulation

To motivate the mobile users and the data requesters to participate in the auction, the total utility of the participants, known as the social welfare, should be maximized. The social welfare is defined as follows:

$$S^W = \sum_i u_i^s + \sum_j u_j^b$$

$$= \sum_i p_i - \sum_i C_i q^i + \sum_j v_j d_j - \sum_j y_j \qquad (8.5)$$

Moreover, the auction needs to guarantee important properties including incentive compatibility (i.e., truthfulness) and individual rationality. The definitions of the properties can be found in Section 6.2.5.

Incentive Mechanism Design

This section presents how to design the single-round double auction to maximize the social welfare while guaranteeing the incentive compatibility and individual rationality. Similar to the standard single-round double auction as described in Section 8.2, after receiving the bids from the data requesters and the asks from the mobile users, the cloud platform performs the winner selection and payment determination. In particular, the winner selection process is implemented so that social welfare is maximized. As expressed in (8.5), social welfare is inversely proportional to the total cost of the mobile users. Thus, the mobile users with the lowest cost or the lowest asks should be selected first. The winner selection and the payment determination are described next.

Winner selection: The cloud platform determines the winner.

- The cloud platform sorts the mobile users in ascending order of their asking prices and sorts the data requesters in descending order of their bidding prices. Assume that the order of asking prices is $a_1 \leq \cdots \leq a_N$, and the order of bidding prices is $b_1 \geq \cdots \geq b_M$. Thus, the order of the mobile users is $\mathcal{N}' = \{1, \ldots, N\}$, and the order of the data requesters is $\mathcal{M}' = \{1, \ldots, M\}$.
- After sorting, the cloud platform finds the largest index n in $\mathcal{N}'$ and the largest index m in $\mathcal{M}'$ that still satisfy the following constraints:

$$\sum_{j=1}^{m} d_j \leq \sum_{i=1}^{n} Q_i \qquad (8.6)$$

$$b_{m+1} \sum_{j=1}^{m} d_j \geq a_{n+1} \sum_{i=1}^{m} q_i \qquad (8.7)$$

The constraint in (8.6) guarantees that the first n mobile users have sufficient sensing time to satisfy the total demand of the first m data requesters. The constraint in (8.7) is defined based on the payment determination described in the next section. In general, this constraint implies that the total price that the winning data requesters need to pay is not less than the total price paid to the mobile users. This ensures that the auction maintains a budget balance.

- The first n mobile users in the list $\mathcal{N}'$ are the winning buyers, and the first m data requesters in the list $\mathcal{M}'$ are the winning sellers.

The algorithm for the winner selection is presented in [252]. The main idea of the algorithm is that the cloud platform greedily allocates the sensing time units of the mobile users to the data requesters. Accordingly, the cloud platform sets $n = 1$ and $m = 1$, meaning that it selects the first mobile user in $\mathcal{N}'$ and the first data requester in $\mathcal{M}'$. Then, the cloud platform verifies the constraints in (8.6) and (8.7). If both the constraints are satisfied, the cloud platform allocates sensing time units of the first mobile user to the first data requester. If the first mobile user still has available sensing time, $q_1 < Q_1$, then the sensing time units left, $Q_1 - q_1$, are allocated to the second data requester. This ensures that all of the sensing time of the mobile users with the low cost is used, which maximizes social welfare as defined in (8.5). This process is repeated until one of the constraints in (8.6) and (8.7) is not satisfied. The running time of the algorithm is $O(m \log m + n \log n)$, which is polynomial time.

Payment determination: To guarantee the truthfulness of the single-round double auction, the critical payment policy described in Section 7.5.1 is used for the payment determination. The critical payment is essentially similar to the Vickrey pricing policy in which the winning buyer pays the price equal to the highest bid of the losing buyers and the winning seller receives the payment equal to the lowest ask of the losing sellers. In particular, the highest bid of the losing buyers is b_{m+1}, and the lowest ask of the lowing sellers is a_{n+1}. Therefore, winning data requester j pays a price y_j that is determined by

$$y_j = b_{m+1} d_j \tag{8.8}$$

Winning mobile user i receives the payment p_i that is determined by

$$p_i = a_{n+1} q_i \tag{8.9}$$

Intuitively, from (8.8), the winning data requesters pay the prices that do not depend on their bidding prices, so the data requesters have an incentive to truthfully submit their bidding prices, such that these prices reflect the true values v_j of the sensing time units. Similarly, the mobile users have an incentive to truthfully submit their asking prices, such that they are equivalent to the real cost C_i for executing the task, since the prices that the mobile users receive do not depend on their asking prices (see (8.9)). The proof of these statements can be found in [231] and [252]. Therefore, the proposed double auction exhibits truthfulness or incentive compatibility. However, the data requesters can cheat on the sensing time demand d_j, and the mobile users can cheat on the available sensing time Q_i. This issue needs to be considered.

Based on the truthfulness property, $b_j = v_j, \forall j \in M$ and $a_i = C_i, \forall i \in N$, the proposed double auction can be proved to support individual rationality. This means that the utility of both the data requesters and mobile users is non-negative, $u_j^b \geq 0, \forall j \in M$ and $u_j^s \geq 0, \forall i \in N$, when they join the auction. In particular for the data requesters, we have

$$u_j^b = v_j d_j - y_j$$
$$= v_j d_j - b_{m+1} d_j$$
$$= d_j (v_j - b_{m+1})$$
$$= d_j (b_j - b_{m+1}) \geq 0$$

Similarly, the utility of the mobile user u_i^s is proved to be non-negative.

In summary, the winner selection and the price determination enable the double auction to improve the social welfare delivered while achieving both individual rationality and truthfulness. The simulation results in [252] show that given a fixed number of sensing tasks, the social welfare obtained through the proposed double auction almost increases as the number of mobile users increases. The reason is that as the number of mobile users increases, the cloud platform can select more mobile users with low costs to perform the sensing tasks. Thus, according to (8.5), the social welfare increases. The results also show that given a fixed number of mobile users, the data requesters achieve a high satisfaction ratio as the number of data requesters is small. The reason is that the ratio of data requesters' demands satisfied is high. Future work can consider how to design a double auction that is able to protect the location privacy of the mobile users.

8.4.2 Location Privacy in Participatory Sensing

In the participatory sensing applications just described, the selected mobile users perform the sensing tasks and then send sensing data to the data requesters. The sensing data may include location information that reveals sensitive information of the mobile users, such as their home addresses and other identifiers. As a result, the mobile users face the risks of physical attacks. To protect the location privacy of mobile users, cryptographic algorithms such as time lapse cryptography [256] and keyed-hash message authentication code [257] can be used. In particular, the time lapse cryptography algorithm uses private keys to encrypt the mobile users' information. However, the cryptographic algorithms often require centralized authorities and computation power for encryption and decryption, requirements that are hardly fulfilled in dynamic and decentralized network environments such as participatory sensing.

Another popular approach that has been widely used to protect the location privacy of mobile users is k-anonymity [258]. With this method, the mobile user's location information is k-anonymous, meaning that the location information of the mobile user cannot be distinguished by an attacker from that of $k - 1$ other mobile users. However, to achieve k-anonymity, there must be at least k mobile users in an *anonymity set*. One major problem is that not all mobile users are sensitive about their own location information. Indeed, when the authors in [259] surveyed mobile users about location information privacy issues, they found that only a few mobile users had privacy concerns (*privacy-sensitive users*), while the majority of the mobile users were not worried about privacy issues (*non-privacy-sensitive users*). The non-privacy-sensitive users often have no incentive to participate in the anonymity set. To provide enough number of mobile users in the anonymity set, dummy users (i.e., virtual users) can be created as proposed in [260]. However, this technique causes significant resources and communication overhead. The double auction can achieve important economic properties such as individual rationality, budget balance, computational efficiency, and truthfulness. Thus, it can be used to provide both the non-privacy-sensitive users and the privacy-sensitive users with incentives to participate in the anonymity set. The approach using the single-round double auction is proposed in [253] and is described next.

Preliminaries

The auction is formulated as follows:

- There are a set $\mathcal{N}$ of N privacy-sensitive users (i.e., mobile users that have location privacy concerns) and a set $\mathcal{M}$ of M non-privacy-sensitive users (i.e., mobile users that have no location privacy issues).
- The privacy-sensitive users in the set $\mathcal{N}$ are interested in joining the anonymity set. They are assumed to have the same privacy degree requirements, meaning that every privacy-sensitive user i in the set $\mathcal{N}$ desires k-anonymity. k is called *location privacy degree requirement* of the mobile user and is also the size of the anonymity set. The privacy-sensitive users invite the non-privacy-sensitive users to join the anonymity set to achieve a certain location privacy degree, and thus the privacy-sensitive users act as buyers. Let b_i denote the bid of privacy-sensitive user i. Here, b_i is the price for the desired k-anonymity privacy that is offered by privacy-sensitive user j. Also, let v_i denote the value of location privacy protection to privacy-sensitive user i.
- When joining the anonymity set, each non-privacy-sensitive user j incurs an extra cost C_j. C_j may include energy consumption and CPU cost. Thus, the non-privacy-sensitive user as a seller is paid a price y_j for joining the set. Let a_j denote that ask of non-privacy-sensitive user j. a_j is the price that the non-privacy-sensitive user is willing to receive. Here, C_j, y_j, and a_j may be different.
- The privacy-sensitive users and the non-privacy-sensitive users submit their bids and asks, respectively, to the auctioneer. Here, the auctioneer can be the middleware or a mobile network operator.
- After receiving the bids and the asks, the auctioneer determines a set $\mathcal{N}^w$ of winning buyers, or privacy-sensitive users, and a set $\mathcal{M}^w$ of winning sellers, or non-privacy-sensitive users. Moreover, the auctioneer calculates price p_i charged to each winning buyer i and price y_j paid to each winning seller j.
- The utility of buyer i is defined as follows:

$$u_i^b = \begin{cases} v_i - p_i, & \text{if } i \in \mathcal{N}^w \\ 0, & \text{otherwise} \end{cases} \quad (8.10)$$

- The utility of seller j is given by

$$u_j^s = \begin{cases} y_j - C_j, & \text{if } j \in \mathcal{M}^w \\ 0, & \text{otherwise} \end{cases} \quad (8.11)$$

To determine the winners and prices, the auctioneer uses a single-round double auction – namely the *anonymity double auction*, which is presented in the next section.

Anonymity Double Auction

The auctioneer first checks whether the number of buyers in the anonymity set is more than k, $N \geq k$. If this is the case, the buyers achieve the k-anonymity themselves without the help of the sellers. If $N < k$, the auctioneer executes the double auction algorithm

---
Algorithm 6 k-Anonymity double auction algorithm [253].

Input: $\mathcal{M}, \mathcal{N}, k$
Output: $\mathcal{M}^w, \mathcal{N}^w, \mathcal{P}, \mathcal{Y}$
1: Sort $\mathcal{M}$ to $\mathcal{M}'$ such that $b_1 \geq b_2 \geq \cdots$
2: Sort $\mathcal{N}$ to $\mathcal{N}'$ such that $a_1 \leq a_2 \geq \cdots$
3: Add the first $N - 1$ buyers into $\mathcal{M}'$ to $\mathcal{M}^w$
4: Add the first $k - N + 1$ sellers into $\mathcal{N}'$ to $\mathcal{N}^w$
5: $p_i = b_N, \forall i \in \mathcal{M}^w$
6: **if** $a_{k-N+2} \leq \frac{(N-1)b_N}{(k-N+1)}$ **then**
7: $\quad y_j = a_{k-N+2}, \forall j \in \mathcal{N}^w$
8: **else**
9: $\quad y_j = \frac{(N-1)b_N}{(k-N+1)}, \forall j \in \mathcal{N}^w$
10: **end if**
---

to find the winning sellers and the winning buyers to form the anonymity set. This algorithm is shown in Algorithm 6 and described as follows:

- The auctioneer sorts the buyers in descending order of their bids and the sellers in ascending order of their asks. Without loss of generality, we assume that the order of the bids is $b_1 \geq b_2 \geq \cdots \geq b_N$ and the order of the asks is $a_1 \leq a_2 \leq \cdots \leq b_M$. Let $\mathcal{N}'$ and $\mathcal{M}'$ be the ordered lists of the buyers and sellers, respectively.

- Then, the auctioneer selects the first $N - 1$ privacy-sensitive users in $\mathcal{N}'$ as the winning buyers. Here, the last buyer with b_N is not selected as the winning buyer. Instead, it is used for the payment determination to guarantee the truthfulness. This is similar to the pricing policy of the Vickrey auction. In particular, each winning buyer pays a price of b_N. Thus, the total price that the winning buyers pay is $(N-1)b_N$.

- To guarantee that the number of mobile users in the set is k, the auctioneer selects the first $k - N + 1$ non-privacy-sensitive users as winning sellers.

- The auctioneer calculates the prices that the winning sellers receive to guarantee truthfulness and budget balance. In particular, each winning seller receives the price that is equal to the ask a_{k-N+2} of seller $k - N + 2$ in the list $\mathcal{M}'$ if $(N - 1)b_N \geq (k - N + 1)a_{k-N+2}$ or $a_{k-N+2} \leq \frac{(N-1)b_N}{(k-N+1)}$. Otherwise, the price paid to each winning seller is $\frac{(N-1)b_N}{k-N+1}$.

The proposed double auction satisfies the following properties (the proof of these properties is presented in [253]):

- *Computation efficiency:* The proposed double auction algorithm consists of sorting algorithms. Thus, the time complexity of the proposed double auction algorithm is $O(N \log N + M \log M)$, which is polynomial.
- *Budget balance:* The difference between the total price that $N - 1$ winning buyers pay and the total price that $k - N + 1$ winning sellers receive is $(N-1)b_N - (k - N + 1)a_{k-N+2} \geq (N-1)b_N - (k - N + 1)\frac{(N-1)b_N}{k-N+1} = 0$.
- *Individual rationality:* Each winning buyer i pays the price $p_i = b_N$. Note that b_N is the bid of the last buyer in $\mathcal{N}'$ and is the minimum bid in the set. Assume the

winning buyer bids truthfully, $b_i = v_i$; then $b_i \geq p_i$ and its utility $u_i^b = v_i - p_i \geq v_i - b_i = 0$. Similarly, assume that each winning seller asks truthfully, $a_j = C_j$; then its utility is $u_j^s = a_{k-n+2} - C_j \geq a_j - C_j = 0$.

- *Truthfulness:* The proposed double auction demonstrates truthfulness since it is truthful for both the buyers and the sellers.

  – For the buyers: Assume that when buyer i bids truthfully, $b_i = v_i$, its payment and utility are p_i and u_i^b, respectively. When the buyer bids untruthfully, $b_i \neq v_i$, its payment and utility are $\hat{p}_i$ and $\hat{u}_i^b$, respectively. We need to show $u_i^b \geq \hat{u}_i^b$ for any $b_i \neq v_i$. There are following cases.

    ○ Case 1: The buyer wins the auction when it bids truthfully, $b_i = v_i$, and bids untruthfully, $b_i \neq v_i$. The buyer pays the same price, p_N, when it bids truthfully and untruthfully. Thus, the buyer achieves the same utility, $u_i^b = \hat{u}_i^b = v_i - p_i = v_i - b_N \geq 0$, as discussed for the individual rationality property.

    ○ Case 2: The buyer wins the auction when it bids truthfully and loses the auction when it bids untruthfully. The utility of the buyer when bidding truthfully is $u_i^b \geq 0$, and that of the buyer when bidding untruthfully is $\hat{u}_i^b = 0$. In this case, $u_i^b \geq \hat{u}_i^b$.

    ○ Case 3: The buyer loses the auction by bidding truthfully and untruthfully. Clearly, the buyer achieves the same zero utility for both situations.

    ○ Case 4: The buyer loses the auction by bidding truthfully and wins the auction by bidding untruthfully. The buyer loses the auction by bidding truthfully, meaning that $v_i < b_N$, and its utility is $u_i^b = 0$. The buyer wins the auction by bidding untruthfully, meaning that it bids $b_i = v_i > b_N$ and achieves the utility $\hat{u}_i^b = v_i - p_i = v_i - b_N < 0$. As such, in this case, $u_i^b > \hat{u}_i^b$.

  – For the sellers: By using the same argument, the proposed double auction is proved to be truthful for the sellers.

In summary, the proposed double auction achieves computational efficiency, individual rationality, budget balance, and truthfulness. The simulation results in [253] show that given $N \geq k$, the auction terminates immediately, with a running time of 0 millisecond, since the auctioneer does not need to consider the sellers to satisfy the buyers' decree requirements. This is consistent with the preceding theoretical analysis. Moreover, the number of winning buyers increases as the number of sellers participating in the auction increases. The reason is the higher probability that the budget balance condition, $a_{k-N+2} \leq \frac{(N-1)b_N}{(k-N+1)}$, is satisfied.

8.4.3 Spectrum Allocation in Heterogeneous Networks

5G heterogeneous networks include multiple tiers, and those tiers can leverage unused spectrum from each other to enhance the spectrum efficiency. Such a situation is shown

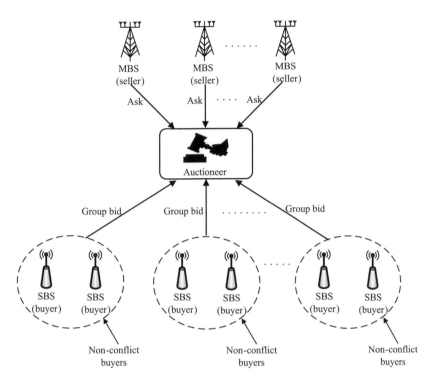

Figure 8.4 Double auction for spectrum trading in heterogeneous networks. SBS and MBS stand for small cell base station and macro base station, respectively.

in Figure 8.4, in which small cell base stations, such as femto access points and pico cell base stations, leverage unused spectrum from macro base stations. The small cell base stations and macro base stations may belong to different mobile network operators. The mobile network operators are rational, so the main issue is how to stimulate them to participate in the trading market. A double auction both incorporates individual rationality and guarantees non-negative utility, or profit, for the participants in the multi-seller/multi-buyer market. Therefore, it can be efficiently used to solve the issue. The use of the single-round double auction for spectrum trading in 5G heterogeneous networks is proposed in [261] and described next.

System Model

The system model shown in Figure 8.4 consists of the following entities:

- There are M macro base stations, N small cell base stations, and one third party as an auctioneer.
- The owners of small cell base stations as buyers purchase spectrum from the owners of macro base stations as sellers.
- Each owner is assumed to own a base station, either a macro base station or a small cell base station. Thus, for convenience, the small cell base stations can be considered to be the buyers, and the macro base stations are the sellers.

- The macro base station may have multiple homogeneous channels for trading, and each channel is potentially reused by *non-conflicting* small cell base stations. The non-conflicting small cell base stations or non-interfering small cell base stations are small cell base stations that can reuse the same channel without causing interference with each other. This setup improves spectrum utilization.
- Let a_j denote the ask of macro base station j and let v_j^s denote the true value of the channel to the macro base station. Here, v_j^s can be the cost for maintaining the channel. Then, the utility of the macro base station for trading the channel is defined as follows:

$$u_j = \begin{cases} y_j - v_j^s, & \text{if macro base station } j \text{ wins} \\ 0, & \text{otherwise} \end{cases} \quad (8.12)$$

where y_j is the price that the macro base station receives for trading its channel.

- Let b_i denote the bid of small cell base station i; b_i is the maximum price that the small cell base station is willing to pay for the channel. Also, let v_i^b denote the true value of the channel to the small cell base station. Then, the utility of the small cell base station can be defined as follows:

$$u_i = \begin{cases} v_i^b - p_i, & \text{if small cell base station } i \text{ wins} \\ 0, & \text{otherwise} \end{cases} \quad (8.13)$$

where p_i is the actual price that the small cell base station needs to pay if it wins the channel.

Typically, the small cell base stations and the macro base stations simultaneously submit their bids and asks, respectively, to the auctioneer for the winner and price determination. However, since the spectrum reusability issue is considered, the auctioneer receives *group bids* rather than individual bids of the small cell base stations. The means by which the group bids are generated is explained in the next section. Given the group bids and the asks, the auctioneer conducts a single-round double auction to (i) select the winning macro base stations and the winning small cell base stations, (ii) allocate the channels of the winning macro base stations to the winning small cell base stations in a conflict-free manner, and (iii) determine the prices to pay/charge the winners. The objective of the auction is to achieve economic properties including truthfulness, individual rationality, and budget balance. In addition, the auction aims to improve spectrum utilization, by ensuring the maximum number of transactions between the small cell base stations and macro base stations.

The double auction is presented in the following section. It is essentially the single-round double auction as described in Section 8.2.

Double Auction Design

Since the proposed double auction considers the spectrum reusability, the buyer group formation needs to be performed before the winner and the price determination steps.

Buyer group formation: The buyer group formation is to group non-conflicting small cell base stations (i.e., non-conflicting buyers) that can be assigned with the same

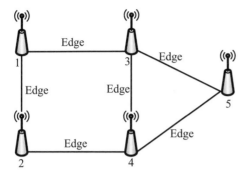

Figure 8.5 An illustration of the conflict graph.

channel without causing interference with each other. The reusability-driven spectrum allocation algorithms such as VERITAS [262] can be used to form the buyer groups based on interference conditions among the small cell base stations. The algorithm is implemented as presented in [262]. The main idea is to build a conflict graph based on the interference conditions. Then, the bidder group formation is equivalent to finding the independent sets in the conflict graph. The independent set is a set of non-conflicting small cell base stations.

To further understand how to form the buyer groups, we consider an example. Assume that the auction has five small cell base stations/buyers. Based on the interference conditions among the small cell base stations, a conflict graph is built as shown in Figure 8.5. In the conflict graph, if two of any small cell base stations do not share an edge, then they do not cause interference with each other and can be grouped into an independent set, a buyer group. Thus, there are three buyer groups: $\mathcal{G}_1 = \{1,5\}$, $\mathcal{G}_2 = \{2,3\}$, $\mathcal{G}_3 = \{4\}$.

For general cases, there are L buyer groups, and each group $\mathcal{G}_l$ has N_l small cell base stations. In the group, a small cell base station calculates a group bid on behalf of other small cell base stations in the group as follows:

$$b_{\mathcal{G}_l} = \min\{b_i | i \in \mathcal{G}_l\} N_l \qquad (8.14)$$

Then, the small cell base station submits the group bid to the auctioneer. Thus, the auctioneer receives L group bids from L buyer groups and M asks from the macro base stations. The third party then determines the winners and the corresponding charges and payments.

Winner selection: The auctioneer uses the winner selection of the single-round double auction to determine the auction winners as follows:

- The auctioneer sorts the buyer groups in descending order of their group bids and sorts the macro base stations in ascending order of their asks. Without loss of generality, we assume that $b_{\mathcal{G}_1} \leq \cdots \leq b_{\mathcal{G}_L}$ and $a_1 \leq \cdots \leq a_M$. Thus, the order of the buyer groups is $\mathcal{L} = \{1, \ldots, L\}$, and the order of the macro base stations is $\mathcal{M} = \{1, \ldots, M\}$.

- The auctioneer finds the largest index m at which the group bid is still greater than the ask:

$$m = \arg \max_{k \leq \min\{L, M\}} b_{\mathcal{G}_k} \geq a_k \qquad (8.15)$$

- The first $(m-1)$ macro base stations in $\mathcal{M}$ are the winning sellers, and the first $(m-1)$ buyer groups in $\mathcal{L}$ are the winning buyer groups.

Price determination: The auctioneer adopts the uniform price policy, presented in Section 8.2.1, for the price determination. With this policy, the winning buyers pay the same price, and the winning sellers receive the same price. However, to guarantee truthfulness, the uniform price policy uses the pricing policy of McAfee's mechanism [251]. In McAfee's mechanism, the winning buyer pays the price equal to the highest bid of the losing buyers and the winning seller receives the payment equal to the lowest ask of the losing sellers. Here, the highest bid of the losing buyer is $b_{\mathcal{G}_m}$, and the lowest ask of the losing sellers is a_m. Thus, the winning buyer groups pay the same price $b_{\mathcal{G}_m}$, and the winning sellers receive the same price a_m.

The question now is, how much will the small cell base stations in each winning buyer group $\mathcal{G}_l$ pay? Note that the small cell base stations in the same group win and use the same channel. Thus, to guarantee fairness, the small cell base stations in the same group should be charged equally. Therefore, each small cell base station i in winning buyer group $\mathcal{G}_l$ pays a price that is determined as follows:

$$p_i = b_{\mathcal{G}_m}/N_l \qquad (8.16)$$

The total price that the $(m-1)$ winning buyer groups pay is $(m-1)b_{\mathcal{G}_m}$, and the total price that the $(m-1)$ winning macro base stations receive is $(m-1)a_m$. Thus, the profit of the auctioneer is

$$\pi = (m-1)(b_{\mathcal{G}_m} - a_m) \qquad (8.17)$$

Properties of the proposed double auction: The proposed double auction demonstrates the following properties:

- *Budget balance:* According to the winner selection scheme, $b_{\mathcal{G}_m} \geq a_m$, and thus $\pi \geq 0$.
- *Individual rationality:* It is easy to show that the utility of the macro base stations (the sellers) is non-negative when participating in the auction. For the small cell base stations (the buyers), we need to show that they are charged less than their bids. From (8.14) and (8.16), the price p_i that small cell base station i in the winning buyer group $\mathcal{G}_l$ pays is the lowest bid in the group. This means that the small cell base station pays the price less than or equal to its bid. Thus, the utility of the small cell base stations is non-negative. As a result, the proposed double auction guarantees the individual rationality.
- *Truthfulness:* We need to show that no participant, whether a small cell base station or a macro base station, gains utility by untruthfully bidding its true value of the channel. First, for the small cell base stations, we prove that any small

cell base station i cannot improve its utility by bidding $b_i \neq v_i^b$. There are following cases:

- Case 1: The small cell base station loses the auction with both bids, the truthful bid and the untruthful bid. In this case, the small cell base station achieves zero utility.
- Case 2: The small cell base station wins the auction with $b_i = v_i^b$ and loses with $b_i \neq v_i^b$. As proved in [261], the winner determination of the proposed double auction exhibits monotonicity; that is, if small cell base station wins the auction by bidding b_i, then it also wins by bidding $b_i' > b_i$. Thus, in this case, $b_i < v_i^b$. The utility of the small cell base station is zero if it loses and is greater than zero if it wins. Thus, it is better for the small cell base station to bid truthfully, $b_i = v_i^b$, such that it wins the auction.
- Case 3: The small cell base station loses the auction with $b_i = v_i^b$ and wins with $b_i \neq v_i^b$. Based on the monotonicity of the winner determination, the utility of small cell base station i is proved to be less than or equal to zero if it submits $b_i \neq v_i^b$ that is no more than the utility when it bids truthfully, $b_i = v_i^b$.
- Case 4: The small cell base station wins the auction with both bids. In this case, the small cell base station pays the same price, and thus it achieves the same utility.

The truthfulness property for the macro base stations can be proved in the same way.

The simulation results in [261] show a trade-off between the economic robustness and the spectrum utilization (i.e., the spectrum efficiency). In particular, the proposed double auction suffers from significant degradation of spectrum utilization of as much as 50% compared with the optimal allocation algorithm. Here, the optimal allocation algorithm is similar to the proposed double auction, but it chooses the buyer groups with the larger sizes without considering the group bids or the bids of the small cell base stations. The optimal allocation algorithm thus improves spectrum utilization, but it may not guarantee properties that are achieved by the proposed double auction. Future work can consider general scenarios with more complex bidding formats, such as small cell base stations bidding for multiple channels.

8.4.4 Cloud Resource Allocation in Edge Computing

Edge computing is an emerging computing paradigm proposed to address issues of cloud computing such as high operational costs, bandwidth bottlenecks, and peak usage. Edge computing exploits edge devices such as cloudlets, including mobile micro-clouds and volunteered computers, to provide cloud resources as well as cloud services. As such, it is able to provide edge cloud services to mobile users with low latency. However, both the service providers, which are the owners of edge cloud services, and the mobile users are rational. Thus, the main issue is how to encourage the service providers to

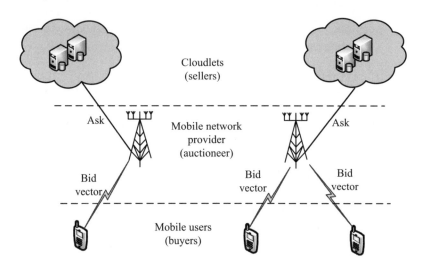

Figure 8.6 Double auction for cloud resource trading in edge computing.

deploy and contribute their edge computing services as well as how to attract the mobile users to buy the edge computing services. The double auction can guarantee a non-negative utility for the participants, and thus it can be used to efficiently design incentive mechanisms for both the service providers and the mobile users. The authors in [255] propose to use a single-round double auction for edge cloud resource allocation. Apart from maintaining individual rationality, which guarantees the non-negative utility of the participants, the proposed approach demonstrates other desirable properties such as truthfulness, system efficiency, and computational efficiency. The proposed approach is described next.

System Model
The system model is an edge computing model, as shown in Figure 8.6, that includes the following entities:

- There are M edge computing service providers, the sellers; N mobile users, the buyers; and one trusted third party, the auctioneer. As shown in Figure 8.6, the mobile network provider provides communication connections to both the cloudlets and the mobile users, and thus the mobile network provider can act as the auctioneer.
- Each service provider is assumed to own a cloudlet. Thus, for convenience, the cloudlet is considered to be a seller that provides cloud services to the mobile users.
- Each mobile user can achieve different utilities when using different cloudlets. For example, by using a cloudlet closer to the mobile user, the bandwidth cost and communication latency of the mobile user are lower, and the utility of the mobile user is higher. As such, the mobile user has different preferences for the cloudlets. Let $\mathbf{b}_i = (b_i^1, \ldots, b_i^M)$ denote the bid vector of mobile user i, where b_i^j

is the bid of mobile user i. b_i^j is the price that mobile user i is willing to pay for receiving a cloud service of cloudlet j. We have $\mathbf{b} = (\mathbf{b}_1, \ldots, \mathbf{b}_N)$.

- Unlike the mobile users, the cloudlets only aim to maximize prices that they receive from trading the cloud services. Thus, the asks of the cloudlet should not differentiate over the mobile users. Let a_j be the ask of cloudlet j, which is the price that the cloudlet is willing to receive from trading the cloud service.
- Assume that there is a buyer–seller mapping. This means that one cloudlet serves only one mobile user, and one mobile user is assigned to only one cloudlet.
- Let v_i^j be the value of the cloud service to mobile user i if it wins the service from cloudlet j. Also, let p_i be the price that the mobile user pays. With the buyer–seller mapping, the utility of the mobile user is given by

$$u_i^b = \begin{cases} v_i^j - p_i, & \text{if mobile user } i \text{ wins the auction} \\ 0, & \text{otherwise} \end{cases} \quad (8.18)$$

Note that in some places in this section, $u_{i,j}^b$ can be used to refer to the utility achieved by mobile user i when it is assigned to cloudlet j.

- Let y_j be the price paid to cloudlet j and let C_j be the cost of the cloudlet for providing the cloud service. The utility, consisting of the surplus or profit, of the cloudlet for trading its cloud service is defined as follows:

$$u_j^s = \begin{cases} y_j - C_j, & \text{if cloudlet } j \text{ wins the auction} \\ 0, & \text{otherwise} \end{cases} \quad (8.19)$$

The cloudlets and mobile users simultaneously submit their bid vectors and asks, respectively, to the auctioneer. The auctioneer adopts the single-round double auction for the winner and payment determination as described in the next sections.

Double Auction Design

The objective of the double auction design is to maintain four desired properties: individual rationality, budget balance, truthfulness, and computational efficiency.

The proposed double auction is performed by two stages: the *candidate-determination and pricing*, and the *candidate-elimination*. In the candidate-determination and pricing stage, the auctioneer determines (i) buyer candidates for each cloudlet, (ii) the prices that the buyer candidates will pay, and (iii) the price paid to the cloudlet. After the candidate-determination and pricing stage, one buyer candidate may win two or more cloudlets. To ensure a seller–buyer mapping, the candidate-elimination stage is then implemented to choose the best cloudlet for the mobile user. The two stages are described in the following subsections.

Candidate-determination and pricing: The algorithm for the candidate-determination and pricing stage is shown in Algorithm 7. Specifically, the auctioneer sorts the cloudlets in ascending order of their asks. Assume that $a_1 \leq a_2 \leq \cdots \leq a_M$. Then the set of ordered cloudlets is $\mathcal{M} = \{1, 2, \ldots, M\}$. Let $\mathbf{a} = \{a_1, \ldots, a_M\}$, and $\mathbf{a}_{-j}$ denote the vector of asks excluding the ask of cloudlet j. Let a_{-j}^{mid} be the median of $\mathbf{a}_{-j}$; then a_{-j}^{mid} is the middle element in $\mathbf{a}_{-j}$. In the case that $\mathbf{a}_{-j}$ has an even number of

Algorithm 7 The algorithm for candidate-determination and pricing [255].

Input: $\mathcal{M}, \mathcal{N}, \mathbf{b}, \mathbf{a}$
Output: $\mathcal{M}_c, \mathcal{N}_c, \mathcal{P}_c, \mathcal{Y}_c$
1: **for** $j \in \mathcal{M}$ **do**
2:     Find the median a_{-j}^{mid} of $\mathbf{a}_{-j}$
3:     $\mathcal{N}^j = \{i : b_i^j \geq a_j, \forall i \in \mathcal{N}\}$
4:     **if** $|\mathcal{N}^j| = 1$ **then**
5:         **if** $b_i^j \geq a_{-j}^{\text{mid}}$ and $a_j \leq a_{-j}^{\text{mid}}$ **then**
6:             $\hat{\sigma}(j) = i$, mobile user i is added into $\mathcal{N}_c$, and cloudlet j is added into $\mathcal{M}_c$
7:             $p_i = y_j = a_{-j}^{\text{mid}}$
8:             p_i and y_j are added into $\mathcal{P}_c$ and $\mathcal{Y}_c$, respectively
9:         **end if**
10:     **else**$|\mathcal{N}^j| > 1$
11:         Sort $\mathcal{N}^j$ to $\mathcal{N}'^j$ such that $b_{i_{(1)}}^j \geq b_{i_{(2)}}^j \geq \cdots$
12:         **if** $b_{i_{(1)}}^j \geq a_{-j}^{\text{mid}}$ **then**
13:             Select first i of $\mathcal{N}'^j$ with the highest bid
14:             $\hat{\sigma}(j) = i$, mobile user i is added into $\mathcal{N}_c$, and cloudlet j is added into $\mathcal{M}_c$
15:             $p_i = y_j = \max\{a_{-j}^{\text{mid}}, b_{i_{(2)}}^j\}$
16:             p_i and y_j are added into $\mathcal{P}_c$ and $\mathcal{Y}_c$, respectively
17:         **end if**
18:     **end if**
19: **end for**

elements, a_{-j}^{mid} is the mean of the two middle elements in $\mathbf{a}_{-j}$. The median is used as a threshold that guarantees that the payment to the cloudlets is not too low. The auctioneer considers two possible cases:

- Case 1: There is only one mobile user i with $b_i^j \geq a_{-j}^{\text{mid}}$. Then, if $a_{-j}^{\text{mid}} \geq a_j$, mobile user i is added to the set of buyer candidates $\mathcal{N}_c$ with price a_{-j}^{mid} and cloudlet j is added to the set of seller candidates $\mathcal{M}_c$ with payment a_{-j}^{mid}. Otherwise, mobile user i does not win the cloud service from cloudlet j.
- Case 2: There are multiple mobile users with bids greater than or equal to a_j. Then, if all the bids of the mobile users are lower than a_{-j}^{mid}, no mobile user wins the cloud service from cloudlet j. Otherwise, the mobile user with the highest bid is added to the buyer candidate set $\mathcal{N}_c$, and cloudlet j is added to the seller candidate set $\mathcal{M}_c$. The price that the mobile user pays cloudlet j is the maximum of the second highest bid and a_{-j}^{mid}. Such a payment scheme is similar to the pricing policy of the Vickrey auction with reserve pricing a_{-j}^{mid}. The payment mechanism guarantees truthfulness and avoids a zero payment, such as in the situation that there is no second highest bid. To further understand this case, we consider the following example. There are two mobile users i and i' with $b_i^j \geq b_{i'}^j \geq a_{-j}^{\text{mid}} \geq a_j$. Mobile user i is then added to the buyer candidate set. The price that the mobile user pays cloudlet j is $b_{i'}^j$.

Candidate-elimination: After the candidate-determination and pricing terminates, each buyer candidate may win more than one cloudlet. Thus, the candidate-elimination

8.4 Development of Double Auction for Computer Networks

Algorithm 8 The algorithm for candidate-elimination [255].
Input: $\mathcal{M}_c, \mathcal{N}_c, \mathcal{P}_c, \mathcal{Y}_c, \mathbf{b}$
Output: $\mathcal{M}_w, \mathcal{N}_w, \mathcal{P}_w, \mathcal{Y}_w$
1:   $\mathcal{M}_w \leftarrow \mathcal{M}_c, \mathcal{N}_w \leftarrow \mathcal{N}_c, \mathcal{P}_w \leftarrow \mathcal{P}_c, \mathcal{Y}_w \leftarrow \mathcal{Y}_c$
2:   **for** each buyer $i \in \mathcal{N}_w$ **do**
3:     **for** any two sellers $j, j' \in \mathcal{M}_w, j \neq j'$ **do**
4:       **if** $u_{i,j}^b = u_{i,j'}^b$ **then**
5:         $j^* \leftarrow$ randomly selected from $\{j, j'\}$
6:       **else**
7:         $j^* \leftarrow \arg\min_{k \in \{j, j'\}} u_{i,k}^b$
8:       **end if**
9:       $\mathcal{M}_w \leftarrow \mathcal{M}_w \setminus \{j^*\}$
10:      $\mathcal{P}_w \leftarrow \mathcal{P}_w \setminus \{p_{j^*}\}$
11:      $\mathcal{Y}_w \leftarrow \mathcal{Y}_w \setminus \{y_{j^*}\}$
12:     **end for**
13: **end for**

stage is performed to select the best cloudlet for the buyer candidate. The algorithm for candidate-elimination is illustrated in Algorithm 8. This algorithm simply selects the best cloudlet that enables the corresponding mobile user to achieve the highest utility. If more than one cloudlet yields the highest utility for the mobile user, the best cloudlet is randomly selected.

The proposed double auction demonstrates the following properties:

- *Computational efficiency:* As shown in [255], the time complexity of Algorithm 7 is $O(M \log M + MN \log N)$, and that of Algorithm 8 is $O(M^2)$. The overall time complexity of the two algorithms is $O(M^2 + MN \log N)$, which is polynomial. Thus, the proposed double auction is computationally efficient.
- *Individual rationality:* In the candidate-determination and pricing stage, each buyer candidate pays either a_{-j}^{mid} or the second highest bid. The prices are lower than or equal to the bid of the buyer candidate that guarantees the non-utility for the buyer candidate. Also, the seller candidate receives either a_{-j}^{mid} or the second highest bid. The prices are higher than the ask that the seller candidate submits. This ensures individual rationality for both buyers and sellers.
- *Budget balance:* The total price that the winning mobile users pay is transferred to the winning cloudlets. Thus, the proposed double auction is budget-balanced.
- *Truthfulness:* As proved in [255], the proposed double auction is truthful for both the mobile users/buyers and the cloudlets/sellers. Thus, the proposed double auction guarantees truthfulness.

To validate the aforementioned properties of the proposed double auction scheme, a simulation is implemented in [255]. In regard to truthfulness, the results show that if the mobile user wins the cloudlet and achieves a positive utility by bidding truthfully, then the mobile user cannot improve the utility by submitting other bids. Similarly, the utility of the cloudlet with a truthful ask is the highest among those with all possible asks. Moreover, the simulation results show that the proposed double auction outperforms the Vickrey auction–based double auction [263] in terms of the number of successful trades or transactions. This implies that the system efficiency obtained by the proposed double auction scheme is improved.

8.5 Summary

In this chapter, we introduce the double auction and its applications for computer networks. Specifically, we first describe the single-round double auction with two pricing policies: uniform pricing and discriminatory pricing. In addition, we show a well-known economic model, the supply and demand model, that works similarly to the single-round double auction. Some specific examples from computer networks are then given and analyzed to show how to determine the winners and the prices in a single-round double auction. The single-round double auction is a sealed-bid double auction in which the participants, including both the buyers and the sellers, cannot learn the bidding strategy of their rivals and discover the real values of the resources. To address this issue, we introduce the continuous double auction, which allows the participants to trade the resources in multiple rounds. An example is provided to further clarify the operation of the continuous double auction. Finally, we review applications of the double auction for emerging issues in computer networks. These issues include sensing task allocation in participatory sensing, location privacy in participatory sensing, spectrum allocation in 5G heterogeneous networks, and cloud resource allocation in edge computing. Important properties achieved by each proposed double auction approach are also discussed.

9 Other Auctions

Future wireless networks will be decentralized, ad hoc, and dynamic: there may not be a centralized authority, and various types of mobile devices can join and leave the network. In such a network, the auctions such as the VCG auction (see Section 6.2) and the combinatorial auction (see Chapter 7) face many challenges. The main reason is that these auctions often require performing centralized computations and obtaining global information from the network. To achieve the target performance with low complexity, distributed auction approaches can be used. In this chapter, we present such distributed auctions and their applications in computer networks. The distributed auctions include the ascending clock auction and the share auction. Moreover, we introduce two auctions recently used in computer networks: the online auction and the wait-line auction. For each auction, we discuss how to apply it to address emerging issues. For convenience, the following summarizes auctions that we present in this chapter.

- *Ascending clock auction:* This is a type of multiple-round auction in which the auctioneer increases the prices of resources until the demand for the resources equals the supply of the resources. The winners are the bidders that participate in the last round of the auction.
- *Share auction:* In this auction, each bidder receives resources proportionally to the bid that the bidder submits to the auctioneer.
- *Online auction:* This auction allows the bidders to submit their bids at any time, and the market is cleared immediately.
- *Wait-line auction:* In this auction, the bidders join a queue and submit their waiting times to the auctioneer. Then, the auctioneer determines the winners based on the waiting times. Specifically, the bidders come the queue earlier – that is, the bidders at the first queue – become the winners of the auction.

Apart from the aforementioned auctions, the sequential auction [264], [265], [266], a powerful method for allocating multiple homogeneous items, has been recently used. In the sequential auction, the items are auctioned off one by one in different rounds. In particular, in each round, the bidder with the highest bid obtains the current item, but the bidder pays the price of the second highest bid. By using the sequential auction, resource allocation achieves both efficiency and fairness. Further details of the sequential auction are found in [266]. The application of the sequential auction for power allocation in two-way relay systems is discussed in [267], the application of the sequential auction for spectrum sharing is presented in [268], and the application of the sequential auction for spectrum allocation in cognitive radio two-way relay networks is discussed in [269].

9.1 Ascending Clock Auction

In Section 7.4.2, we discuss the ascending clock auction. However, the ascending clock auction discussed in that chapter was used as the first phase for the price discovery in the clock-proxy auction. How to determine the winners and the payments in the ascending clock auction was not presented. In this section, we further discuss the ascending clock auction as well as its application for to deal with security issues in computer networks.

9.1.1 Auction Process

To better understand how the ascending clock auction works, we consider a resource trading market in a 5G heterogeneous network. The market consists of a service provider (the seller) and multiple users (the buyers). In the heterogeneous resource market, the service provider can sell different types of resources such as spectrum, power, caching, and antenna. Assume that the service provider is willing to sell two types of resources, bandwidth and antenna, which are denoted by resource types W and A, respectively. The number of bandwidth units for trading is $\overline{Q}_W$, and the number of antennas is $\overline{Q}_A$. To sell the resources, the service provider can adopt an ascending clock auction, which proceeds as follows:

- Before the auction starts, the service provider sets the initial prices of resources – namely, the price per bandwidth unit and the price per antenna.
- Given the prices of the resources, in the first auction round, each user bids for the resources by submitting its demand for the resources. The demand specifies the number of bandwidth units and the number of antennas that the user is willing to buy.
- Upon receiving the demands for the resources from the users, the service provider increases the prices of those resources with excess demands. For example, if the total demand for antennas is greater than $\overline{Q}_A$, the service provider increases the price per antenna. Then, the service provider announces the new prices to the users.
- Given the new prices, the users submit their demands again in the next round.
- This process continues until the total demand is equal to the supply of resources of all types.
- The users that participate in the last round of auction are the winning users (the winners) of the auction.
- The winning users pay the service provider the prices that the service provider announced in the last round.

It can be seen that the ascending clock auction has some similarities to the English auction as presented in Section 4.1. First, both auctions are ascending auctions in which the auctioneer, the service provider in this context, increases the prices of the resources over rounds. This may improve the service provider's revenue. Second, the winner determination and the payment mechanism of the ascending clock auction are the same as those in the English auction. In particular, for the payment mechanism, the winners pay the prices that are set by the service provider in the last round of the auction. As the users only participate in any round of auction if the users' utilities in that round are non-negative, the ascending clock auction supports individual rationality.

However, the ascending clock auction and the English auction have one key difference. In the ascending clock auction, the bidders, the users in the example context, submit their demands for resources rather than the prices that they are willing to pay. This means that the prices that the bidders are willing to pay are not revealed during the auction. We can say that the ascending clock auction preserves the privacy of the bidders. This feature enables the ascending clock auction to be an effective solution for addressing privacy issues as well as security issues in the future wireless network [201], [270]. In the next section, we describe how to apply the ascending clock auction to ensure physical layer security in cooperative wireless networks.

9.1.2 Application of Ascending Clock Auction for Physical Layer Security

As presented in Section 2.3, physical layer security is proposed as an alternative solution that exploits physical characteristics of the wireless channels, such as channel gains and noise. In particular, a legitimate source employs friendly jamming power from friendly jammers to maximize the secrecy rate of reliable information transmitted from the source to an intended destination at which eavesdroppers are unable to decode the information. This problem can be regarded as a power allocation issue that can be efficiently solved by auction theory [148], [271], [272]. In this section, we describe the use of the ascending clock auction to provide physical layer security in a cooperative wireless network to maximize the secrecy rate of the network. Note that this scheme is proposed in [201].

System Model

The system model is shown in Figure 9.1 and described here:

- There is a set $\mathcal{N}$ of N sources. The sources are the bidders or the buyers. Let S_i denote source i.
- There are N destinations corresponding to N sources. Let D_i denote the destination of source i.
- There is one malicious eavesdropper, denoted by E, that tries to eavesdrop on the information transmitted from the sources to their destinations.
- Let $h_{S_i D_i}$ and $h_{S_i E}$ be the power channel gains from source S_i to destination D_i and eavesdropper E, respectively.
- There is one friendly jammer, denoted by J, that transmits power P_i^J to help improve the secrecy rate of information transmission from source S_i to its destination. The friendly jammer is thus the seller or auctioneer.

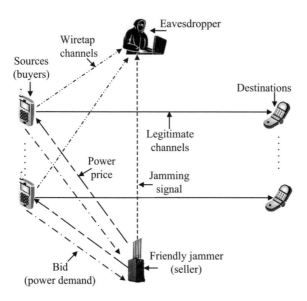

Figure 9.1 Physical layer security in cognitive wireless networks based on ascending clock auction [273].

- Let h_{JD_i} and h_{JE} be the power channel gains from the friendly jammer to destination D_i and eavesdropper E, respectively.
- Let σ^2 denote the independent additive white Gaussian noise variance at the destination and eavesdropper devices. The channel bandwidth that each source accesses for the information transmission is α. Here, the channels that the sources use are assumed to be orthogonal.

As discussed in Section 6.1.4, the objective when using the friendly jammer is to reduce the data rate that is "leaking" from the sources to the eavesdropper. However, this also reduces the useful data rate from the sources to the destinations. Thus, the approaches designed for physical layer security should guarantee that the information rate from the sources to the eavesdropper decreases faster than the data rate from the sources to the destinations. In other words, the friendly jamming power is determined to improve the secrecy rate R_S^i of each source S_i, which is defined as follows:

$$R_S^i = \left[\alpha \log_2\left(1 + \frac{h_{S_i D_i} P_i}{\sigma^2 + h_{JD_i} P_i^J}\right) - \alpha \log_2\left(1 + \frac{h_{S_i E} P_i}{\sigma^2 + h_{JE} P_i^J}\right)\right]^+ \quad (9.1)$$

where $[a]^+$ represents $\max(0, a)$, and P_i is the transmit power of source S_i.

Note that the first term in (9.1) is the channel capacity of source S_i to destination D_i, and the second term is the channel capacity for source S_i to malicious eavesdropper E. Also, both the first term and the second term are decreasing and convex functions of jamming power P_i^J.

Next, we present how the ascending clock auction is used for power allocation to improve R_S^i. We first define the utility functions of the sources and the friendly jammer.

- *Sources' utility functions:* Each source S_i, if it wins the auction, needs to pay the cost to the friendly jammer for wining the jamming power. Thus, the utility function of source S_i can be defined as follows:

$$U_i(P_i^J, c) = g(P_i^J) - p(P_i^J, c) \qquad (9.2)$$

where $g(P_i^J)$ is the performance gain with jamming power P_i^J, $p(P_i^J, c)$ is the cost paid to the friendly jammer, and c is the unit price of jamming power that the friendly jammer sets during the auction. Note that although c is a constant for all units of power, it may change in different auction rounds according to the resource demand and supply rule. The performance gain and the cost functions are defined as follows:

  - Performance gain is the difference between the secrecy rate with the jamming power and the secrecy rate without the jamming power. Thus, $g(P_i^J)$ is defined as $R_S^i - \hat{R}_S^i$, where $\hat{R}_S^i$ is the secrecy rate R_S^i of source S_i without the jamming power, $P_i^J = 0$.
  - In general, the source must pay a high cost if it uses a large amount of jamming power. Thus, the cost function $p(P_i^J, c)$ is monotonically increasing with $p(P_i^J)$. The cost function can be simply defined as $p(P_i^J, c) = cP_i^J$.

- *Jammer's utility function:* By providing the jamming service, the friendly jammer receives the payments from the sources. Thus, the utility of the friendly jammer is defined as follows:

$$U_J = c \sum_i P_i^J \qquad (9.3)$$

$$\text{s.t. } 0 \leq \sum_i P_i^J \leq P_{max}$$

where P_{max} is the maximum power that the friendly jammer can provide. To guarantee a non-negative benefit for the friendly jammer, the cost should be greater than the minimum cost or the reserve price c^0. The reserve price can be determined based on the cost of transmitting a unit of jamming power.

Ascending Clock Auction Process

The problem of the friendly jammer is to allocate the jamming power to each source S_i under the power constraint P_{max} so as to maximize the secrecy rate. To solve this problem, the friendly jammer can adopt a centralized approach. In general, the centralized approach uses private information, such as the transmit power and the channel information, of each pair of source and destination to maximize the global secrecy rate. However, the sources and destinations are geographically distributed, and it may be impossible for the friendly jammer to collect all the private information. The ascending clock auction can be used to develop the distributed power allocation scheme proposed in [201].

The algorithm is shown in Algorithm 9 and summarized as follows:

Algorithm 9 Ascending clock auction for the friendly jamming power allocation [201].
1: The friendly jammer initializes the power price with c^0 in round $t = 0$.
2: The friendly jammer announces c^0 to all the sources.
3: Each source S_i computes its optimal bid $P^J_{i,0}$ according to (9.4).
4: The sources submit the optimal bids to the friendly jammer.
5: The friendly jammer computes $P^J_{sum,0} = \sum_i P^J_{i,0}$ and compares $P^J_{sum,0}$ with P_{max}.
6: **if** $P^J_{sum,0} \leq P_{max}$ **then**
7:     The friendly jammer concludes the auction.
8: **else**
9:     **repeat**
10:         The friendly jammer sets $c^{t+1} = c^t + \delta, t = t + 1$.
11:         The friendly jammer announces c^t to all the sources.
12:         Source S_i computes its optimal bid $P^J_{i,t} = \arg\max_{(P^J_i, P_i)} U_i(P^J_i, c^t)$.
13:         The sources submit the optimal bids to the friendly jammer.
14:         The friendly jammer computes $P^J_{sum,t} = \sum_i P^J_{i,t}$.
15:     **until** $P^J_{sum,t} \leq P_{max}$.
16:     The friendly jammer sets $T = t$ and allocates P^{*J}_i to each source S_i, where

$$P^{*J}_i = P^J_{i,T} + \frac{P^J_{i,T-1} - P^J_{i,T}}{\sum_i P^J_{i,T-1} - \sum_i P^J_{i,T}} \left(P_{max} - \sum_i P^J_{i,T} \right)$$

17: **end if**

- Before the auction starts, at round $t = 0$, the friendly jammer sets the price of jamming power to the reserve price c^0. Then, the friendly jammer announces the price to all the sources.
- Given the price, in the first round $t = 1$, each source S_i determines its optimal bid. The optimal bid is the optimal power demand $P^J_{i,0}$ that is determined as follows:

$$P^J_{i,0} = \arg\max_{(P^J_i, P_i)} U_i(P^J_i, c^0) \qquad (9.4)$$

- Upon receiving the bids from the sources, the friendly jammer calculates the total demand of the sources as $P^J_{sum,0} = \sum_i P^J_{i,0}$. The friendly jammer compares $P^J_{sum,0}$ with its maximum power P_{max}. If $P^J_{sum,0} \leq P_{max}$, the friendly jammer concludes the auction, and there is no resource trading due to the low trading price, c^0. Otherwise, the friendly jammer increases the price by δ and announces the new price to all the sources.
- Given the new price, in the next round $t = 2$, the sources compute again their optimal bids and submit the bids to the friendly jammer. The friendly jammer compares the total demand and its maximum power to adjust the price.
- This process continues until $P^J_{sum,t=T} < P_{max}$ or $P^J_{sum,t=T} < P_{max}$.

Note that when the auction terminates at $P^J_{sum,T} < P_{max}$, the whole power of the friendly jammer is not fully utilized. To make sure that $P^J_{sum,T} = P_{max}$, $P^J_{sum,T}$ can be modified by using the proportional rationing rule [274]. According to the proportional rationing rule, if each source S_i submits its final bid $P_{i,T}^J < P_{i,T-1}^J$, the final power allocated to the source, denoted by P_i^{*J}, must be greater than the final bid: $P_i^{*J} > P_{i,T}^J$. This guarantees that even if the source submits a low final bid (demand), a high amount of power is still allocated to the source that enables the power of the friendly jammer to be fully utilized. Therefore, the power allocated to source S_i is given by

$$P_i^{*J} = P_{i,T}^J + \frac{P_{i,T-1}^J - P_{i,T}^J}{\sum_i P_{i,T-1}^J - \sum_i P_{i,T}^J} \left(P_{max} - \sum_i P_{i,T}^J \right) \quad (9.5)$$

where $\sum_i P_i^{*J} = P_{max}$.

As proven in [201], with a sufficiently large t, the value of T is finite. Thus, the proposed auction can conclude in a finite number of rounds; that is, the proposed algorithm converges.

The simulation results in [201] show that the proposed auction obtains a positive performance gain in terms of the system secrecy rate compared with the no-jamming case. Here, the system secrecy rate is defined as the total secrecy rate of all the source–destination links. The results imply that the proposed auction achieves efficient jamming power allocation. However, the proposed auction still incurs a considerable amount of communication overhead for exchanging information between the sources and the friendly jammer. Thus, one open issue is how to reduce the communication overhead while maintaining the efficiency of the proposed auction scheme.

9.2 Share Auction

In traditional auctions such as the first-price and second-price sealed-bid auction, bidders submit their bids, indicating the prices that the bidders are willing to pay for a single item, to the seller or the auctioneer. The bidder with the highest bid is selected as the winner of the auction. The winner receives the item from the seller, while the losers receive nothing. In resource trading markets, the item can be a divisible item. For example, in the energy and spectrum trading markets, power and spectrum can be considered to be divisible goods. In such a market, the item, such as an amount of power or a range of spectrum for trading, may not be fully utilized by the winner, and the item should be shared among several bidders. The issue is how to share the item efficiently and fairly among the bidders; the share auction can be applied to ensure this allocation. With the share auction, the item can be split among all the bidders rather than allocated to only the winner. The share auction is thus sometimes called a *divisible auction*. Unlike in the traditional auctions, there is no winner determination in the share auction. In other words, any bidder participating in the share auction becomes the winner as long as its bid is a positive scalar. The share auction process is illustrated in Figure. 9.2. The allocation and the payment of the share auction is implemented as follows:

- For the resource allocation, each bidder receives a portion of the item or an amount of resource that is proportional to the bid that the bidder submits. Consider a share auction with N bidders and a bidding profile of $\mathbf{b} = (b_1, \ldots, b_N)$.

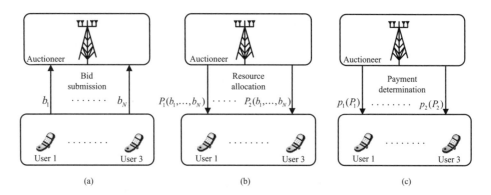

Figure 9.2 An illustration of share auction process: (a) bid submission, (b) resource allocation, and (c) payment determination.

Then, each bidder i receives an amount of resource defined as follows:

$$P_i = \frac{b_i}{\beta + \sum_{j}^{N} b_j} P_0 \tag{9.6}$$

where P_i is the amount of resource allocated to bidder i, P_0 is the maximum amount of resource that the seller can provide, and β is called the *reserve bid*. The reserve bid is used to guarantee a unique desirable outcome, such as Nash equilibrium, of the auction. This is different from a general reserve bid (price), which is used to guarantee a revenue gain for the seller. In general, the value of β does not affect the resource allocation at the Nash equilibrium, and its value is not important as long as it is a positive value. For simplicity, β is set to 1.

- For the payment, the price that the bidder pays is generally a function of the price that the seller announces and the allocated resources. Thus, the price that the bidder pays may not be related to the prices that the bidders submit. This is different from some traditional auctions such as the first-price or the second-price sealed-bid auctions.

Except for the winner determination, the share auction is simple and has low computational complexity. The resource allocation and the payment for the bidders can be further determined depending on a specific context. For convenience, we consider again the physical layer security issue as described in Section 9.1.2 and discuss how the resource allocation and the payment are performed. The share auction approach is actually proposed in [145]. The model consists of N sources, N corresponding destinations, one eavesdropper, and one friendly jammer. The sources are the bidders, and the friendly jammer is the seller or the auctioneer. Further description of the model is provided in Section 9.1.2. The friendly jammer conducts the share auction for the jamming power allocation as follows:

- The friendly jammer announces a positive reserve bid β and a price per power unit c to all the sources before the auction starts.
- Each source S_i computes a bid b_i and submits it to the friendly jammer.

- Upon receiving the bids from the sources, the friendly jammer assigns jamming power to the sources according to (9.6). Specifically, the jamming power allocated to each source S_i is

$$P_i^J = \frac{b_i}{\beta + \sum_j^N b_j} P_{max} \qquad (9.7)$$

where P_{max} is the maximum power that the friendly jammer can provide.
- Given the allocated power P_i^J, source S_i pays the friendly jammer a price p_i. The price p_i can be simply determined similarly to that in Section 9.1.2:

$$p_i = c P_i^J \qquad (9.8)$$

The question now is how the sources determine the optimal bids, which will maximize their utility. Here, the utility of a particular bidder depends on its bid and the bids of the other sources. The share auction can be thus regarded as a noncooperative game in which the utility of source S_i can be expressed by

$$U_i(b_i; \mathbf{b}_{-i}, c) = g(b_i; \mathbf{b}_{-i}) - c P_i^J \qquad (9.9)$$

where $\mathbf{b}_{-i}$ is the bidding profile excluding the bid of source S_i, $g(b_i; \mathbf{b}_{-i})$ is the performance gain with jamming power P_i^J that is defined in (9.2), and $p(b_i; \mathbf{b}_{-i}, c)$ is the price paid to the friendly jammer. Thus, the source can determine its optimal bid, or best response, as follows:

$$b_i^*(\mathbf{b}_{-i}, c) = \{b_i | b_i = \arg\max_{b_i \geq 0} U_i(b_i; \mathbf{b}_{-i}, c)\} \qquad (9.10)$$

Intuitively, the value of $b_i^*(\mathbf{b}_{-i}, c)$ depends on the price per power unit c as follows:

- Case I: c is too small. Then, the utility function U_i is an increasing function in the allocated power, and thus the source tries to bid a high price such that its utility is at is maximum. In this case, b_i^* should be ∞.
- Case II: c is too large. Then, cP_i^J may increase faster than $g(b_i; \mathbf{b}_{-i})$, and thus the utility function U_i is a decreasing function. Therefore, the source can set $b_i^* = 0$.
- Case III: c is set to a value such that the utility function U_i increases first and then decreases within the feasible region. Then, the utility function is a quasi-concave function, and there is an optimal bid b_i^* that maximizes the source's utility as well as its secrecy rate.

The bidding profile $\mathbf{b}^* = (b_1^*, \ldots, b_N^*)$ constitutes the Nash equilibrium of the auction at which no source wants to change unilaterally its bid. In other words, for every source S_i, we have $U_i(b_i^*; \mathbf{b}_{-i}^*, c) \geq U_i(b_i; \mathbf{b}_{-i}^*, c), \forall b_i$.

The simulation results in [145] show that the secrecy capacity obtained by the proposed auction is greatly improved compared with the non-jammer case. Moreover, the proposed auction scheme can attain the same secrecy capacity as does the centralized solution. Here, the centralized solution is a traditional approach that aims to maximize only the secrecy capacity. The centralized solution is thus considered to be an upper-bound approach. However, the centralized solution requires global information and suffers from high computational complexity.

Apart from improving physical layer security, the share auction is used as a distributed solution to determine relay selection and relay power allocation for cooperative communications as proposed in [275].

9.3 Online Auction

The aforementioned auctions can be considered to be offline auctions in which bidders and sellers can make their bids and asks at any time, but the auction market is cleared at certain specified time points. For example, the auctioneer receives bids from the buyers and asks from the sellers for a predefined period. After this period, the market is cleared, as matching between the bidders and the sellers is done. However, in digital markets such as eBay, whenever a bid and an ask arrive, the market is cleared immediately. Such an auction is called an *online auction*, or sometimes an *Internet auction* because it is typically used for trading digital goods via the Internet.

Since the auctioneer needs to make decisions about allocation and payments in real time, the online auction is generally more complicated than the offline auction. However, the online auction is suitable to some practical wireless systems in which the auction requests could be generated randomly and need to be handled as soon as possible. Consider spectrum allocation: when offline auctions are used, the auctioneer collects bids from the users (bidders) and clears the market only at the certain time. However, the user may need the resource urgently before that time. Thus, online auctions are more practical for such scenarios. These auctions allow the users to submit their bids for the resources at any time, with the auctioneer then allocating the requested spectrum immediately to satisfy the users' needs.

The main theoretical, experimental, and empirical work describing online auctions is well presented in [276]. In this section, we provide some basic terminologies commonly used in the online auction. We then present a specific application of the online auction in computer networks to understand how the online auction works.

9.3.1 Basic Terminologies

- *Sellers in online auctions:* A potential seller needs to be registered to offer its items. A secure online payment system is recommended for the seller. In the context of computer networks, the sellers can be service providers that provide network services or network resources. The sellers can also be mobile users that provide their sensing data or under-utilized resources. For example, in mobile crowdsensing networks [31], [277], the sellers are mobile users that arrive sequentially in a geographic area and provide sensing data as shown in Figure 9.3. When the seller chooses to offer an item through an online auction, it needs to specify the following information:

9.3 Online Auction

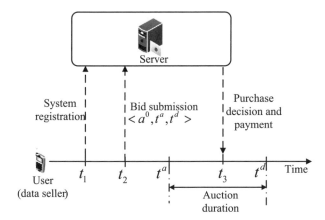

Figure 9.3 Online auction process in mobile crowdsensing networks with one mobile user. t_1 is the system registration time of the mobile user, t_2 is the bid submission time, and t_3 is the time of purchase decision and payment of the server. a^0 is the starting bid, and t^a and t^d are the arrival time and the departure time, respectively, of the mobile user. t_3 must be between t^a and t^d.

- *Starting bid:* The seller needs to set the price at which the seller offers the item. The seller can determine the price based on the same or a similar item sold in the past.
- *Starting time:* The seller needs to specify the time when the item is available. In the computer network environment, the starting time is often the *arrival time* – that is, the time that the network resources become available.
- *Auction duration:* The seller needs to specify the duration time, which is how long the item will be available. In general markets, such as the digital goods market, the duration may be up to 20 or 30 days. For computer network environments that typically operate in real time, the auction duration is much shorter, in the range of milliseconds. Also, the term *departure time*, referring to the time that the network resources are no longer available, is often used rather than "auction duration."

- *Buyers in online auctions:* Similar to the sellers, the buyers need to register as clients in the system to participate in online auctions. Before bidding on an item, the buyers can analyze and compare prices as well as the reputations of different sellers. The buyers can be required to specify a maximum price that they are willing to pay for the item. This information is not made publicly to either other bidders or the seller. Typically, the bidder is guaranteed to be the winner if its maximum bid is higher than those of its rivals. In wireless network environments such as cognitive radio networks [278], [279], the buyers can be secondary spectrum users that can submit bids for resources at any time. In cloud computing [280], the buyers can be cloud users that bid for a pool of different types of resources including CPU, storage, and bandwidth.

9.3.2 Development of Online Auction for Cloud Resource Pooling

Cloud service providers face two big challenges. First, the cloud services heavily depend on centralized data centers in the cloud, and accidents and natural disasters such as fires and earthquakes can have significant impacts on service performance. Second, the service providers have to pay substantial costs for purchasing and renting cloud hardware infrastructure, power, and network bandwidth. To address these challenges, client-assisted or user-assisted cloud models have recently been developed. Examples of client-assisted cloud models include FS2You [281] and Triton [282]. The client-assisted cloud models are decentralized cloud paradigms that form resource pools by exploiting the under-utilized storage, computing, and network resources of clients or users. The clients are edge devices located in the external network environment of the data centers. Thus, they guarantee high data availability and reliability while reducing network costs.

One critical issue with the client-assisted cloud models is how to incentivize the clients to contribute their local resources while minimizing the service provider's cost for pooling the resources. The incentive mechanism can be designed using offline auctions such as the VCG auction. However, due to the asynchronous arrivals of clients, meaning that the clients may arrive and depart the system at any time, an offline auction may not be the best suitable option. Alternatively, the online auction can be adopted as an efficient solution. This section presents such an approach as proposed in [283].

System Model and Problem Formulation

The system model shown in Figure 9.4 includes the following entities:

- There is one service provider that is willing to buy resources from clients that can provide cloud services. Here, the resource is a combination of storage and network bandwidth. Thus, the service provider is the buyer or the auctioneer, and the clients are the sellers.
- Consider the duration $[0, T]$, during which the service provider wants to buy capacity denoted by S from the clients. Note that at each time $t \in [0, T]$, the set of clients is unknown to the service provider in advance.
- At each time t, the service provider receives a set of asks $\mathcal{A} = (a_1, \ldots, a_j, \ldots)$ from clients. Here, each ask a_j refers to the available resource that client j can contribute and the price that the client is willing to receive. The ask can be expressed by a tuple $a_j = (\hat{t}_j^a, \hat{t}_j^d, \hat{Q}_j, \hat{c}_j)$, where $\hat{t}_j^a$ is the announced arrival time of available resources, $\hat{t}_j^d$ is the announced departure time of available resources, $\hat{Q}_j$ is the announced available resource capacity, and $\hat{c}_j$ is the announced marginal cost. In particular, $\hat{c}_j$ is the cost for the client to provide one more unit of resource for one time unit.
- The type of the client is denoted by $v_j = (t_j^a, t_j^d, Q_j, c_j)$; it is private information known to the client only.
- Let $\mathcal{A}_{-j}$ be the set of all asks except the ask of client j.
- Upon receiving ask a_j, the service provider needs to determine the amount of resource to buy $q_j(t, a_j)$, where $0 \leq q_j \leq \hat{Q}_j$ and time $t \in [\hat{t}_j^a, \hat{t}_j^d]$, and the

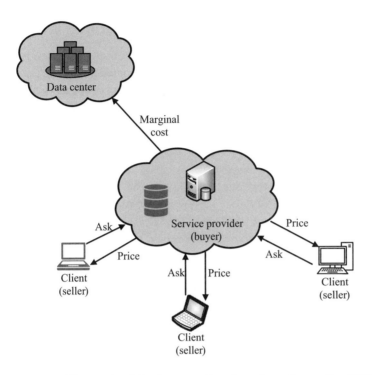

Figure 9.4 Client-assisted cloud storage allocation using online auction [201].

corresponding payment $p_j(t, a_j)$ to minimize the resource pooling cost of the service provider. To incentivize the clients to contribute their resources, the resource trading also guarantees a non-negative utility for the clients.

- The utility of client j at time t can be defined as $u_j(t, q_j, p_j) = p_j(t, a_j) - c_j q_j(t, a_j)$. Thus, the utility of the client during time $[\hat{t}_j^a, \hat{t}_j^d]$ is defined by

$$U_j(q_j, p_j) = \int_{\hat{t}_j^a}^{\hat{t}_j^d} u_j(t, q_j, p_j) dt \qquad (9.11)$$

To achieve the objectives of the service provider, the following properties are important:

- *Individual rationality:* The online auction needs to support individual rationality to guarantee the non-negative utility for the clients. Individual rationality is guaranteed if for any client j, we have

$$u_j(t, q_j, p_j) = p_j(t, a_j) - c_j q_j(t, a_j) \geq 0 \qquad (9.12)$$

- *Truthfulness:* The online auction also maintains truthfulness such that the clients reveal resource availability and resource cost truthfully. The online auction is truthful if for any client j with type $v_j = (t_j^a, t_j^d, Q_j, c_j)$, and for every ask $a_j = (\hat{t}_j^a, \hat{t}_j^d, \hat{Q}_j, \hat{c}_j)$ of the client satisfying $\hat{t}_j^a \geq t_j^a, \hat{t}_j^d \leq t_j^d, \hat{Q}_j \leq Q_j$, and $\hat{c}_j \neq c_j$, we have

$$U_j(q_j(t,v_j),p_j(t,v_j)) \geq U_j(q_j(t,a_j),p_j(t,a_j)) \quad (9.13)$$

- *Resource pooling cost:* The resource pooling cost of the service provider is defined based on the *completeness ratio* [284], denoted by $R(t',t)$. Let $\mathcal{A}_t$ be the set of stream asks received by time t. Given $\mathcal{A}_t$, we can define the completeness ratio of the service provider at any time $t \leq t' \leq T$ as follows:

$$R(t',t) = \frac{\sum_j q_j(t',a_j)}{S} \in [0,1] \quad (9.14)$$

$R(t',t)$ refers to the completeness ratio of the service provider at time t' after receiving the streaming asks $\mathcal{A}_t$ at time t. The value of $R(t',t)$ implies that the required capacity S of the service provider at time t is achieved, $R(t',t) = 1$, or not achieved, $R(t',t) < 1$. In the case of $R(t',t) < 1$, the service provider needs to buy the remaining resources from servers in the data centers with the marginal resource cost of c_s. Thus, the resource pooling cost $C(\mathcal{A})$ of the service provider includes the cost that it pays the clients and the cost that it pays the servers. $C(\mathcal{A})$ is defined as follows:

$$C(\mathcal{A}) = \sum_j \int_{\hat{t}_j^a}^{\hat{t}_j^d} p_j(t,a_j)dt + \int_0^T c_s S[1 - R(t,t)]dt$$

$$= \int_0^T \sum_j \left[p_j(t,a_j) + c_s S[1 - R(t,t)] \right] dt \quad (9.15)$$

The problem of the service provider can be defined as follows:

$$\min_{\{p_j,q_j\}} C(\mathcal{A}) \quad (9.16)$$

$$\text{s.t.} (9.12), (9.13), (9.14) \quad (9.17)$$

Note that q_j and p_j can be decoupled at different time points. Thus, the optimization problem in (9.17) is equivalent to the following problem for any $0 \leq t \leq T$:

$$\min_{\{p_j,q_j\}} \sum_j p_j(t,a_j) + c_s S[1 - R(t,t)] \quad (9.18)$$

$$\text{s.t.} (9.12), (9.13), 0 \leq R(t,t) \leq 1$$

Algorithm for Online Auction

One of the challenges when solving (9.18) is the need to guarantee the truthfulness of the auction. To address this challenge, Myerson's well-known characterization is used. Myerson's principle is applied for the forward auction with a single indivisible good. We present the principle as follows:

LEMMA 9.1 *[194] Consider a forward auction including one seller and multiple bidders. Let $P_i(b_i)$ be the probability of buyer i with bid b_i winning an auction. The auction is truthful if and only if it satisfies the two following properties:*

- $P_i(b_i)$ *is monotonically nondecreasing in b_i.*
- *Buyer i submitting b_i is charged with $b_i P_i(b_i) - \int_0^{b_i} P_i(b)db$.*

The first property refers to the monotonicity of the allocation rule. Accordingly, if seller i wins the auction with ask a_i, then it also wins with any higher ask $a_j > a_i$. Given the allocation rule, the second property determines the conditional payment for each bidder i. As proved in [285], the price that the bidder pays is independent of its bid b_i, which guarantees the truthfulness of the auction.

Myerson's principle is proposed for the forward auction with a single indivisible good. The auction that we consider is a type of reverse auction with divisible goods. Thus, Myerson's principle has to be modified. In particular, allocation monotonicity can be defined as follows:

DEFINITION 9.2 *[283] Consider two asks $a_j = (\hat{t}_j^a, \hat{t}_j^d, \hat{Q}_j, \hat{c}_j)$ and $a'_j = (\hat{t}_j'^a, \hat{t}_j'^d, \hat{Q}'_j, \hat{c}'_j)$. An ask a_j dominates a'_j, denoted by $a_j \succeq a'_j$, if the resource marginal cost of ask a_j is less than or equal to that of ask a'_j, $\hat{c}_j \geq \hat{c}'_j$, and the resource availability of ask a'_j is a subset of that of ask a_j, $\hat{t}_j^a \leq \hat{t}_j'^a$, $\hat{t}_j^d \geq \hat{t}_j'^d$, $\hat{Q}_j \geq \hat{Q}'_j$. Then, an allocation mechanism is monotone if*

$$\int_{\hat{t}_j^a}^{\hat{t}_j^d} p_j(t, a_j) dt \geq \int_{\hat{t}_j'^a}^{\hat{t}_j'^d} p_j(t, a'_j) dt, \forall j, if\ a_j \succeq a'_j \quad (9.19)$$

As the allocation rule is determined, the payment scheme that guarantees the truthfulness of the online auction is determined by the following theorem.

THEOREM 9.3 *There exists a payment rule such that the online auction is truthful if the payment scheme is defined as follows:*

$$p_j(t, a_j) = \hat{c}_j q_j(t, a_j) + \int_{\hat{c}_j}^{+\infty} q_j\bigl(t, (\hat{t}_j^a, \hat{t}_j^d, \hat{Q}_j, c)\bigr) dc \quad (9.20)$$

The proof of Theorem (9.3) can be found in [283].

In summary, if any reverse auction with divisible goods performs the allocation rule according to (9.19) and the payment rule according to (9.20), then the auction is truthful. However, similar to Meyerson's principle, the allocation and payment rules are only applied to offline auctions in which the asks of the sellers arrive at the same time. To apply the online auctions, the allocation and payment rules can be modified based on a *marginal pricing function* [286].

Let $\rho(t, R)$ denote the marginal pricing function of the service provider for buying resources from clients, as the completeness ratio of resource pooling at time t is R. Based on the marginal pricing function, the allocation rule is determined as follows:

$$q_j(t, a_j) = \begin{cases} 0, & \text{if } \hat{c}_j \geq \rho(t, R) \\ \min\{\hat{Q}_j, S[\rho^{-1}(t, \hat{c}_j)] - R(t, \hat{t}_j^{a(-)})]\}, & \text{if } \hat{c}_j < \rho(t, R) \end{cases} \quad (9.21)$$

where $R(t, \hat{t}_j^{a(-)})$ is the completeness ratio for resource pooling at time t after receiving streaming asks at time $\hat{t}_j^{a(-)}$, the time just before receiving ask a_j.

The payment rule is defined as follows:

$$p_j(t, a_j) = \int_{R(t, \hat{t}_j^{a(-)})}^{R(t, \hat{a}_j)} S\rho(t, R) dR \quad (9.22)$$

As proved in [283], if the allocation rule follows (9.21) and the payment rule follows (9.22), with $\rho(t, R)$ being a non-increasing in completeness ratio R, the online auction is truthful. Moreover, by using the allocation and payment rules, the online auction problem in (9.22) can be converted into an online algorithm design problem that has the marginal pricing function $\rho(t, R)$ as its variable. Specifically, by substituting the payment rule (9.22) into the problem (9.18), the online algorithm design problem is expressed by

$$\min_{\{p_j, q_j\}} \sum_j \int_{R(t, \hat{r}_j^{a(-)})}^{R(t, \hat{a}_j)} S\rho(t, R) dR + c_s S[1 - R(t, t)] \quad (9.23)$$

$$\text{s.t. } \rho(t, R) \text{ is a nonincreasing function} \quad (9.24)$$

$$\rho(t, 1) \leq c_{min} \quad (9.25)$$

The constraints of (9.23) are equivalent to those of (9.24). Specifically, the constraint in (9.24) guarantees the truthfulness of the auction. The constraint in (9.25) guarantees that $0 \leq R(t, t) \leq 1$. Moreover, the price determined in (9.22) is proved to be higher than the resource marginal cost that the client submits. Thus, for every client j, $U_j(q_j, p_j) \geq 0$.

The algorithm of the online auction can be summarized as shown in Algorithm 10. The algorithm is implemented at the service provider. The input of Algorithm 10 includes the marginal pricing function $\rho(t, R)$. Thus, the marginal pricing function needs to be specified. By using the social cost of the offline VCG auction as a lower bound of the resource pooling cost of the offline optimal auction, the marginal pricing function that minimizes the resource pooling cost of the online auction is determined by $\rho(t, R) = c_s - c_s(1 - \frac{1}{\lambda})e^{\frac{1}{\lambda}}$, where λ is the competitive ratio. The competitive ratio is defined as the ratio of the resource pooling cost of the online auction to that of the offline optimal auction. Thus, λ is expected to be small. Given the pricing function definition, we have the following theorem.

THEOREM 9.4 *By using the marginal pricing function of $\rho(t, R) = c_s - c_s(1 - \frac{1}{\lambda})e^{R/\lambda}$, the online auction can achieve a resource pooling cost that is not greater than λ times that of the offline optimal auction, $C(\mathcal{A}) \leq \lambda C_{opt}(\mathcal{A})$, where $C_{opt}(\mathcal{A})$ is the service provider's resource pooling cost under the offline optimal auction, and λ is within $[\lambda_{min}, \lambda_{max}]$, where λ_{min} and λ_{max} are the solutions to equations $(1 - \frac{1}{\lambda})e^{1/\lambda} = 1 - \frac{c_{max}}{c_s}$ and $(1 - \frac{1}{\lambda})e^{1/\lambda} = 1 - \frac{c_{min}}{c_s}$, respectively.*

The simulation results in [283] show that the resource polling cost in the online auction is always less than that of the offline VCG auction when varying the resource pooling demand, the number of asks, and the ratio between the server cost and the average ask. In future work, the cloud resource trading among the clients (the users) can be considered. This context can be considered to be a *self-organization cloud*.

Algorithm 10 Algorithm for the online auction [283].

Input: $\rho(t, R), S, T$
Output: $q_j(t, a_j), p_j(t, a_j)$
1: Initialize the completeness ratio $R_t = 0$ at time $t \in [0, T]$
2: **While** collect ask a_j **do**
3:     **for** $t \in [\hat{t}_j^a, \hat{t}_j^d]$ **do**
4:         **if** $\hat{c}_j < \rho(t, R_t)$ **then**
5:             **if** $\rho^{-1}(t, \hat{c}_j) > 1$ **then**
6:                 Set $\rho^{-1}(t, \hat{c}_j) = 1$
7:             **end if**
8:         Service provider buys $q_j(t, a_j)$ resource from seller j according to (9.21)
9:         Service provider pays $p_j(t, a_j)$ resource from seller j according to (9.22)
10:         Update $R_t = R_t + \frac{q_j(t, a_j)}{S}$
11:         **end if**
12:     **end for**
13: **end While**

The online double auction can be adopted for the dynamic cloud resource trading between the demand clients and the supply clients.

9.4 Waiting-Line Auction

In this section, we introduce a "non-money auction," known as the waiting-line auction, and discuss the Nash equilibrium of this auction.

The previously described auctions can be regarded as "money auctions" in which bidders submit prices to the seller or the auctioneer. The bidders with the high prices often win the auction and receive resources from the seller. This means that "poor bidders" with limited budgets may never win the auction since they typically bid low prices. To improve the winning probabilities of the poor bidders, the waiting-line auction [287] can be used. Unlike the traditional auctions, such as the sealed-bid auctions, the waiting-line auction has no explicit process of bid submission. Instead, the bidders join a queue (a line), and the seller allocates the resources sequentially to the bidders in the queue at some unknown time on a first-come-first-serve basis. As such, the waiting-line auction determines the winners based on the arrival times of the bidders rather than their prices. In other words, the bidders that come earlier and join the first queue become the winners of the auction. The waiting-line auction thus enhances the winning probabilities of the poor bidders if they arrive in the queue early. This is similar to the situation in which poorer people, rather than richer people, are willing to waiting for a long time for discounted sales. The reason is that the opportunity cost of the time spent

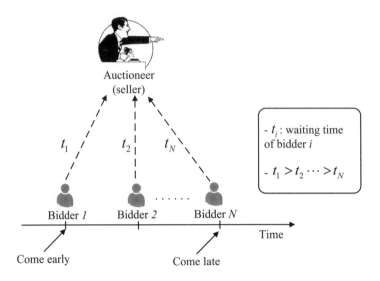

Figure 9.5 Waiting-line auction for spectrum allocation.

for waiting is lower for the poorer people; in other words, the poorer people can tolerate long waiting times. Since the waiting-line auction determines the winners according to the arrival times of the bidders, it is also called a *waiting-time auction*. The waiting-line auction has been widely used for allocating items such as World Series game tickets, theater tickets, theater seats for ticket holders, and certain classes of airline tickets.

The following terminology is used in the waiting-line auction:

- *Bid:* Bidders compete for items by submitting their bids to the auctioneer. The bid specifies the time that the bidder can wait for receiving the items. Thus, the bid refers to the waiting time of the bidder (see Figure 9.5). A large bid means that a bidder willing to come the queue early has a high probability of winning the items.
- *Opportunity cost:* When the bidder decides to join the waiting line, the bidder can lose or sacrifice its profit/utility from other alternatives. We say that the bidder pays an opportunity cost for joining the waiting line. In the context of computer networks, when one mobile user decides to join the waiting line, that user may need to stop its data transmission, which results in a loss of its utility.
- *Private value:* The private value of the bidder refers to the value of the item to the bidder. As the bidder assigns a higher value to the item, it is willing to come the queue earlier and pays a high opportunity cost. Thus, the private value can be considered to be a function of the opportunity cost.

Similar to traditional auctions, such as the first-or second sealed-bid price auctions, the waiting-line auction is expected to have a Nash equilibrium. Next, we discuss how to determine the Nash equilibrium in the waiting-line auction. For convenience, we consider a scenario of cognitive radio networks. We can use the waiting-line auction in cognitive radio network when we want to improve the spectrum access opportunities of

9.4 Waiting-Line Auction

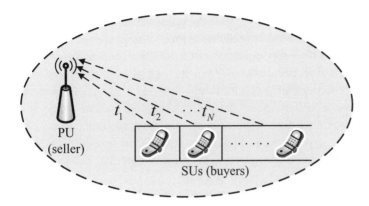

Figure 9.6 Waiting-line auction for spectrum allocation [288]. PU and SU stand for primary user and secondary user, respectively.

secondary users with poor cognitive ability. The secondary users with poor cognitive ability are the mobile devices that have low resources, such as low communication rate and low energy. In this scenario, the unused spectrum of the primary users can be regarded as discounted sales, and the waiting-line auction is used to allocate the unused spectrum to the secondary users with poor cognitive ability if those users are willing to wait for a longer time. The waiting-line auction approach in the cognitive radio network is proposed in [288].

The cognitive radio network model is shown in Figure 9.6, and the entities in the model can be described as follows:

- There are N secondary users that are willing to use unused spectrum units from one primary user. Thus, the secondary users are the bidders or buyers, and the primary user is the seller.
- The primary user has M unused spectrum units, where M is assumed to be less than N: $M < N$. Thus, the secondary users need to compete to obtain the spectrum units. Each winner can receive at most one spectrum unit.
- The primary user allocates M spectrum units to the secondary users at known time t_a on a first-come-first-serve basis. This means that the earliest M secondary users are the winners for the spectrum units. Each secondary user i needs to determine its waiting time t_i before t_a, and then submits the waiting time to the primary user. t_i indicates how long the secondary user can wait until receiving the spectrum unit. For example, if $t_i = 3$, the secondary user is willing to wait for 3 hours to receive the spectrum unit. Thus, a large t_i means that secondary user i is willing to wait for longer time. The secondary user has a higher winning probability since it arrives at the queue earlier. The lowest value of t_i may be zero, and in this case the secondary user is not willing to wait to receive the spectrum unit.
- If secondary user i decides to join the waiting line, that user is assumed to stop its current data transmission during the waiting time. This assumption is reasonable since the secondary user may be equipped with a single radio.

- The secondary user can increase the waiting time t_i to improve its winning probability, but it also has to spend more time waiting. We can say that the secondary user pays a high opportunity cost of time. Let w_i denote the opportunity cost per time unit of secondary user i. Then, the waiting cost of this user is $w_i t_i$.
- The secondary user knows its waiting time, but it does not know the waiting times of other secondary users. Thus, the waiting-line auction can be formulated as a noncooperative game with incomplete information. In the game, each secondary user as a player seeks an equilibrium strategy, the optimal waiting time, to maximize its utility. For this, the private value of spectrum unit to the secondary user should be defined.
- The private value can be considered to be a function of the opportunity cost of time w_i of the secondary user. If w_i is smaller, the secondary user is willing to wait for longer time, such that t_i is larger. Since the secondary user is willing to come earlier, we say that the private value of the secondary user is higher. We can assume that $v(w_i)$ is a monotonic decreasing function of w_i [288].
- If the secondary user wins the auction, it receives the spectrum unit with private value $v(w_i)$. Therefore, the utility of the secondary user can be defined by

$$U_i^W = v(w_i) - w_i t_i - k w_i \quad (9.26)$$

where k is a constant time period that can be seen as the time required to reach the queue.

- If the secondary user loses the auction, its utility is given by

$$U_i^L = -w_i t_i \quad (9.27)$$

- For simplification, the term "unified utility" can be defined as the ratio of the utility function to the opportunity cost of time. If secondary user i wins, the unified utility, denoted by π_i^W, is defined by

$$\pi_i^W = \frac{U_i^W}{w_i} = a_i - t_i - k \quad (9.28)$$

where $a_i = v(w_i)/w_i$ is called the *time value* of a spectrum unit for user i. a_i is also a decreasing function of w_i. π_i^W is the utility in time units for the winner that waits t_i time units. The utility function in (9.28) is quite similar to that of the sealed-bid auction (see (5.2)), in which the private value is a_i and the bid is t_i.

- Similarly, as secondary user i loses, its unified utility, denoted by π_i^L, is defined by

$$\pi_i^L = \frac{U_i^L}{w_i} = -t_i - k \quad (9.29)$$

This utility function in (9.29) is different from that in (5.2) since the loser needs to pay an amount of $t_i + k$. In the case that $k = 0$, secondary user i pays the amount equal to its bid, t_i.

The optimal waiting time t_i of secondary user i is determined depending on its time value a_i and the expected waiting times of other secondary users. We consider a symmetric model in which the time values of a spectrum unit to the secondary users are drawn from the same probability distribution. Then, the subscript i can be removed from a_i. We need to find a common strategy function $\sigma(a)$ that determines the best waiting times for all secondary users as a function of their own time values. $\sigma(a)$ is called the equilibrium strategy function and is assumed to be positive, strictly increasing, and differentiable in a. $\sigma(a)$ can be explained as follows.

Given M spectrum units, secondary users with M largest waiting times receive the spectrum units. Thus, secondary user i will win one spectrum unit if t_i exceeds the Mth largest of the other secondary users' waiting times. Since $\sigma(a)$ is strictly increasing in a, secondary user i will win if a_i exceeds the Mth largest of the other secondary users' time values. Let $F(a)$ be the probability that a exceeds the Mth largest of the other secondary users' time values, and let $f(a)$ be the corresponding density function. The authors in [287] derive the equilibrium strategy function $\sigma(a)$ for the utility functions given in (9.28) and (9.29) as follows:

$$\sigma(a) = \int_{a^*}^{a} y f(y) dy \qquad (9.30)$$

where a^* is called the *cutoff value*. It is the lowest time value at which a secondary user is indifferent between arriving at the allocation time and not participating in the auction at all. a^* can be determined by solving the equation $F(a^*)a^* = k$.

It is proven in [287] that if $N - 1$ competitors of secondary user i select their waiting times $t_j = \sigma(a_j), \forall j \neq i$, then the expected utility of secondary user i is globally maximized by selecting $t_i = \sigma(a_i)$. The simulation results in [288] show that a small deviation of t_i reduces the winning probability of the corresponding secondary user by up to 60% and the utility by up to 40%. Therefore, the secondary user cannot deviate from the unique equilibrium strategy as given in (9.30).

9.5 Summary

In this chapter, we introduce four special auctions and their applications in computer networks. First, we present the ascending clock auction, which has some similarities with the English auction. To understand how this auction works, we discuss the use of the ascending clock auction for improving physical layer security in a cognitive wireless network. Second, we introduce the share auction, which has low computational complexity and is typically used in markets with divisible goods. Third, we introduce the online auction, which allows the seller to make decisions about allocation and payments in real time. Some basic terminologies that are commonly used in the online auction are defined as well. An online auction approach for cloud resource pooling in a client-assisted cloud model is discussed. Fourth, we introduce the waiting-line auction, which can be formulated as a noncooperative game and its Nash equilibrium. The waiting-line auction is a non-money auction in which the winners are determined based on waiting times submitted by the bidders rather than their bidding prices. The terminologies and the Nash equilibrium of the waiting-line auction are defined for computer networks. In future work, researchers can consider how to preserve the privacy of the buyers and the sellers in online auctions.

10 Optimal Auction Using Machine Learning

In the previous chapter, we discussed various auctions and desired properties of those auctions. However, the means of designing an optimal auction in terms of maximizing revenue for the seller and simultaneously guaranteeing the desired properties, such incentive compatibility and individual rationality, remains an open question. For example, the first-price sealed-bid auction using the pay-what-you-bid rule (see Chapter 5) can achieve revenue maximization, but it cannot guarantee incentive compatibility. This means that bidders always have an incentive to submit untruthful bids to improve their utility. Also, the second-price sealed-bid auctions, including the Vickrey auction and the VCG auction, as discussed in Chapter 6 can ensure incentive compatibility by using the second-price rule, but may not achieve revenue gains.

In this chapter, we discuss how to design optimal auctions. These optimal auctions are based on deep learning techniques that are proposed by the authors in [192]. Specifically, we first introduce the optimal auction design problem. Then, we introduce the deep learning technique and the motivations of applying the deep learning technique to optimal auctions. After that, we describe neural networks that are used to derive optimal auctions. We further provide examples to show that the deep learning-based optimal auction approaches can be efficiently used to solve emerging issues in computer networks.

10.1 Optimal Auction

In this section, we discuss the optimal auction problems. We discuss these problems in the context of a fog resource trading market that is described as follows [192]:

- The market consists of one service provider, the seller or the auctioneer, that has M computing resource units for trading.
- There are N users, the bidders. Each user i has a valuation or a value v_i of the computing resource unit. Assume that the values of different computing resource units to the user are similar, and let $\mathbf{v} = (v_1, \ldots, v_N)$ denote the valuation profile of the users. v_i is drawn independently from distribution F_i over a possible valuation profile $\mathbf{V}_i$.
- The service provider may not know the realized values of the users, but it is assumed to know the distribution functions.

10.1 Optimal Auction

- The users submit bids, representing the prices that they are willing to pay, to the service provider. Let $b_i \in V_i$ denote the bid of user i, and $\mathbf{b} = (b_1, \ldots, b_N)$ denote the bid profile of the users. Also, let $\mathbf{b}_{-i}$ denote the bid profile $\mathbf{b}$ excluding the bid of user i.

Upon receiving the bid profile $\mathbf{b}$ from the users, the service provider determines an allocation rule and a pricing rule. The allocation rule includes the winning probabilities $g_i, i = 1, \ldots, N$, of the users, and the pricing rule includes the conditioned prices p_i, $i = 1, \ldots, N$, for the users. The allocation and pricing rules are both functions of the bid profile. Let $\mathbf{g}(\mathbf{b})$ and $\mathbf{p}(\mathbf{b})$ denote the allocation rule and the pricing rule, respectively.

The key objective is to maximize the expected revenue of the service provider while guaranteeing two important properties: incentive compatibility and individual rationality. The definitions of these properties are presented in Section 6.2.5. In general, incentive compatibility and individual rationality are related to the utility functions of the users. The utility function of the users depends on the values of computing resource units to the users as well as the winning probabilities and the conditioned prices of the users. Specifically, the utility of each user i can be defined as follows [192]:

$$u_i(v_i, \mathbf{b}) = v_i(g_i(\mathbf{b})) - p_i(\mathbf{b}) \tag{10.1}$$

We have the following definitions [192]:

- The expected revenue is the total price that the service provider receives from the users. The expected revenue can be thus defined as $\sum_i p_i(\mathbf{v})$.
- The auction provides *incentive compatibility* if $u_i(v_i, (v_i, \mathbf{b}_{-i})) \geq u_i(v_i, (b_i, \mathbf{b}_{-i}))$, $\forall i, \forall v_i, \forall b_i$, and $\forall \mathbf{b}_{-i}$. In other words, the auction demonstrates incentive compatibility if the utility of each user i is maximized by submitting truthfully its bid regardless of the other users' bids.
- The auction provides individual rationality if $u_i(v_i, (v_i, \mathbf{b}_{-i})) \geq 0$ for $\forall i, \forall v_i$, $\forall \mathbf{b}_{-i}$. This means that every user i has a non-negative utility when participating in the auction.

The problem of designing the optimal auction entails determining the allocation rule and the pricing rule by solving the following optimization problem [192]:

$$\max_{\mathbf{g}, \mathbf{p}} \left[\sum_i p_i(\mathbf{v}) \right] \tag{10.2}$$

$$\text{s.t. [IC], [IR]}$$

where [IC] and [IR] are the incentive compatibility and individual rationality constraints, respectively. The problem in (10.2) is the constrained optimization. In general, solving the constrained optimization problem to derive the optimal auction is difficult [192]. Myerson's optimal mechanism [194] can be adopted by using the concept of *virtual values*. However, Myerson's optimal mechanism is constrained to a single item. For multi-item cases, we can find the optimal auction designs proposed in [289], [290], and [291]. However, these optimal auction approaches are constrained to one bidder or user.

In recent years, a deep learning technique that has the ability to automatically identify relevant features of data has gained considerable attention. Recent theoretical results in [195] show that deep learning using stochastic gradient descent can successfully find globally optimal solutions for complex problems such as non-convex and non-concave optimization problems. This motivates the authors in [192] to adopt the deep learning technique to solve the constrained optimization problem in (10.2).

10.2 Machine Learning

A simple deep learning model uses a multi-layer neural network [292]. As shown in Figure 10.1, the feed-forward neural network consists of multiple layers: the input layer, the hidden layers, and the output layer. The information moves in one direction from the input layer, through the hidden layers, and to the output layer. In each layer, there are a number of neural nodes. The neural nodes use activation functions such as softmax, sigmoid, and rectifier to proceed to the deeper level of inception. The neural nodes are connected with each other through weights. These weights refer to the strength of the connections among the neural nodes, and the weights imply how much the output changes as the input varies. As such, the feed-forward neural networks used in deep learning map from inputs to outputs [192].

There are three types of learning: supervised learning, semi-supervised learning, and unsupervised learning. In particular, supervised learning finds weights of the neural network to minimize a *loss function* or *cost function* based on the training data or data set. The training data contains a set of training examples. Let $\mathcal{L}(\cdot)$ denote the loss function, $\mathbf{w}$ be the set of the weights of the neural network, and $\mathbf{x}$ and $\mathbf{y}$ be the input and the output of the neural network, respectively. Here, $\mathbf{y}$ is also the target value or

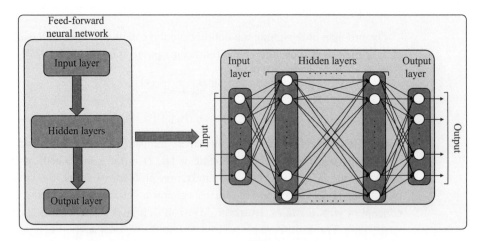

Figure 10.1 A general deep learning model.

reference value. The $(\mathbf{x}, \mathbf{y})$ pairs are sampled from the training data. Then, the problem of the supervised learning can be expressed by

$$\min_{\mathbf{w}} \mathcal{L}(\mathbf{x}, \mathbf{y}; \mathbf{w}) \tag{10.3}$$

Stochastic gradient descent is typically used to find the optimal weights $\mathbf{w}$ to optimize (i.e., minimize) the loss function.

From (10.3), we see that the optimization problem (10.2) can be solved by deep learning [192].

- We define the negative expected revenue $-\left[\sum_i p_i(\mathbf{v})\right]$ as the loss function $\mathcal{L}(\cdot)$.
- The values of computing resource units to the bidders are the input $\mathbf{x}$ of the neural network.
- The allocation rule $\mathbf{g}$ and the pricing rule $\mathbf{p}$ are functions of the weights $\mathbf{w}$ of the neural networks. Also, $\mathbf{g}$ and $\mathbf{p}$ are the outputs of the neural networks.

Note that the optimization problem in (10.2) has both individual rationality constraints [IR] and incentive compatibility constraints [IC]. However, the individual rationality and incentive compatibility constraints can be introduced into the loss function $\mathcal{L}()$ – for example, by using the augmented Lagrangian method [293]. The next section discusses how to design the optimal auction.

10.3 Machine Learning for Optimal Auction

In this section, we present how to construct the loss function and the neural networks for the optimal auction. Furthermore, we provide an example to show the application of the optimal auction to handle the resource management issue in computer networks.

10.3.1 Design

We consider again the fog cloud resource trading market as presented in Section 10.1, but in a general case. Specifically, each user submits different bids (prices) for different computing resource units.

- Let b_{ij} be the bid of user i for computing resource unit j, and let $\mathbf{b}$ denote the bid profit that includes all bids of the users.
- Given the bid profile $\mathbf{b}$ as the input, let $g_{ij}(\mathbf{b})$ denote the probability that user i wins resource unit j.
- Also, let $p_i(\mathbf{b})$ denote the expected price for each user i. $p_i(\mathbf{b})$ is the total price that the user pays the service provider for winning the resource units.

Loss Function

To construct the loss function, we define the following terms [192]:

- *Individual rationality violation:* The auction guarantees individual rationality, which means that the utility or payoff of any user should not be negative when it participates in the auction. If the utility of a user is negative, we say that its individual rationality is violated or there is an individual rationality violation for that user. Thus, the auction should be designed such that the individual rationality violation for any user is as small as possible. The individual rationality violation is preferably zero. Let u_i denote the utility of user i; then the individual rationality violation for user i is defined as $IR_i = \max\{0, -u_i\}$.

- *Incentive compatibility violation:* Incentive compatibility guarantees that every user achieves the highest utility just by submitting its truthful bid or true value. In turn, an incentive compatibility violation is defined as the maximum gain in utility that the user can receive if the user submits a non-truthful bid knowing the bids of others. Let IC_i denote the incentive compatibility violation for user i and let v_i denote the truthful bid, or true value, of the user. Then, the incentive compatibility violation for user i is defined as follows [192]:

$$IC_i = \max_{v'_i \neq v_i}(u_i(v'_i, \mathbf{b}_{-i})) - u_i(v_i, \mathbf{b}_{-i}) \tag{10.4}$$

When IC_i is small, the utility gain that the user receives when it submits a non-truthful bid is small. Thus, IC_i is expected to be zero such that the user has no incentive to submit a non-truthful bid.

- *Negative expected revenue:* The expected revenue is the total price that the service provider receives from the users, $\left[\sum_i p_i(\mathbf{v})\right]$. Thus, the negative expected revenue is defined as $-\left[\sum_i p_i(\mathbf{v})\right]$.

Again, the aim of the auction design is to maximize the expected revenue and minimize the individual rationality and incentive compatibility violations. If we consider the revenue function as the objective function and the individual rationality and incentive compatibility violations as the constraints, then the augmented Lagrangian method can be used to formulate the loss function. In particular, the loss function, denoted by $\hat{\mathcal{L}}()$, is defined as follows [192]:

$$\hat{\mathcal{L}}(\lambda, \lambda', \mathbf{w}) = -\left[\sum_i p_i(\mathbf{v}, \mathbf{w})\right] + \sum_{i=1}^{N} \lambda_i IC_i(\mathbf{w}) + \sum_{i=1}^{N} \lambda'_i IR_i(\mathbf{w}) \tag{10.5}$$

$$+ \frac{\rho}{2}\sum_{i=1}^{N}\left((IC_i(\mathbf{w}))^2 + (IR_i(\mathbf{w}))^2\right)$$

where $\lambda = [\lambda_1, \ldots, \lambda_N]$ and $\lambda = [\lambda_1, \ldots, \lambda_N]$ are the vectors of Lagrange multipliers associated with the incentive compatibility constraints and the individual rationality constraints, respectively. ρ is the parameter that is used to adjust the weight on the penalty terms for the constraint violations.

Neural Networks for Auctions

We need to construct neural networks and train those networks to design the optimal auction. The optimal auction outputs (i) the winning probabilities of the users and (ii) the price decisions for the users. Thus, we need two neural networks: an *allocation*

10.3 Machine Learning for Optimal Auction

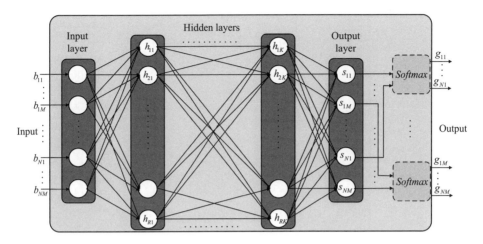

Figure 10.2 Allocation system.

system and a *pricing system* [192]. The allocation system and the pricing system are shown in Figures 10.2 and 10.3, respectively.

Allocation system: The allocation system determines the winning probabilities of the users for the computing resource units. The allocation system is proposed in [192] and described here (see Figure 10.2):

- The allocation system uses a neural network including one input layer, K hidden layers, and one output layer. For simplicity, we assume that the hidden layers have the same number of neural nodes, designated as R neural nodes.
- The input layer receives bid profile **b** of the users.
- The hidden layers use sigmoid activation functions to transform the bid profile to the output layer.
- The output layer uses softmax activation functions that represent the competition among the users for the computing resource units.
- The allocation system produces M allocation vectors. Each allocation vector includes the allocation probabilities $g_{1j}(\mathbf{b}), \ldots, g_{Nj}(\mathbf{b})$ for resource unit j through a softmax activation function. Therefore, we have $\sum_{i=1}^{N} g_{ij}(\mathbf{b}) \leq 1$, where $g_{ij}(\mathbf{b})$ refers to the probability that user i wins resource unit j given the input profile **b**.
- Let h_{jk} denote the linear function of neural node j in layer k, and let $\mathbf{h}_k$ denote the vector of linear functions in layer k: $\mathbf{h}_k = (h_{1k}, \ldots, h_{Rk})$.
- The linear functions in hidden layer 1 of the allocation system are defined as follows:

$$h_{j1} = \sigma(\mathbf{w}_{j1}^g \mathbf{b}), \forall j = 1, \ldots, R \quad (10.6)$$

where $\mathbf{w}_{j1}^g$ contains weights corresponding to layer 1, and $\sigma(x) = \frac{1}{(1+e^{-z})}$ is the sigmoid function.

- The linear functions in each hidden layer $k, k = 2, \ldots, K$, in the allocation system are defined as follows:

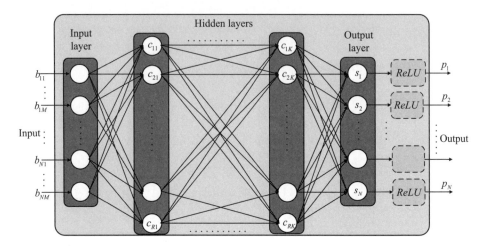

Figure 10.3 Pricing system.

$$h_{jk} = \sigma(\mathbf{w}^g_{jk}\mathbf{h}_{k-1}), \forall k = 2, \ldots, K, j = 1, \ldots, R \quad (10.7)$$

where $\mathbf{w}^g_{jk}$ consists of weights corresponding to hidden layer k.
- In the output layer, let $\mathbf{s} = \{s_{ij}, \forall i = 1, \ldots, N, j = 1, \ldots, M\}$. $\mathbf{s}$ is determined as follows:

$$\mathbf{s} = \mathbf{w}^g_{K+1}\mathbf{h}_K \quad (10.8)$$

where $\mathbf{w}_{K+1}$ consists of weights corresponding to the output layer. Let $\mathbf{w}^g$ denote the matrix that contains all the weights of the allocation system. Then, the output of the allocation system is determined by [192]

$$g_{ij}(\mathbf{b}, \mathbf{w}^g) = softmax_i(s_{1j}, \ldots, s_{Nj}) \quad (10.9)$$

$$= \frac{e^{s_{ij}}}{\sum_{n=1}^{N} e^{s_{nj}}}, \forall i = 1, \ldots, N, j = 1, \ldots, M$$

where g_{ij} represents the probability that user i wins resource unit j.

Pricing system: The pricing system uses a neural network with multiple layers, as illustrated in Figure 10.3. Note that the neural network in the allocation system is independent from that in the pricing system. Also, the number of hidden layers and the number of neural nodes in each hidden layer in the pricing system may differ from those in the allocation system. However, for expression simplicity, we assume that the number of hidden layers in the pricing system is K, and the number of neural nodes in each hidden layer is R. The output of the pricing system is a vector of price decisions. The vector is represented by $\mathbf{p} = (p_1, \ldots, p_N)$, where p_i is the price that user i pays the service provider if it wins the computing resource units. To ensure non-negative prices, the output layer in the pricing system uses rectifier activation functions [192], indicated by *ReLU*s in Figure 10.3. The parameters of the pricing system are determined as follows [192]:

$$c_{j1} = \sigma(\mathbf{w}^p_{j1}\mathbf{b}), \forall j = 1, \ldots, R \quad (10.10)$$

10.3 Machine Learning for Optimal Auction

$$c_{jk} = \sigma(\mathbf{w}_{jk}^p \mathbf{c}_{k-1}), \forall k = 2, \ldots, K, j = 1, \ldots, R$$

$$\mathbf{s} = \mathbf{w}_{K+1}^p \mathbf{c}_K$$

where $\mathbf{w}_{jk}^p$ contains weights corresponding to layer k. Let $\mathbf{w}^p$ denote the matrix that contains all the weights of the neural network in the pricing system. The output of the pricing system is determined by [192]

$$p_i(\mathbf{b}, \mathbf{w}^p) = ReLU(s_i), \forall i = 1, \ldots, N \tag{10.11}$$

where $ReLU(x) = \max(0, x)$ that guarantees the non-negative prices.

Training process: The training process uses a data set to find the weight parameters of the neural networks that optimize the loss function. In particular, the training process determines parameters $(\mathbf{w}^g, \mathbf{w}^p)$ of the allocation and pricing systems, respectively, to minimize the loss function.

We use the loss function $\hat{\mathcal{L}}(\cdot)$ for both the allocation and pricing systems as defined in (10.5). However, since there are two neural networks, the loss function is written as [192]

$$\hat{\mathcal{L}}(\lambda, \lambda', \mathbf{w}^g, \mathbf{w}^p) = -\left[\sum_i p_i(\mathbf{v}, \mathbf{w}^g, \mathbf{w}^p)\right] + \sum_{i=1}^N \lambda_i IC_i(\mathbf{w}^g, \mathbf{w}^p) \tag{10.12}$$

$$+ \sum_{i=1}^N \lambda_i' IR_i(\mathbf{w}^g, \mathbf{w}^p) + \frac{\rho}{2}\sum_{i=1}^N \left((IC_i(\mathbf{w}^g, \mathbf{w}^p))^2 + (IR_i(\mathbf{w}^g, \mathbf{w}^p))^2\right)$$

where $\mathbf{v}$ is the valuation profile that is a vector of values (i.e., truthful bids) of the computing resource units to the users. The valuation profiles are used as the data set to train the allocation and pricing systems.

The training algorithm for the optimal multi-item auction is shown in Algorithm 11. Note that the service provider generally has no information about the values of the users. However, the service provider can know the distributions of the values of the users, perhaps by observation, in a specific application of fog computing, presented in the next subsection. In particular, we introduce the fog computing resource market that supports the mobile blockchain network. Then, we discuss how to obtain the values of the computing resource units to the users. Finally, we provide and discuss numerical results to show the performance improvement obtained through the use of the optimal auction.

Algorithm 11 Deep learning algorithm for multi-item auction.

Input: $N, M, K, R, \mathbf{v}, \rho$
Output: Allocation probabilities ($\mathbf{g}$) and conditional prices ($\mathbf{p}$)
 1: Initialize: $\mathbf{w}^g, \mathbf{w}^p, \lambda, \lambda'$
 2: **repeat**
 3:    Compute $g_{ij}(\mathbf{v}, \mathbf{w}^g), \forall i = 1, \ldots, N, j = 1, \ldots, M$, according to (10.9)
 4:    Compute $p_i(\mathbf{v}, \mathbf{w}^p), \forall i = 1, \ldots, N$ according to (10.11)
 5:    Compute $\hat{\mathcal{L}}(\lambda, \lambda', \mathbf{w}^g, \mathbf{w}^p)$ according to (10.5)
 6:    Update $(\mathbf{w}^g, \mathbf{w}^p)$ using the stochastic gradient descent solver
 7: **until** The loss function $\hat{\mathcal{L}}(\lambda, \lambda', \mathbf{w}^g, \mathbf{w}^p)$ reaches its minimum

10.3.2 Example

In this section, we consider a specific application of fog computing to the mobile blockchain network [294]. We consider the application due to the following reasons:

- The blockchain network is a promising application that has gained enormous popularity within business, government, and academia. Built for Bitcoin, blockchain is a decentralized, peer-to-peer (P2P) data management framework [295] that can maintain high data security, integrity, scalability, and extensibility.
- Currently, deployment of blockchain in mobile environments faces a series of critical challenges. The reason is that the mining process, which entails solving the proof-of-work (PoW) puzzle, requires high computing power and energy from mobile devices [296].
- Fog computing leverages resources of devices, called *fog* nodes, at the edge of the network. This enables fog computing to provide computing, storage, and offloading of services with low latency, low bandwidth consumption cost, flexibility, scalability, security, and mobility. In essence, fog computing is a promising solution that allows the mining task of mobile devices to be offloaded to the service provider [193], [297].

Fog Computing Resource Market

The fog computing resource market supporting the blockchain network is shown in Figure 10.4 and described here:

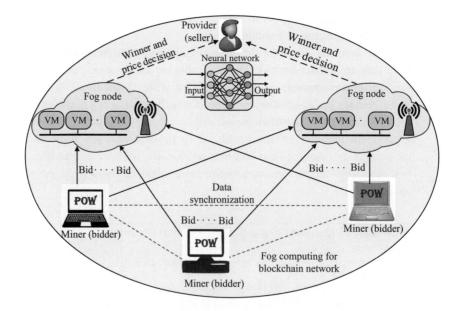

Figure 10.4 Fog computing for blockchain networks [193].

10.3 Machine Learning for Optimal Auction

- There are one service provider and N mobile devices. Here, the mobile devices can be laptops or any lightweight devices. The mobile devices are miners in the blockchain network.
- The service provider deploys near-to-end computing devices, known as fog nodes, across the blockchain network to provide nearby computing resource units to the miners.
- Each miner can buy multiple computing resource units at the fog nodes to support its solving of PoW puzzles. The miners have to compete for a limited number of computing resource units in the market.
- The service provider conducts a multi-item auction for trading the computing resource units. The service provider is the seller and auctioneer, and the miners are the bidders.
- Each miner submits bids to the service provider. The bids refer to prices that the miner is willing to pay for the computing resource units. The price is zero, meaning that the miner is not willing to buy the computing resource unit. For example, miner i submits bid b_{ij} for computing resource unit j. $b_{ij} = 0$ implies that the miner is not willing to buy computing resource unit j.
- Upon receiving the bids, the service provider determines the winners and the prices that the winners need to pay.
- The objective is to maximize the service provider's revenue while guaranteeing individual rationality and incentive compatibility.
- To achieve the objective, the service provider adopts the allocation system and pricing system described in Section 10.3.1 to determine the winners and the prices for the winning miners. For this purpose, the allocation and pricing systems need to be trained based on the data set that can be constructed from the miners' values for the resource computing units.
- To estimate the values of the resource computing units to the miners and then construct the data set, the service provider can use some information such as the mining rewards and the size of the block. The details are well presented in [297], [298], and [299]. In general, the value of the resource computing unit to miner i is obtained as follows:

$$v_i = (T + rs_i)e^{-\frac{1}{\tau}\epsilon s_i} \quad (10.13)$$

where

- T is the fixed reward for mining a new block.
- r is the transaction fee rate.
- s_i is the size of the block, equivalent to the number of transactions in the block.
- ϵ is a constant that indicates the impact of the size of block on the miner's block propagation time.
- τ is the average time of mining a block.

Typically, the parameters T, r, ϵ, and τ are publicly known in the network. The size of the block is privately chosen by the miners. However, the service provider can know its distribution, for example, by observation. Note that we consider the symmetric auction model in which the size of block in (10.13) for all miners is sampled independently and identically distributed from a known distribution function. The distribution function may be uniform $U[s_{min}; s_{max}]$, where s_{min} and s_{max} are the minimum and maximum values of the size of block, respectively.

Numerical Examples

We provide some experimental results to demonstrate that the deep learning–based auction can be efficiently used for resource allocation in fog computing to support blockchain networks. To implement the deep learning Algorithm 11, the TensorFlow deep learning library [300] is used. The performance of the deep learning–based auction scheme is measured in terms of (i) expected revenue of the service provider, (ii) expected individual rationality violation, and (iii) incentive compatibility violation. For comparison purposes, the greedy algorithm [301] is used as the baseline scheme. In the greedy algorithm, the service provider first sorts the miners in descending order of their bids. Then, it iteratively selects the miners with the highest bids as the winners and allocates computing resource units to the winners. The selection and allocation continue until no computing resource units are left. The winners pay the service provider prices equal to their bids.

The parameters of the fog computing system and the deep learning model are set as follows:

- For ease of presenting the findings, we consider a blockchain network with $N = 5$ miners and $M = 3$ computing resource units.
- The neural networks in the allocation and pricing systems have $K = 2$ hidden layers. The number of neural nodes in each hidden layer is 20.
- The L_2 regularization can be used in the training process to ensure that the weight parameters of the neural networks are bounded.
- The learning rate is set to 0.001. The Adam optimizer is used, as it allows us to adjust the learning rate during the training process.
- The data set consists of 5,000 valuation profiles. Each valuation profile is a vector of values of miners. The element, or valuation, of the vector is determined according to the expression given in (10.13) where
    - The fixed reward for mining a new block is $T = 2.0$.
    - The distribution of size s_i of the block follows the uniform distribution between 0 and 500, denoted by $U[0; 500]$.
    - The average time of mining a block is $\tau = 600$ seconds.
    - The transaction fee rate is $r = 0.007$.
    - The constant ϵ is 1.

It is important to highlight that the values of λ_i and λ'_i in the loss function given in (10.5) affect the revenue and the individual rationality and incentive compatibility

10.3 Machine Learning for Optimal Auction

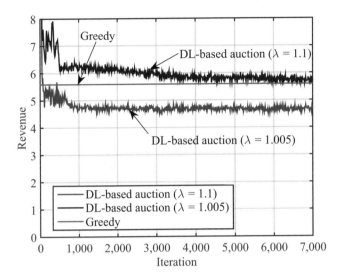

Figure 10.5 Test for revenue. DL stands for deep learning.

violations obtained by the deep learning-based auction. Here, we let $\lambda_i = \lambda'_i = \lambda$. In general, the small values of λ allow the deep learning–based auction to deliver high revenue for the provider, but the individual rationality and incentive compatibility violations may be large. Also, the large values of λ yield low revenue, but small individual rationality and incentive compatibility violations. λ is thus considered to be the trade-off parameter to be varied in the performance evaluation. Specifically, we choose two trade-off parameters of 1.1 and 1.005. The simulation results for the revenue versus the number of iterations are shown in Figure 10.5, and those for individual rationality and incentive compatibility violations versus the number of iterations are illustrated in Figures 10.6 and 10.7, respectively.

As shown in Figure 10.5, the deep learning–based auction converges to different values of revenue. Specifically, when the trade-off parameter is 1.1, the revenue reaches 4.7 while that obtained by the greedy scheme is 5.58. There is the gap between the deep learning–based auction and the greedy scheme since the deep learning–based scheme guarantees small individual rationality and incentive compatibility violations while the greedy scheme incurs large individual rationality and incentive compatibility violations. As we reduce the trade-off parameter to 1.005, the deep learning–based auction significantly improves the revenue of the provider from 4.7 to 5.75. The improved revenue of the deep learning–based auction is higher than that of the greedy scheme. However, this increases the individual rationality and incentive compatibility violations of the deep learning–based auction, as shown in Figures 10.6 and 10.7.

As shown in Figure 10.6, the incentive compatibility violation of the deep learning–based auction is not significantly affected when the trade-off parameters change. The reason can be that the incentive compatibility violation of a miner is defined as the difference between two utility functions of the miner. As the trade-off parameters change, both utility functions change (i.e., increase or decrease) together, and thus the difference

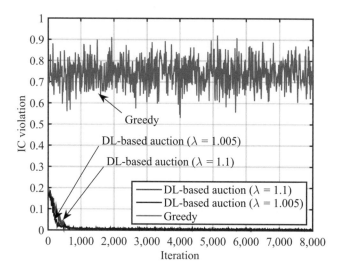

Figure 10.6 IC violation. IC stands for incentive compatibility.

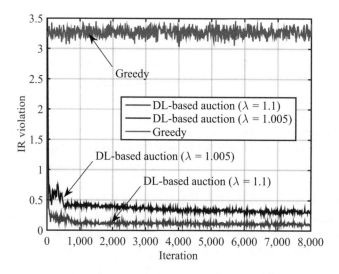

Figure 10.7 IR violation. IR stands for individual rationality.

between the utility functions (i.e., the incentive compatibility violation) seems not to change. Specifically, the incentive compatibility violation of the deep learning–based scheme only increases from 0.0055 to 0.006 as the trade-off parameter decreases from 1.1 to 1.005. For the greedy scheme, the incentive compatibility violation is around 0.77. Clearly, the incentive compatibility violation of the proposed deep learning–based auction is much lower than that of the greedy scheme. Recall that an incentive compatibility violation that is close to zero means that the incentive compatibility, or truthfulness, is guaranteed; that is, the miners have no incentive to offer bids that deviate from their truthful valuations. Therefore, the proposed deep learning–based auction outperforms the greedy scheme in terms of guaranteeing truthfulness.

Next, we evaluate the performance in terms of individual rationality violations. For the greedy scheme, even if the miners win the auction, such that their winning probabilities are 1, they are charged equal to their bids, meaning that their utilities are zero. However, the winning probabilities are generally less than 1, so the individual rationality violation of the greedy scheme is expected to be large. As shown in Figure 10.7, the individual rationality violation for the greedy scheme is around 3.2, while the corresponding violations for the deep learning–based auction reach 0.09 and 0.3 for the trade-off parameters of 1.1 and 1.005, respectively. Apparently, the individual rationality violation of our proposed scheme is much lower than that of the greedy scheme. Note that the small value of the individual rationality violation means that the utilities of the miners have a lower chance to be negative, and thus the miners have more incentive to participate in the auction; in other words, the individual rationality is guaranteed.

From Figures 10.6 and 10.7, we can also observe that the individual rationality violation of the deep learning–based auction is more sensitive to a change in the trade-off parameters compared with the incentive compatibility violation. The reason comes from the definitions of the revenue and the individual rationality violation. Both revenue and individual rationality violations are proportional to the prices that the provider receives from the miners. While we vary the trade-off parameters to increase the prices, thereby improving the revenue, the individual rationality violation also increases. This may offer the miners less incentive to participate in the auction. Thus, choosing the trade-off parameters to guarantee an acceptable individual rationality violation is important. In fact, the provider can first determine the acceptable individual rationality violation and then choose the trade-off parameters. The acceptable individual rationality violation is determined by, for example, measuring how much negative utility that the miners can "suffer."

The aforementioned results confirm that deep learning can be used for the multi-item auction to solve the optimization problem (i.e., revenue maximization) with multiple constraints, such as the individual rationality and incentive compatibility guarantees. In the next section, we show that deep learning is also able to recover the well-known Myerson auction, referring to the optimal auction based on the Myerson theorem [194]. In particular, we can use Myerson's characterization results to construct neural networks for the optimal auction. This approach is proposed in [192].

10.4 Machine Learning for Myerson Auction

The Myerson auction is used as an analytical solution of the optimal single-item auction that maximizes the revenue and guarantees two desired properties, individual rationality and incentive compatibility. In this section, we discuss how to construct a neural network architecture based on Myerson's characterization results for the optimal single-item auction. First, we present an overview of the Myerson auction. Based on Myerson's characterization results, we explain how to construct the neural networks for (i) the monotone transform functions, (ii) the allocation rule, and (iii) the pricing rule. Finally, we show experimental results in the scenario of a blockchain network to confirm the efficiency of using deep learning for deriving the optimal auction.

10.4.1 Design

Myerson Auction

We consider a fog computing resource market that includes one service provider and N users. Since the Myerson auction is applicable in the single-item setting, we consider a market with one computing resource unit. This means that N users (i.e., bidders) need to compete for the computing resource unit of the service provider.

The key idea of the Myerson auction is to determine the allocation and pricing rules by using monotone transform functions. Therefore, we need to determine the monotone transform functions first. In general, the monotone transform functions convert the input bids into *transformed bids* as follows:

- Let b_i denote the bid of user i for the computing resource unit.
- Let $\phi_i(b_i)$ denote the monotone transform function of the user.
- Let $\overline{b}_i$ be the *transformed bid* of the user. Then, $\overline{b}_i$ is defined as follows [192]:

$$\overline{b}_i = \phi_i(b_i), i = 1, \ldots, N \quad (10.14)$$

Based on the transformed bids $\overline{b}_i, i = 1, \ldots, N$, the Myerson auction adopts the Vickrey auction with a reserve price [302] to determine the allocation and conditional pricing rules for the users. In particular, the reserve price refers to the lowest price that is considered acceptable by the service provider. The reason for using the Vickrey auction with a reserve price is to guarantee that (i) the auction provides incentive compatibility, and (ii) the revenue of the service provider is not too low. Let g_i^0 denote the winning probability of user i based on the Vickrey auction with a reserve price. Also, let p_i^0 denote the price decision for the user based on the Vickrey auction with a reserve price. Note that g_i^0 and p_i^0 are functions of the transformed bids $\overline{b}_i, i = 1, \ldots, N$. Then, the Myerson auction determines the allocation rule and the conditional pricing rule as follows:

- The allocation rule for user i, denoted by g_i, is given by

$$g_i = g_i^0 \circ \phi_i \quad (10.15)$$

- The conditional pricing rule for user i, denoted by p_i, is

$$p_i = \phi_i^{-1} \circ p_i^0 \quad (10.16)$$

We have the Myerson theorem as follows:

THEOREM 10.1 *[194] For any set of strictly monotonically increasing functions $\phi_1, \ldots, \phi_N : \mathbb{R}_{\geq 0} \mapsto \mathbb{R}_{\geq 0}$, an auction that is defined by the allocation rule given in*

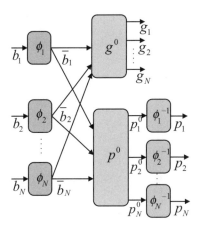

Figure 10.8 Neural network architecture for Myerson auction [193].

(10.15) and the conditional pricing rule given in (10.16) is incentive compatible and individual rational.

Theorem 10.1 means that if we design a mechanism with allocation and conditional pricing rules, g_i and p_i, then the mechanism guarantees both incentive compatibility and individual rationality [192]. This means that we can use Theorem 10.1 to design the neural network architecture to learn the auction. As such, the auction learned by the neural network is incentive compatible and individual rational. Figure 10.8 shows the neural network architecture. The neural network architecture performs (i) the monotone transform functions ϕ_i, (ii) the allocation rule g_i, and (iii) the conditional pricing rule p_i. The next sections describe the necessary steps.

Monotone Transform Functions

Monotone transform functions $\phi_i, i = 1, \ldots, N$, map bids $b_i, i = 1, \ldots, N$, of the users to their transformed bids $\bar{b}_i, i = 1, \ldots, N$. Figure 10.9 shows a feed-forward neural network that constructs the transform function ϕ_i. The parameters of the neural network are determined as follows [192]:

- The neural network consists of K groups.
- Each group k has J linear functions.
- The linear function j in group k is denoted as $h_{kj}(b_i)$ and is defined as

$$h_{kj}(b_i) = w_{kj}^i b_i + \beta_{kj}^i \qquad (10.17)$$

where $k = 1, \ldots, K$, $j = 1, \ldots, J$, $w_{kj}^i \in R_{>0}$ and β_{kj}^i are the weights and bias, respectively.

- The transform function ϕ_i is defined as follows:

$$\phi_i(b_i) = \min_{k=1,\ldots,K} \max_{j=1,\ldots,J} (w_{kj}^i b_i + \beta_{kj}^i) \qquad (10.18)$$

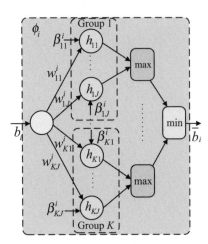

Figure 10.9 Neural network for monotone transformation ϕ_i [193].

- The inverse transform ϕ_i^{-1} can be deduced from the parameters for the forward transform as follows:

$$\phi_i^{-1}(y) = \max_{k=1,\ldots,K} \min_{j=1,\ldots,J} (w_{kj}^i)^{-1}(y - \beta_{kj}^i) \qquad (10.19)$$

In fact, the neural network architecture shown in Figure 10.8 is similar to a general auto-encoder neural network including the encoder and the decoder [192]. The encoder transforms the input bids to a different representation, the transformed bids, by using (10.18); the decoder then inverts the transformed bids by using (10.19).

Allocation Rule

After the transformed bids are determined, the allocation rule is implemented to determine the winning probabilities of the users. In general, the allocation rule is based on the allocation rule of the second-price auction. In particular, the user with the highest transformed bid has the highest probability of winning the computing resource unit. Thus, the neural network for the allocation rule maps the input, consisting of the transformed bids $\bar{\mathbf{b}} = (\bar{b}_1, \ldots, \bar{b}_N)$, to the output, which comprises the allocation probabilities $\mathbf{g} = (g_1, \ldots, g_N)$. Therefore, we construct the neural network as shown in Figure 10.10. Since the market has one computing resource unit, there is competition among the users. Thus, the softmax function should be used at the output of the neural network. The allocation probabilities of the users are determined as follows [192]:

$$g_i(\bar{\mathbf{b}}) = softmax_i(\bar{b}_1, \ldots, \bar{b}_{N+1}; \kappa) = \frac{e^{\kappa \bar{b}_i}}{\sum_{j=1}^{N+1} e^{\kappa \bar{b}_j}}, \forall i \in N \qquad (10.20)$$

where $\bar{b}_{N+1} = 0$ is an additional dummy input, and $\kappa > 0$ is the parameter that determines the quality of the approximation. In particular, the accuracy of the approximation increases as the value of κ becomes higher. However, the allocation function may be less smooth and becomes more difficult to optimize [192].

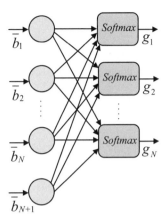

Figure 10.10 Illustration of allocation rule **g** [193].

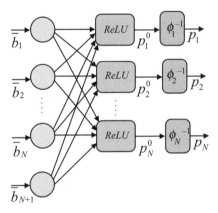

Figure 10.11 Illustration of pricing rule **p** [193].

In summary, the neural network for the allocation rule consists of (i) a layer of linear functions, (ii) a layer of max–min operations, and (iii) a layer of softmax activation functions.

Pricing Rule

The pricing rule is to determine price p_i to user i given the winning probability of the user. This pricing rule is sometimes called a *conditional pricing rule*. The neural network for the pricing rule is shown in Figure 10.11. This rule uses the transformed bids to determine the prices for the users. In particular, it is implemented as follows [192]:

- First, the pricing rule calculates the price p_i^0 to each user i by using the Vickrey auction with a reserve price. Here, the reserve price is zero, so p_i^0 is determined as the maximum of the transformed bids from the other users and zero. Therefore, p_i^0 can be determined using a $ReLU$ activation unit as follows:

$$p_i^0(\bar{\mathbf{b}}) = ReLU(\max_{j \neq i} \bar{b}_j), \forall i \in N \quad (10.21)$$

where $ReLU(z) = \max(z, 0)$ that ensures the non-negative payments.

- Second, the pricing rule determines price p_i using Theorem 10.1 as follows:

$$p_i = \phi_i^{-1} \circ p_i^0(\bar{\mathbf{b}}) = \phi_i^{-1}(p_i^0(\bar{\mathbf{b}})) \quad (10.22)$$

where $\phi_i^{-1}(y)$ is determined according to (10.19).

As such, the neural network for the pricing rule includes (i) a layer of linear functions, (ii) a layer of max–min operations, (iii) a layer of $ReLU$ activation functions, (iv) another layer of linear functions, and (v) another layer of min–max operations.

Neural Network Training

We need to train the neural networks constructed for the allocation and pricing rules to derive the optimal auction. For this purpose, we optimize the weights and bias of the neural networks to minimize the loss function based on data training. The weights and the bias are w_{kj}^i and β_{kj}^i, where $i = 1, \ldots, N$, $j = 1, \ldots, J$, and $k = 1, \ldots, K$. The training data and loss function are defined as follows [192]:

- *Training data:* The training data consists of bidder valuation profiles of the users. Similar to the general auction design case, the bidder values of the users are sampled independently and identically distributed from a distribution function. The distribution function can be determined depending on a specific scenario that is discussed in the next section. Let $\mathbf{v}^s = (v_1^s, \ldots, v_N^s)$ be the bidder valuation profile s of the miners, where $s = 1, \ldots, S$; S is the size of the training data; and v_i^s is the value of the computing resource unit to user i, $v_i^s \sim f_V(v)$.
- *Loss function:* Let $g_i^{(\mathbf{w},\boldsymbol{\beta})}(\mathbf{v}^s)$ and $p_i^{(\mathbf{w},\boldsymbol{\beta})}(\mathbf{v}^s)$ denote the allocation probability and conditional pricing of miner i, respectively. $\mathbf{w}$ and $\boldsymbol{\beta}$ are the matrices containing the weights of the neural network. The objective is to maximize the expected revenue of the service provider. Note that we do not mention the individual rationality and incentive compatibility as the constraints of the objective, since the definitions of the allocation rule $\mathbf{g}$ and the pricing rule $\mathbf{p}$ include the constraints. The loss function is defined as the expected, negated revenue [192]:

$$\hat{L}(\mathbf{w}, \boldsymbol{\beta}) = -\sum_{i=1}^{N} g_i^{(\mathbf{w},\boldsymbol{\beta})}(\mathbf{v}^s) p_i^{(\mathbf{w},\boldsymbol{\beta})}(\mathbf{v}^s) \quad (10.23)$$

We optimize the loss function $\hat{L}(\mathbf{w}, \boldsymbol{\beta})$ in (10.23) over parameters $(\mathbf{w}, \boldsymbol{\beta})$ using the stochastic gradient descent solver. To do so, we need the training data discussed in the next section. The algorithm is shown in Algorithm 12.

10.4.2 Example

In this section, we present how to obtain the bidder valuation profiles, for the purposes of the data training, in the scenario of a blockchain network. Then, we show experimental

10.4 Machine Learning for Myerson Auction

Algorithm 12 Deep learning algorithm for single-item auction based on Myerson's results [193].

Input: $N, S, J, K, \kappa, f_T(t), f_C(c), \mathbf{v}^s = (v_1^s, \ldots, v_N^s), \bar{v}_{N+1}^s$
Output: Allocation probabilities $(g_1, \ldots, g_N)$ and conditional prices $(p_1, \ldots, p_N)$
1: Initialize: weights **w** and bias β
2: **repeat**
3:     Determine $\bar{v}_i^s = \phi_i(v_i^s)$ according to (10.18)
4:     Calculate g_i according to (10.20)
5:     Determine p_i^0 according to (10.21)
6:     Determine $p_i = \phi_i^{-1}(p_i^0)$ according to (10.19)
7:     Compute $\hat{L}(\mathbf{w}, \beta)$ according to (10.23)
8:     Update $(\mathbf{w}, \beta)$ using the stochastic gradient descent solver.
9: **until** The loss function $\hat{L}(\mathbf{w}, \beta)$ minimizes

results of applying the deep learning–based auction to resource management in the blockchain network.

Bidder Valuation Profile

We consider again the fog computing system for blockchain networks as shown in Figure 10.4 with one computing resource unit, $M = 1$. In general, we can determine the value of each miner as shown in (10.13). However, each miner has an initial capacity that may impact its willingness-to-buy and the value it assigns to the unit. Thus, the value of the miner should be changed. To determine the bidder valuation profile, $\mathbf{v}^s = (v_1^s, \ldots, v_N^s)$, we need to find the distribution function of value v, denoted by $f_V(v)$ [193]:

- Let c_i denote the initial computing capacity of miner i. As c_i is large, the miner can complete its block mining with the initial computing capacity. Thus, the miner may have no incentive or less incentive to buy the computing resource unit. As a result, the miner assigns a lower value to the computing resource unit, and it is willing to pay a lower price.
- Let s_i denote the size of a block; s_i is the number of transactions to be included in the block chosen by the miner. As s_i is large, the miner needs more computing resources to complete its block mining. Thus, the miner has more incentive to buy the computing resource unit of the service provider. As a result, the miner assigns a higher value to the computing resource unit, and it is willing to pay a higher price.
- The relation among the value v_i, the initial computing capacity c_i, and the size of the block s_i can be expressed as follows:

$$v_i^s = t_i/c_i \qquad (10.24)$$

- Based on (10.24), the distribution of value $f_V(v)$ can be determined according to the distribution of the size block t_i, denoted as $f_T(t)$, and the distribution of

Table 10.1 Simulation parameters [193].

Parameters	Values
Number of miners (N)	10 and 20
Training set size (S)	1,000 valuation profiles
Number of groups (K)	5
Number of linear functions (J)	10
Number of iterations	4,000
Approximate quality κ	1 and 2
Distribution of size of block $f_T(t)$	$\sim U[0;1]$
Distribution of initial capacity $f_C(c)$	$\sim U[0.2;0.5], U[0.4;0.7]$

the initial computing capacity c_i, denoted by $f_C(c)$. Here, $f_T(t)$ and $f_C(c)$ are available, as they are based on past observations.

- Assume that variables t and c are independent from each other and follow uniform distributions [108]: $t \sim U[t_{min};t_{max}]$ and $c \sim U[c_{min};c_{max}]$, $c_{min} > 0$.
- Then, by using the Jacobian determinant [303], the distribution function $f_V(v)$ is determined as follows [193]:

$$f_V(v) = \frac{c_{min} + c_{max}}{2(t_{max} - t_{min})} \qquad (10.25)$$

where v is within $[t_{min}/c_{max}; t_{max}/c_{min}]$.

Performance Evaluation

This section provides experimental results to confirm the benefits of the deep learning–based auction as presented in Section 10.4. For comparison purposes, the second price auction or Vickrey auction [221] is adopted as a baseline scheme. The deep learning–based auction is implemented by using the TensorFlow deep learning library. The simulation parameters for the mobile blockchain network and the deep learning model are shown in Table 10.1.

To evaluate the performance of the deep learning–based auction, different scenarios are considered by varying the number of miners and the distribution of the miners' initial capacity. Figures 10.12–10.14 show the performance in terms of revenue, and Figure 10.15 shows the winning probability of the miners as their initial capacities vary.

Revenue comparison: As shown in Figures 10.12 and 10.13, the deep learning–based auction is able to converge quickly to the stable values of revenue. Also, this auction outperforms the second-price auction scheme in terms of revenue for a given number of miners and distribution of initial capacity. For instance, given $N = 20$, $c_i \sim U[0.2;0.5]$, and $\kappa = 1$, the revenue obtained by the deep learning–based auction is 3.38, while the revenue obtained by the second-price auction scheme is 3.11. Also, given $N = 10$, $c_i \sim U[0.2;0.5]$, and $\kappa = 1$, the revenue obtained by the deep learning–based auction is 2.75, while the revenue obtained by the second-price auction scheme is 2.55. This confirms the effectiveness of the deep learning–based auction.

10.4 Machine Learning for Myerson Auction

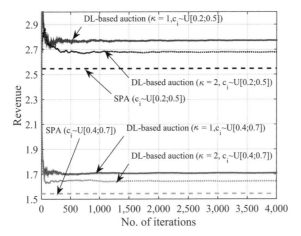

Figure 10.12 Test revenue for 10 miners [193]. DL stands for deep learning, and SPA stands for second-price auction or Vickrey auction.

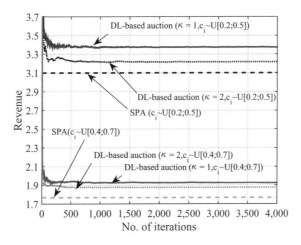

Figure 10.13 Test revenue for 20 miners [193]. DL stands for deep learning, and SPA stands for second-price auction or Vickrey auction.

Impact of the number of miners on the revenue: In the general case, as the number of buyers increases, the competition among the buyers increases. Thus, the prices that the buyers are willing to pay the seller are higher. As a result, the revenue that the seller receives is higher. This is confirmed by the simulation results shown in Figure 10.14. Specifically, the revenue values that the service provider receives are 2.8, 3.17, and 3.4 when the number of miners in the auction is 10, 15, and 20, respectively.

Impact of the initial capacity on the revenue: As shown in Figure 10.12 for the case of $N = 10$ miners, when c_i is within a small range, $c_i \sim U[0.2; 0.5]$, the expected revenue that the service provider receives increases compared with the large range,

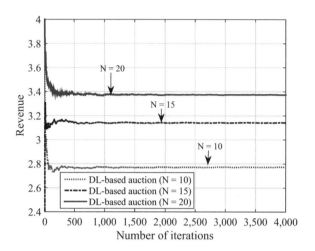

Figure 10.14 Test revenue for $c_i \sim U[0.2; 0.5]$ and $\kappa = 1$ [193].

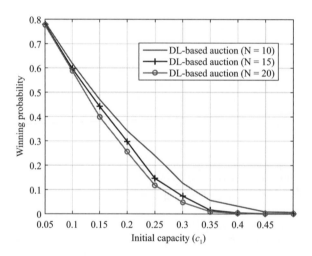

Figure 10.15 Winning probability of miner 1 versus initial capacity [193].

$c_i \sim U[0.4; 0.7]$. This occurs because the bids of miners are inversely proportional to the initial capacity. Thus, the bids are higher when the initial capacity is lower, which in turn improves the expected revenue of the service provider.

Impact of the initial capacity on the winning probability: Without loss of generality, we consider the winning probability of miner 1 as its initial capacity c_1 varies from 0.05 to 0.5. Note that the initial capacity of other miners is $c_i \sim U[0.2; 0.5]$, $i = 2, \ldots, N$. As shown in Figure 10.15, the winning probability of miner 1 decreases as c_1 increases. This occurs because the bid of the miner decreases as c_1 increases, which results in a reduced winning probability.

10.5 Summary

In this chapter, we introduce the design of optimal auctions using the deep learning technique. In particular, we first introduce the optimal auction design problems. Then, we present the deep learning technique and the motivations for its use for designing the optimal auctions. After that, we describe neural networks that are used to derive the optimal multi-item auction. In addition, we present the neural network architecture for recovering the Myerson auction. To demonstrate the efficiency of the deep learning–based auctions, we apply them to the resource management problem in the blockchain network. As illustrated in the simulation results, the deep learning–based auction approaches clearly outperform the baseline approaches.

References

[1] V. Krishna, *Auction Theory*. San Diego, CA: Academic Press, 2002.

[2] M. Shubik, *The Theory of Money and Financial Institutions: Volume 1*. Cambridge, MA: MIT Press, 2004.

[3] O. Hekster, *Rome and Its Empire, AD 193–284*. Edinburgh: Edinburgh University Press, 2008.

[4] About auctions. Federal Communications Commission [Online]. (Dec. 2017). Available: www.fcc.gov/auctions/about-auctions

[5] K. Binmore and P. Klemperer, "The biggest auction ever: The sale of the British 3G telecom licences," *Economic Journal*, vol. 112, no. 478, pp. 74–96, Apr. 2002.

[6] Results of auction. Ofcom [Online]. (Apr. 2018). Available: www.ofcom.org.uk

[7] The advantages of eBay for online sellers [Online]. (Jan. 2019). Available: www.channelreply.com/blog/view/advantages-of-ebay

[8] J. Gubbi, R. Buyya, S. Marusic, and M. Palaniswami, "Internet of Things (IoT): A vision, architectural elements, and future directions," *Future Generation Computer Systems*, vol. 29, no. 7, pp. 1645–1660, Sept. 2013.

[9] L. Brock, A naming scheme for physical objects [Online]. (2001). Available: http://cocoa.ethz.ch/downloads/2014/06/None-MIT-AUTOID-WH-002.pdf

[10] M. Roberto, B. Abyi, and R. Domenico, "Towards a definition of the Internet of Things (IoT)," *Technical Report*, 2015.

[11] L. Atzori, A. Iera, and G. Morabito, "The Internet of Things: A survey," *Computer Networks*, vol. 54, no. 15, pp. 2787–2805, Oct. 2010.

[12] Ibid.

[13] L. Zhou, A. Chong, and E. Ngai, "Supply chain management in the era of the Internet of Things," *International Journal of Production Economics*, vol. 159, no. 1, pp. 1–3, Oct. 2015.

[14] C. Sarkar, S. N. A. U. Nambi, R. V. Prasad, and A. Rahim, "A scalable distributed architecture towards unifying IoT applications," in *2014 IEEE World Forum on Internet of Things (WF-IoT)*, Seoul, Korea, Mar. 2014, pp. 508–513.

[15] E. Borgia, "The Internet of Things vision: Key features, applications and open issues," *Computer Communications*, vol. 54, pp. 1–31, Dec. 2014.

[16] N. C. Luong, D. T. Hoang, P. Wang, D. Niyato, D. I. Kim, and Z. Han, "Data collection and wireless communication in Internet of Things (IoT) using economic analysis and pricing models: A survey," *IEEE Communications Surveys & Tutorials*, vol. 18, no. 4, pp. 2546–2590, 2016.

[17] D. Uckelmann, M. Harrison, and F. Michahelles, "An architectural approach towards the future Internet of Things," in *Architecting the Internet of Things*. Berlin/Heidelberg: Springer, 2011, pp. 1–24.

[18] A. Gluhak, S. Krco, M. Nati, D. Pfisterer, N. Mitton, and T. Razafindralambo, "A survey on facilities for experimental Internet of Things research," *IEEE Communications Magazine*, vol. 49, no. 11, pp. 58–67, Nov. 2011.

[19] P. J. Nesse, S. Svaet, D. Strasunskas, and A. A. Gaivoronski, "Assessment and optimisation of business opportunities for telecom operators in the cloud value network," *Transactions on Emerging Telecommunications Technologies*, vol. 24, no. 5, pp. 503–516, June 2013.

[20] M. Sarvary and P. M. Parker, "Marketing information: A competitive analysis," *Marketing Science*, vol. 16, no. 1, pp. 24–38, Feb. 1997.

[21] Y. Feng, B. Li, and B. Li, "Price competition in an oligopoly market with multiple iaas cloud providers," *IEEE Transactions on Computers*, vol. 63, no. 1, pp. 59–73, Jan. 2014.

[22] D. Niyato, X. Lu, P. Wang, D. I. Kim, and Z. Han, "Economics of Internet of Things (IoT): An information market approach," *IEEE Wireless Communications*, vol. 23, no. 4, pp. 136–145, Aug. 2016.

[23] S. Bin, L. Yuan, and W. Xiaoyi, "Research on data mining models for the Internet of Things," in *IEEE International Conference on Image Analysis and Signal Processing (IASP)*, Zhejiang, China, Apr. 2010, pp. 127–132.

[24] Z. Chen, F. Xia, T. Huang, F. Bu, and H. Wang, "A localization method for the Internet of Things," *The Journal of Supercomputing*, vol. 63, no. 3, pp. 657–674, Mar. 2013.

[25] Location-based service revenues will grow to 34.8 billion euros in 2020. *IoT Business News* [Online]. (2015). Available: https://iotbusinessnews.com/2015/08/31/79312-location-based-service-revenues-will-grow-to-34-8-billion-euros-in-2020/

[26] R. Honicky, E. A. Brewer, E. Paulos, and R. White, "N-smarts: Networked suite of mobile atmospheric real-time sensors," in *Proceedings of the Second ACM SIGCOMM Workshop on Networked Systems for Developing Regions*, Seattle, WA, Aug. 2008, pp. 25–30.

[27] M. Ennis and R. K. Rowe, "Hygienic biometric sensors," July 2012, U.S. Patent 8,229,185.

[28] H. Gao, C. H. Liu, W. Wang, J. Zhao, Z. Song, X. Su, J. Crowcroft, and K. K. Leung, "A survey of incentive mechanisms for participatory sensing," *IEEE Communications Surveys & Tutorials*, vol. 17, no. 2, pp. 918–943, 2015.

[29] L. G. Jaimes, I. J. Vergara-Laurens, and A. Raij, "A survey of incentive techniques for mobile crowd sensing," *IEEE Journal Internet of Things*, vol. 2, no. 5, pp. 370–380, Oct. 2015.

[30] X. Zhang, Z. Yang, W. Sun, Y. Liu, S. Tang, K. Xing, and X. Mao, "Incentives for mobile crowd sensing: A survey," *IEEE Communications Surveys & Tutorials*, vol. 18, no. 1, pp. 54–67, 2016.

[31] R. K. Ganti, F. Ye, and H. Lei, "Mobile crowdsensing: Current state and future challenges," *IEEE Communications Magazine*, vol. 49, no. 11, pp. 32–39, Nov. 2011.

[32] J. Liu, H. Shen, H. S. Narman, W. Chung, and Z. Lin, "A survey of mobile crowdsensing techniques: A critical component for the Internet of Things," *ACM Transactions on Cyber-Physical Systems*, vol. 2, no. 3, pp. 18:1–18:26, July 2018.

[33] N. C. Luong, D. T. Hoang, P. Wang, D. Niyato, and Z. Han, "Applications of economic and pricing models for wireless network security: A survey," *IEEE Communications Surveys & Tutorials*, vol. 19, no. 4, pp. 2735–2767, 2017.

[34] N. Sultan, "Cloud computing for education: A new dawn?" *International Journal of Information Management*, vol. 30, no. 2, pp. 109–116, Apr. 2010.

[35] M. R. Rahimi, J. Ren, C. H. Liu, A. V. Vasilakos, and N. Venkatasubramanian, "Mobile cloud computing: A survey, state of art and future directions," *Mobile Networks and Applications*, vol. 19, no. 2, pp. 133–143, Apr. 2014.

References

[36] M.-H. Kuo, "Opportunities and challenges of cloud computing to improve health care services," *Journal of Medical Internet Research*, vol. 13, no. 3, pp. 67–89, Sept. 2011.

[37] Z. Li, C. Chen, and K. Wang, "Cloud computing for agent-based urban transportation systems," *IEEE Intelligent Systems*, vol. 26, no. 1, pp. 73–79, Feb. 2011.

[38] K. Chard, S. Caton, O. F. Rana, and K. Bubendorfer, "Social cloud: Cloud computing in social networks," in *International Conference on Cloud Computing*, Miami, FL, July 2010, pp. 99–106.

[39] G. Lawton, "Developing software online with platform-as-a-service technology," *Computer*, vol. 41, no. 6, pp. 13–15, June 2008.

[40] S. Bhardwaj, L. Jain, and S. Jain, "Cloud computing: A study of infrastructure as a service (IaaS)," *International Journal of Engineering and Information Technology*, vol. 2, no. 1, pp. 60–63, 2010.

[41] A. Benlian and T. Hess, "Opportunities and risks of software-as-a-service: Findings from a survey of IT Executives," *Decision Support Systems*, vol. 52, no. 1, pp. 232–246, Dec. 2011.

[42] N. C. Luong, P. Wang, D. Niyato, Y. Wen, and Z. Han, "Resource management in cloud networking using economic analysis and pricing models: A survey," *IEEE Communications Surveys & Tutorials*, vol. 19, no. 2, pp. 954–1001, 2017.

[43] Q. Duan, Y. Yan, and A. V. Vasilakos, "A survey on service-oriented network virtualization toward convergence of networking and cloud computing," *IEEE Transactions on Network and Service Management*, vol. 9, no. 4, pp. 373–392, Dec. 2012.

[44] P. Murray, A. Sefidcon, R. Steinert, V. Fusenig, and J. Carapinha, "Cloud networking: An infrastructure service architecture for the wide area," in *Future Network & Mobile Summit (FutureNetw)*, Berlin, Germany, July 2012.

[45] X. Xiang, C. Lin, F. Chen, and X. Chen, "Greening geo-distributed data centers by joint optimization of request routing and virtual machine scheduling," in *Proceedings of the IEEE/ACM 7th International Conference on Utility and Cloud Computing*, London, UK, Dec. 2014.

[46] N. Bitar, S. Gringeri, and T. J. Xia, "Technologies and protocols for data center and cloud networking," *IEEE Communications Magazine*, vol. 51, no. 9, pp. 24–31, Sept. 2013.

[47] A. Levin and P. Massonet, "Enabling federated cloud networking," in *Proceedings of the 8th ACM International Systems and Storage Conference*. Haifa, Israel: ACM, May 2015, pp. 23–23.

[48] A. Jamakovic, T. M. Bohnert, and G. Karagiannis, *Mobile Cloud Networking: Mobile Network, Compute, and Storage as One Service On-Demand*. Berlin/Heidelberg: Springer, 2013.

[49] G. Karagiannis, A. Jamakovic, A. Edmonds, C. Parada, T. Metsch, D. Pichon, M. Corici, S. Ruffino, A. Gomes, P. S. Crosta, et al., "Mobile cloud networking: Virtualisation of cellular networks," in *21st International Conference on Telecommunications (ICT)*, Lisbon, Portugal, May 2014, pp. 410–415.

[50] G. Lewis, S. Echeverría, S. Simanta, B. Bradshaw, and J. Root, "Tactical cloudlets: Moving cloud computing to the edge," in *IEEE Military Communications Conference*, Baltimore, MD, Oct. 2014, pp. 1440–1446.

[51] I. Stojmenovic and S. Wen, "The fog computing paradigm: Scenarios and security issues," in *Federated Conference on Computer Science and Information Systems (FedCSIS)*, Warsaw, Poland, Sept. 2014.

[52] B. Ahlgren, P. A. Aranda, P. Chemouil, S. Oueslati, L. M. Correia, H. Karl, M. Söllner, and A. Welin, "Content, connectivity, and cloud: Ingredients for the network of the future," *IEEE Communications Magazine*, vol. 49, no. 7, pp. 62–70, June 2011.

[53] M. T. Beck, M. Werner, S. Feld, and S. Schimper, "Mobile edge computing: A taxonomy," in *Proceedings of the Sixth International Conference on Advances in Future Internet*, Lisbon, Portugal, Nov. 2014, pp. 1–7.

[54] J. E. Smith and R. Nair, "The architecture of virtual machines," *Computer*, vol. 38, no. 5, pp. 32–38, May 2005.

[55] Y. Yuan, C.-r. Wang, and C. Wang, "A game based approach for sharing the data center network," in *International Symposium on Neural Networks*, Shenyang, China, July 2012, pp. 641–649.

[56] G. Carella, A. Edmonds, F. Dudouet, M. Corici, B. Sousa, and Z. Yousaf, "Mobile cloud networking: From cloud, through NFV and beyond," in *IEEE Conference on Network Function Virtualization and Software Defined Network (NFV-SDN)*, San Francisco, CA, Nov. 2015, pp. 7–8.

[57] P. Rost, C. J. Bernardos, A. De Domenico, M. Di Girolamo, M. Lalam, A. Maeder, D. Sabella, and D. Wübben, "Cloud technologies for flexible 5G radio access networks," *IEEE Communications Magazine*, vol. 52, no. 5, pp. 68–76, Sept. 2014.

[58] M. Peng, C. Wang, V. Lau, and H. V. Poor, "Fronthaul-constrained cloud radio access networks: Insights and challenges," *IEEE Wireless Communications*, vol. 22, no. 2, pp. 152–160, Apr. 2015.

[59] A. Checko, H. L. Christiansen, Y. Yan, L. Scolari, G. Kardaras, M. S. Berger, and L. Dittmann, "Cloud ran for mobile networks: A technology overview," *IEEE Communications Surveys & Tutorials*, vol. 17, no. 1, pp. 405–426, 2015.

[60] A. Greenberg, J. Hamilton, D. A. Maltz, and P. Patel, "The cost of a cloud: Research problems in data center networks," *ACM SIGCOMM Computer Communication Review*, vol. 39, no. 1, pp. 68–73, Jan. 2008.

[61] P. Garcia Lopez, A. Montresor, D. Epema, A. Datta, T. Higashino, A. Iamnitchi, M. Barcellos, P. Felber, and E. Riviere, "Edge-centric computing: Vision and challenges," *ACM SIGCOMM Computer Communication Review*, vol. 45, no. 5, pp. 37–42, Oct. 2015.

[62] A. Ahmed and E. Ahmed, "A survey on mobile edge computing," in *Proceedings of the 10th IEEE International Conference on Intelligent Systems and Control (ISCO 2016)*, Coimbatore, India, Jan. 2016, pp. 1–8.

[63] W. Shi, J. Cao, Q. Zhang, Y. Li, and L. Xu, "Edge computing: Vision and challenges," *IEEE Internet of Things Journal*, vol. 3, no. 5, pp. 637–646, Oct. 2016.

[64] P. Mach and Z. Becvar, "Mobile edge computing: A survey on architecture and computation offloading," *IEEE Communications Surveys & Tutorials*, vol. 19, no. 3, pp. 1628–1656, 2017.

[65] Y. Mao, C. You, J. Zhang, K. Huang, and K. B. Letaief, "A survey on mobile edge computing: The communication perspective," *IEEE Communications Surveys & Tutorials*, vol. 19, no. 4, pp. 2322–2358, 2017.

[66] I. Newslog. IPTV standardization on track say industry experts [Online]. (2006). Available: www.itu.int/ITU-T/newslog/IPTV+Standardization+On+Track+Say+Industry+Experts.aspx

[67] K. A. Hua, Y. Cai, and S. Sheu, "Patching: A multicast technique for true video-on-demand services," in *Proceedings of the Sixth ACM International Conference on Multimedia*. Bristol, UK: ACM, Sept. 1998, pp. 191–200.

[68] P. Murray. Cloud networking architecture description [Online]. (2012). Available: www.sail-project.eu/wp-content/uploads/2011/09/SAIL_DD1_final_public.pdf

[69] Z. Liu, Q. Wang, J. Huang, Y. Wu, Y. Wang, X. Jia, and H. Chen, "Cloud-based video-on-demand services for smart TV," in *International Conference on Information Science and Technology (ICIST)*, Da Nang, Vietnam, Apr. 2017, pp. 81–84.

[70] Y. Wu, C. Wu, B. Li, X. Qiu, and F. Lau, "Cloudmedia: When cloud on demand meets video on demand," in *International Conference on Distributed Computing Systems (ICDCS)*, Minneapolis, MN, June 2011, pp. 268–277.

[71] M. Agiwal, A. Roy, and N. Saxena, "Next generation 5G wireless networks: A comprehensive survey," *IEEE Communications Surveys & Tutorials*, vol. 18, no. 3, pp. 1617–1655, 2016.

[72] IMT vision framework and overall objectives of the future development of IMT for 2020 and beyond. ITU [Online]. Available: www.itu.int/

[73] E. Hossain and M. Hasan, "5G cellular: Key enabling technologies and research challenges," *IEEE Instrumentation & Measurement Magazine*, vol. 18, no. 3, pp. 11–21, May 2015.

[74] N. C. Luong, P. Wang, D. Niyato, Y.-C. Liang, Z. Han, and F. Hou, "Applications of economic and pricing models for resource management in 5G wireless networks: A survey," *IEEE Communications Surveys & Tutorials*, vol: 21, no: 4, pp. 3298–3339, 2018.

[75] E. G. Larsson, O. Edfors, F. Tufvesson, and T. L. Marzetta, "Massive MIMO for next generation wireless systems," *IEEE Communications Magazine*, vol. 52, no. 2, pp. 186–195, Feb. 2014.

[76] M. Di Renzo, H. Haas, A. Ghrayeb, S. Sugiura, and L. Hanzo, "Spatial modulation for generalized MIMO: Challenges, opportunities, and implementation," *Proceedings of the IEEE*, vol. 102, no. 1, pp. 56–103, Jan. 2014.

[77] P. C. Cramton, Y. Shoham, R. Steinberg, et al., *Combinatorial Auctions*. Cambridge, MA: MIT Press, 2006.

[78] M. Sawahashi, Y. Kishiyama, A. Morimoto, D. Nishikawa, and M. Tanno, "Coordinated multipoint transmission/reception techniques for LTE-advanced [coordinated and distributed MIMO]," *IEEE Wireless Communications*, vol. 17, no. 3, pp. 26–34, June 2010.

[79] M. Kamel, W. Hamouda, and A. Youssef, "Ultra-dense networks: A survey," *IEEE Communications Surveys & Tutorials*, vol. 18, no. 4, pp. 2522–2545, May 2016.

[80] A. B. Sediq, R. H. Gohary, R. Schoenen, and H. Yanikomeroglu, "Optimal tradeoff between sum-rate efficiency and Jain's fairness index in resource allocation," *IEEE Transactions on Wireless Communications*, vol. 12, no. 7, pp. 3496–3509, June 2013.

[81] *IEEE 802.11ad. Part 11: Wireless LAN medium access control (MAC) and physical layer (PHY) specifications – amendment 3: Enhancements for very high throughput in the 60 GHz band*, Standard, Dec. 2012.

[82] M. Elkashlan, T. Q. Duong, and H.-H. Chen, "Millimeter-wave communications for 5G: Fundamentals: Part I [Guest editorial]," *IEEE Communications Magazine*, vol. 52, no. 9, pp. 52–54, Sept. 2014.

[83] Y. Niu, Y. Li, D. Jin, L. Su, and A. V. Vasilakos, "A survey of millimeter wave communications (mmWave) for 5G: Opportunities and challenges," *Wireless Networks*, vol. 21, no. 8, pp. 2657–2676, Nov. 2015.

[84] Y. Xu, G. Athanasiou, C. Fischione, and L. Tassiulas, "Distributed association control and relaying in millimeter wave wireless networks," in *IEEE ICC*, Kuala Lumpur, Malaysia, May 2016, pp. 1–6.

[85] Y. Xu, H. Shokri-Ghadikolaei, and C. Fischione, "Auction based dynamic distributed association in millimeter wave networks," in *IEEE GLOBECOM*, Washington, DC, Dec. 2016, pp. 1–6.

[86] Y. Zhang, C. Lee, D. Niyato, and P. Wang, "Auction approaches for resource allocation in wireless systems: A survey," *IEEE Communications Surveys & Tutorials*, vol. 15, no. 3, pp. 1020–1041, 2013.

[87] A. Asadi, Q. Wang, and V. Mancuso, "A survey on device-to-device communication in cellular networks," *IEEE Communications Surveys & Tutorials*, vol. 16, no. 4, pp. 1801–1819, 2014.

[88] *3rd Generation Partnership Project; Technical Specification Group Services and System Aspects; Service requirements for Machine-Type Communications (MTC); Stage 1 (Release 10)*, 3GPP TS 22.368 V11.2.0 Standard, 2011.

[89] F. Ghavimi and H.-H. Chen, "M2M communications in 3GPP LTE/LTE-A networks: Architectures, service requirements, challenges, and applications," *IEEE Communications Surveys & Tutorials*, vol. 17, no. 2, pp. 525–549, 2015.

[90] J. Kim, J. Lee, J. Kim, and J. Yun, "M2M service platforms: Survey, issues, and enabling technologies," *IEEE Communications Surveys & Tutorials*, vol. 16, no. 1, pp. 61–76, 2014.

[91] R. Ratasuk, A. Prasad, Z. Li, A. Ghosh, and M. A. Uusitalo, "Recent advancements in M2M communications in 4G networks and evolution towards 5G," in *18th International Conference on Intelligence in Next Generation Networks*, Paris, France, Feb. 2015, pp. 52–57.

[92] B. Pourghebleh and N. J. Navimipour, "Data aggregation mechanisms in the Internet of Things: A systematic review of the literature and recommendations for future research," *Journal of Network and Computer Applications*, vol. 97, pp. 23–34, Nov. 2017.

[93] D. L. Donoho, "Compressed sensing," *IEEE Transactions on Information Theory*, vol. 52, no. 4, pp. 1289–1306, Apr. 2006.

[94] N. Cao, S. Brahma, and P. K. Varshney, "An incentive-based mechanism for location estimation in wireless sensor networks," in *IEEE GlobalSIP*, Austin, TX, Dec. 2013, pp. 157–160.

[95] J.-S. Lee and B. Hoh, "Dynamic pricing incentive for participatory sensing," *Pervasive and Mobile Computing*, vol. 6, no. 6, pp. 693–708, Dec. 2010.

[96] T. Luo, H. P. Tan, and L. Xia, "Profit-maximizing incentive for participatory sensing," in *INFOCOM*, Toronto, ON, Apr. 2014, pp. 127–135.

[97] I. Krontiris and A. Albers, "Monetary incentives in participatory sensing using multi-attributive auctions," *International Journal of Parallel, Emergent and Distributed Systems*, vol. 27, no. 4, pp. 317–336, Aug. 2012.

[98] I. Koutsopoulos, "Optimal incentive-driven design of participatory sensing systems," in *IEEE INFOCOM*, Turin, Italy, Apr. 2013, pp. 1402–1410.

[99] K. Lan and H. Wang, "On providing incentives to collect road traffic information," in *International Wireless Communications & Mobile Computing Conference (IWCMC 13)*, 2013.

[100] T. Luo and C. K. Tham, "Fairness and social welfare in incentivizing participatory sensing," in *IEEE Communications Society Conference on Sensor, Mesh and Ad Hoc Communications and Networks (SECON)*, Seoul, South Korea, June 2012, pp. 425–433.

[101] H. N. Pham, B. S. Sim, and H. Y. Youn, "A novel approach for selecting the participants to collect data in participatory sensing," in *IEEE 11th International Symposium on Applications and the Internet (SAINT)*, Bavaria, Germany, July 2011, pp. 50–55.

[102] W. Nan, B. Guo, S. Huangfu, Z. Yu, H. Chen, and X. Zhou, "A cross-space, multi-interaction-based dynamic incentive mechanism for mobile crowd sensing," in *UUTC-ATC-ScalCom*, Bali, Indonesia, Dec. 2014, pp. 179–186.

[103] D. Schrage, C. Farnham, and P. G. Gonsalves, "A market-based optimization approach to sensor and resource management," in *Defense and Security Symposium*, 2006, p. 62 290I.

[104] N. Edalat, W. Xiao, N. Roy, S. K. Das, and M. Motani, "Combinatorial auction-based task allocation in multi-application wireless sensor networks," in *9th International Conference on Embedded and Ubiquitous Computing (EUC)*, Melbourne, VIC, Australia, Oct. 2011, pp. 174–181.

[105] D. Liu, L. Wang, Y. Chen, M. Elkashlan, K. Wong, R. Schober, and L. Hanzo, "User association in 5G networks: A survey and an outlook," *IEEE Communications Surveys Tutorials*, vol. 18, no. 2, pp. 1018–1044, 2016.

[106] Y. Xu, R. Q. Hu, L. Wei, and G. Wu, "QoE-aware mobile association and resource allocation over wireless heterogeneous networks," in *IEEE GLOBECOM*, Austin, TX, Dec. 2014, pp. 4695–4701.

[107] H. Shokri-Ghadikolaei, Y. Xu, L. Gkatzikis, and C. Fischione, "User association and the alignment-throughput tradeoff in millimeter wave networks," in *International Forum on Research and Technologies for Society and Industry Leveraging*, Turin, Italy, Sept. 2015, pp. 100–105.

[108] S. Maghsudi and E. Hossain, "Distributed downlink user association in small cell networks with energy harvesting," in *IEEE ICC*, Kuala Lumpur, Malaysia, May 2016, pp. 1–6.

[109] R. Sun, M. Hong, and Z.-Q. Luo, "Joint downlink base station association and power control for max–min fairness: Computation and complexity," *IEEE Journal on Selected Areas in Communications*, vol. 33, no. 6, pp. 1040–1054, June 2015.

[110] Z. Mao, G. Nan, and M. Li, "A dynamic pricing scheme for congestion game in wireless machine-to-machine networks," *International Journal of Distributed Sensor Networks*, vol. 8, no. 8, pp. 1–9, Jan. 2012.

[111] B. Romanous, N. Bitar, A. Imran, and H. Refai, "Network densification: Challenges and opportunities in enabling 5G," in *IEEE International Workshop on Computer Aided Modelling and Design of Communication Links and Networks (CAMAD)*, Guildford, UK, Sept. 2015, pp. 129–134.

[112] H. Jo, C. Mun, J. Moon, and J. Yook, "Interference mitigation using uplink power control for two-tier femtocell networks," *IEEE Transactions on Wireless Communications*, vol. 8, no. 10, pp. 4906–4910, Oct. 2009.

[113] S. Kandukuri and S. Boyd, "Optimal power control in interference-limited fading wireless channels with outage-probability specifications," *IEEE Transactions on Wireless Communications*, vol. 1, no. 1, pp. 46–55, Jan. 2002.

[114] C. Kosta, B. Hunt, A. U. Quddus, and R. Tafazolli, "On interference avoidance through inter-cell interference coordination (ICIC) based on OFDMA mobile systems," *IEEE Communications Surveys Tutorials*, vol. 15, no. 3, pp. 973–995, 2013.

[115] M. Hasan and E. Hossain, "Distributed resource allocation in D2D-enabled multi-tier cellular networks: An auction approach," in *IEEE ICC*, London, June 2015, pp. 2949–2954.

[116] M. Lashgari, B. Maham, H. Kebriaei, and W. Saad, "Distributed power allocation and interference mitigation in two-tier femtocell networks: A game-theoretic approach," in *International Wireless Communications and Mobile Computing Conference (IWCMC)*, Dubrovnik, Croatia, Aug. 2015, pp. 55–60.

[117] Y. Chen, W. Li, Y. Hu, and Q. Zhu, "Dormancy mechanism based power allocation in heterogeneous networks: A Stackelberg game approach," *Mobile Networks and Applications*, vol. 22, no. 3, pp. 552–563, June 2017.

[118] N. D. Duong, A. Madhukumar, and D. Niyato, "Stackelberg Bayesian game for power allocation in two-tier networks," *IEEE Transactions on Vehicular Technology*, vol. 65, no. 4, pp. 2341–2354, Apr. 2016.

[119] "Toward cooperation by carrier aggregation in heterogeneous networks: A hierarchical game approach," *IEEE Transactions on Vehicular Technology*, vol. 66, no. 2, pp. 1670–1683, Feb. 2017.

[120] X. Gu, X. Zhang, Z. Zhou, Y. Cheng, and J. Peng, "Game theory based interference control approach in 5G ultra-dense heterogeneous networks," in *Asia-Pacific Services Computing Conference*, Zhangjiajie, China, Nov. 2016, pp. 306–319.

[121] Z. Wang, B. Hu, X. Wang, and S. Chen, "Interference pricing in 5G ultra-dense small cell networks: A Stackelberg game approach," *IET Communications*, vol. 10, no. 15, pp. 1865–1872, Oct. 2016.

[122] L. Li, G. Zhao, and R. S. Blum, "A survey of caching techniques in cellular networks: Research issues and challenges in content placement and delivery strategies," *IEEE Communications Surveys & Tutorials*, vol. 20, no. 3, pp. 1710–1732, 2018.

[123] L. Breslau, P. Cao, L. Fan, G. Phillips, S. Shenker, et al., "Web caching and zipf-like distributions: Evidence and implications," in *IEEE INFOCOM*, vol. 1, no. 1, New York, Mar. 1999, pp. 126–134.

[124] M. Arlitt, L. Cherkasova, J. Dilley, R. Friedrich, and T. Jin, "Evaluating content management techniques for web proxy caches," *ACM SIGMETRICS Performance Evaluation Review*, vol. 27, no. 4, pp. 3–11, Mar. 2000.

[125] M. Abrams, C. R. Standridge, G. Abdulla, S. Williams, and E. A. Fox, "Caching proxies: Limitations and potentials," Department of Computer Science, Virginia Polytechnic Institute & State, Technical, Report, 1995.

[126] H. Zhou, H. Wang, X. Li, and V. C. Leung, "A survey on mobile data offloading technologies," *IEEE Access*, vol. 6, pp. 5101–5111, Jan. 2018.

[127] A. Aijaz, H. Aghvami, and M. Amani, "A survey on mobile data offloading: Technical and business perspectives," *IEEE Wireless Communications*, vol. 20, no. 2, pp. 104–112, Apr. 2013.

[128] A. Bousia, E. Kartsakli, A. Antonopoulos, L. Alonso, and C. Verikoukis, "Auction-based offloading for base station switching off in heterogeneous networks," in *European Conference on Networks and Communications (EuCNC)*, Athens, Greece, June 2016, pp. 335–339.

[129] S. Paris, F. Martisnon, I. Filippini, and L. Clien, "A bandwidth trading marketplace for mobile data offloading," in *IEEE INFOCOM*, Turin, Italy, Apr. 2013, pp. 430–434.

[130] F. Sken, P.-H. Lin, L. Sanguinetti, M. Debbah, and E. A. Jorswieck, "An energy-aware auction for hybrid access in heterogeneous networks under QoS requirements," in *IEEE ICASSP*, Shanghai, China, Mar. 2016, pp. 3606–3610.

[131] N. Zhang, N. Cheng, A. T. Gamage, K. Zhang, J. W. Mark, and X. Shen, "Cloud assisted HetNets toward 5G wireless networks," *IEEE Communications Magazine*, vol. 53, no. 6, pp. 59–65, June 2015.

[132] F. Rebecchi, M. D. De Amorim, V. Conan, A. Passarella, R. Bruno, and M. Conti, "Data offloading techniques in cellular networks: A survey," *IEEE Communications Surveys & Tutorials*, vol. 17, no. 2, pp. 580–603, 2015.

[133] S. Yi, P. Naldurg, and R. Kravets, "Security-aware ad hoc routing for wireless networks," in *Proceedings of the 2nd ACM International Symposium on Mobile Ad Hoc Networking & Computing*. Long Beach, CA: ACM, Oct. 2001, pp. 299–302.

[134] X. Su, G. Peng, and S. Chan, "Multi-path routing and forwarding in non-cooperative wireless networks," *IEEE Transactions on Parallel and Distributed Systems*, vol. 25, no. 10, pp. 2638–2647, Aug. 2014.

[135] H. Delfs and H. Knebl, "Symmetric-key encryption," in *Introduction to Cryptography*. Berlin/Heidelberg: Springer, 2007, pp. 11–31.

[136] A. Salomaa, *Public-Key Cryptography*. Berlin/Heidelberg: Springer Science & Business Media, 2013.

[137] Y. Zou, J. Zhu, X. Wang, and V. C. Leung, "Improving physical-layer security in wireless communications using diversity techniques," *IEEE Network*, vol. 29, no. 1, pp. 42–48, Jan. 2015.

[138] M. Bloch and J. Barros, *Physical-Layer Security: From Information Theory to Security Engineering*. Cambridge: Cambridge University Press, 2011.

[139] Y.-S. Shiu, S. Y. Chang, H.-C. Wu, S. C.-H. Huang, and H.-H. Chen, "Physical layer security in wireless networks: A tutorial," *IEEE Wireless Communications*, vol. 18, no. 2, pp. 66–74, Apr. 2011.

[140] J. Barros and M. R. Rodrigues, "Secrecy capacity of wireless channels," in *IEEE International Symposium on Information Theory*, Seattle, WA, Dec. 2006, pp. 356–360.

[141] E. A. Jorswieck, A. Wolf, and S. Gerbracht, *Secrecy on the Physical Layer in Wireless Networks*. London: INTECH Open Access Publisher, 2010.

[142] L. Dong, Z. Han, A. P. Petropulu, and H. V. Poor, "Improving wireless physical layer security via cooperating relays," *IEEE Transactions on Signal Processing*, vol. 58, no. 3, pp. 1875–1888, Dec. 2010.

[143] K. Seong, M. Mohseni, and J. M. Cioffi, "Optimal resource allocation for OFDMA downlink systems," in *IEEE International Symposium on Information Theory*, Seattle, WA, July 2006, pp. 1394–1398.

[144] M. O. Hasna and M.-S. Alouini, "Optimal power allocation for relayed transmissions over Fayleigh-fading channels," *IEEE Transactions on Wireless Communications*, vol. 3, no. 6, pp. 1999–2004, Nov. 2004.

[145] H. Zhu, M. Ninoslav, M. Debbah, and A. Hjorungnes, "Improved wireless secrecy capacity using distributed auction theory," in *5th International Conference on Mobile Ad-Hoc and Sensor Networks*, Fujian, China, Dec. 2009, pp. 442–447.

[146] Y. Zhang, R. Zhang, L. Song, Z. Han, and B. Jiao, "Ascending clock auction for physical layer security," in *Physical Layer Security in Wireless Communications*. Cambridge: Cambridge University Press, 2013, pp. 209–235.

[147] M. Li, Y. Guo, K. Huang, and F. Guo, "Secure power and subcarrier auction in uplink full-duplex cellular networks," *China Communications*, vol. 12, no. 1, pp. 157–165, Jan. 2015.

[148] T. Wang, L. Song, Z. Han, and B. Jiao, "Improve secure communications in cognitive two-way relay networks using sequential second price auction," in *IEEE Symposium on New Frontiers in Dynamic Spectrum Access Networks (DySPAN)*, Aachen, Germany, May 2011, pp. 308–315.

[149] J. Deng, R. Zhang, L. Song, Z. Han, and B. Jiao, "Truthful mechanisms for secure communication in wireless cooperative system," *IEEE Transactions on Wireless Communications*, vol. 12, no. 9, pp. 4236–4245, Aug. 2013.

[150] S. Liu, R. Zhang, L. Song, Z. Han, and B. Jiao, "Enforce truth-telling in wireless relay networks for secure communication," in *IEEE INFOCOM*, Shanghai, China, Apr. 2011, pp. 1071–1075.

[151] I. Stanojev and A. Yener, "Cooperative jamming via spectrum leasing," in *International Symposium on Modeling and Optimization in Mobile, Ad Hoc and Wireless Networks*, Princeton, NJ, May 2011, pp. 265–272.

[152] X. Wang, Y. Ji, H. Zhou, and J. Li, "Dasi: A truthful double auction mechanism for secure information transfer in cognitive radio networks," in *12th Annual IEEE International Conference on Sensing, Communication, and Networking (SECON)*, Seattle, WA, Jun. 2015, pp. 19–27.

[153] X. Wang, Y. Ji, H. Zhou, and J. Li, "Auction based frameworks for secure communications in static and dynamic cognitive radio networks," *IEEE Transactions on Vehicular Technology*, vol: 66, no: 3, pp. 2658–2673, Mar. 2017.

[154] P. Siyari, M. Krunz, and D. N. Nguyen, "Price-based friendly jamming in a MISO interference wiretap channel," in *IEEE INFOCOM*, San Francisco, CA, Apr. 2016, pp. 1–9.

[155] K. Grover, A. Lim, and Q. Yang, "Jamming and anti-jamming techniques in wireless networks: A survey," *International Journal of Ad Hoc and Ubiquitous Computing*, vol. 17, no. 4, pp. 197–215, Dec. 2014.

[156] H. Liu, Z. Liu, Y. Chen, and W. Xu, "Determining the position of a jammer using a virtual-force iterative approach," *Wireless Networks*, vol. 17, no. 2, pp. 531–547, Feb. 2011.

[157] P. Tague, D. Slater, R. Poovendran, and G. Noubir, "Linear programming models for jamming attacks on network traffic flows," in *International Symposium on Modeling and Optimization in Mobile, Ad Hoc, and Wireless Networks and Workshops*, Berlin, Germany, Apr. 2008, pp. 207–216.

[158] G. Alnifie and R. Simon, "A multi-channel defense against jamming attacks in wireless sensor networks," in *Proceedings of the 3rd ACM Workshop on QoS and Security for Wireless and Mobile Networks*, New York, 2007, pp. 95–104.

[159] A. D. Wood, J. A. Stankovic, and S. H. Son, "Jam: A jammed-area mapping service for sensor networks," in *IEEE Real-Time Systems Symposium*, Cancun, Mexico, Dec. 2003, pp. 286–297.

[160] S. Aasha Nandhini, R. Kishore, and S. Radha, "An efficient anti jamming technique for wireless sensor networks," in *International Conference on Recent Trends in Information Technology*, Tamil Nadu, India, Apr. 2012, pp. 361–366.

[161] S. K. Jain and K. Garg, "A hybrid model of defense techniques against base station jamming attack in wireless sensor networks," in *2009 First International Conference on Computational Intelligence, Communication Systems and Networks*, July 2009, pp. 102–107.

[162] P. Popovski, H. Yomo, and R. Prasad, "Strategies for adaptive frequency hopping in the unlicensed bands," *IEEE Wireless Communications*, vol. 13, no. 6, pp. 60–67, Dec 2006.

[163] C. Popper, M. Strasser, and S. Capkun, "Anti-jamming broadcast communication using uncoordinated spread spectrum techniques," *IEEE Journal on Selected Areas in Communications*, vol. 28, no. 5, pp. 703–715, June 2010.

[164] L. Zhang, Z. Guan, and T. Melodia, "Cooperative anti-jamming for infrastructure-less wireless networks with stochastic relaying," in *IEEE INFOCOM*, Toronto, Canada, Apr. 2014, pp. 549–557.

[165] L. Zhang, Z. Guan, and T. Melodia, "United against the enemy: Anti-jamming based on cross-layer cooperation in wireless networks," *IEEE Transactions on Wireless Communications*, vol. 15, no. 8, pp. 5733–5747, Aug. 2016.

[166] X. Tang, P. Ren, and Z. Han, "Combating full-duplex active eavesdropper: A game-theoretic perspective," in *IEEE ICC*, Kuala Lumpur, Malaysia, Dec. 2016, pp. 1–6.

[167] A. Chorppath, T. Alpcan, and H. Boche, "Bayesian mechanisms and detection methods for wireless network with malicious users," *IEEE Transactions on Mobile Computing*, vol. 15, no. 10, pp. 2452–2465, Oct. 2016.

[168] S. T. Zargar, J. Joshi, and D. Tipper, "A survey of defense mechanisms against distributed denial of service (DDoS) flooding attacks," *IEEE Communications Surveys & Tutorials*, vol. 15, no. 4, pp. 2046–2069, Mar. 2013.

[169] C. Perkins, E. Belding-Royer, and S. Das. Ad hoc on-demand distance vector (AODV) routing [Online]. (2003). Available: www.cs.cornell.edu/people/egs/615/aodv.pdf

[170] A. Agah, K. Basu, and S. K. Das, "Security enforcement in wireless sensor networks: A framework based on non-cooperative games," *Pervasive and Mobile Computing*, vol. 2, no. 2, pp. 137–158, Apr. 2006.

[171] J. B. Joshi, W. G. Aref, A. Ghafoor, and E. H. Spafford, "Security models for web-based applications," *Communications of the ACM*, vol. 44, no. 2, pp. 38–44, Feb. 2001.

[172] *Information processing systems – Open Systems Interconnection – Basic Reference Model – Part 2: Security Architecture*, ISO Standard 7498-2:1989 [Online]. (Feb. 1989). Available: www.iso.org/iso/catalogue-detail.htm?csnumber=14256

[173] I. Bashir, E. Serafini, and K. Wall, "Securing network software applications: Introduction," *Communications of the ACM*, vol. 44, no. 2, pp. 28–30, Feb. 2001.

[174] E. F. Stone, H. G. Gueutal, D. G. Gardner, and S. McClure, "A field experiment comparing information-privacy values, beliefs, and attitudes across several types of organizations," *Journal of Applied Psychology*, vol. 68, no. 3, pp. 459–468, Aug. 1983.

[175] D. R. Stinson, *Cryptography: Theory and Practice*. Boca Raton, FL: CRC Press, 2005.

[176] M. Gruteser and D. Grunwald, "Anonymous usage of location-based services through spatial and temporal cloaking," in *International Conference on Mobile Systems, Applications and Services*, San Francisco, CA, May 2003, pp. 31–42.

[177] Y. Zhang, W. Tong, and S. Zhong, "On designing satisfaction-ratio–aware truthful incentive mechanisms for k-anonymity location privacy," *IEEE Transactions on Information Forensics and Security*, vol. 11, no. 11, pp. 2528–2541, Nov. 2016.

[178] M. Yokoo, Y. Sakurai, and S. Matsubara, "Robust double auction protocol against false-name bids," *Decision Support Systems*, vol. 39, no. 2, pp. 241–252, Apr. 2005.

[179] Q. Wang, B. Ye, B. Tang, T. Xu, S. Guo, S. Lu, and W. Zhuang, "Robust large-scale spectrum auctions against false-name bids," *IEEE Transactions on Mobile Computing*, vol: 16, no: 6, pp. 1730–1743, June 2017.

[180] S. Rathinakumar and M. K. Marina, "Gavel: Strategy-proof ascending bid auction for dynamic licensed shared access," in *Proceedings of the 17th ACM International Symposium on Mobile Ad Hoc Networking and Computing*, Paderborn, Germany, July 2016, pp. 121–130.

[181] B. Wang, Y. Wu, Z. Ji, K. R. Liu, and T. C. Clancy, "Game theoretical mechanism design methods," *IEEE Signal Processing Magazine*, vol. 25, no. 6, pp. 74–84, Nov. 2008.

[182] M. Pan, J. Sun, and Y. Fang, "Purging the back-room dealing: Secure spectrum auction leveraging Paillier cryptosystem," *IEEE Journal on Selected Areas in Communications*, vol. 29, no. 4, pp. 866–876, Mar. 2011.

[183] V. Krishna, *Auction Theory*. San Diego, CA: Academic Press, 2009.

[184] M. J. Osborne et al., *An Introduction to Game Theory*. New York: Oxford University Press, 2004, vol. 3, no. 3.

[185] Z. Han, D. Niyato, W. Saad, T. Başar, and A. Hjørungnes, *Game Theory in Wireless and Communication Networks: Theory, Models, and Applications*. Cambridge: Cambridge University Press, 2012.

[186] D. C. Parkes, "Classic mechanism design," Ph.D. dissertation, 2001.

[187] M. A. Satterthwaite and S. R. Williams, "Bilateral trade with the sealed bid k-double auction: Existence and efficiency," *Journal of Economic Theory*, vol. 48, no. 1, pp. 107–133, June 1989.

[188] J. D. Hartline, *Mechanism design and approximation*, 2013.

[189] G.-Y. Lin and H.-Y. Wei, "A multi-period resource auction scheme for machine-to-machine communications," in *IEEE International Conference on Communication Systems*, Macau, China, Nov. 2014, pp. 177–181.

[190] L. Gao, P. Li, Z. Pan, N. Liu, and X. You, "Virtualization framework and VCG based resource block allocation scheme for LTE virtualization," in *IEEE Vehicular Technology Conference*, Nanjing, China, May 2016, pp. 1–6.

[191] Y. Gui, Z. Zheng, F. Wu, X. Gao, and G. Chen, "Soar: Strategy-proof auction mechanisms for distributed cloud bandwidth reservation," in *IEEE International Conference on Communication Systems*, Macau, China, Nov. 2014, pp. 162–166.

[192] P. Dütting, Z. Feng, H. Narasimhan, and D. C. Parkes, "Optimal auctions through deep learning," arXiv preprint arXiv:1706.03459, 2017.

[193] N. C. Luong, Z. Xiong, P. Wang, and D. Niyato, "Optimal auction for edge computing resource management in mobile blockchain networks: A deep learning approach," in *IEEE International Conference on Communications (ICC)*, Kansas City, MO, May 2018, pp. 1–6.

[194] R. B. Myerson, "Optimal auction design," *Mathematics of Operations Research*, vol. 6, no. 1, pp. 58–73, Feb. 1981.

[195] K. Kawaguchi, "Deep learning without poor local minima," in *Advances in Neural Information Processing Systems*, Barcelona, Spain, Dec. 2016, pp. 586–594.

[196] Y. Chen, "Banking panics: The role of the first-come, first-served rule and information externalities," *Journal of Political Economy*, vol. 107, no. 5, pp. 946–968, Oct. 1999.

[197] R. P. McAfee and J. McMillan, "Auctions and bidding," *Journal of Economic Literature*, vol. 25, no. 2, pp. 699–738, June 1987.

[198] C. A. Gizelis and D. D. Vergados, "A survey of pricing schemes in wireless networks," *IEEE Communications Surveys & Tutorials*, vol. 13, no. 1, pp. 126–145, July 2010.

[199] S. Parsons, J. A. Rodriguez-Aguilar, and M. Klein, "Auctions and bidding: A guide for computer scientists," *ACM Computing Surveys (CSUR)*, vol. 43, no. 2, p. 10, Jan. 2011.

[200] I. Koutsopoulos and G. Iosifidis, "Auction mechanisms for network resource allocation," in *International Symposium on Modeling and Optimization in Mobile, Ad Hoc, and Wireless Networks*, Avignon, France, May 2010, pp. 554–563.

[201] R. Zhang, L. Song, Z. Han, and B. Jiao, "Ascending clock auction for physical layer security," in *Physical Layer Security in Wireless Communications*. Boca Raton, FL: CRC Press, 2016, pp. 223–249.

[202] A. Mochón, Y. Sáez, et al., *Understanding Auctions*. Basel, Switzerland: Springer, 2015.

[203] J. K. Goeree, "Bidding for the future: Signaling in auctions with an aftermarket," *Journal of Economic Theory*, vol. 108, no. 2, pp. 345–364, Feb. 2003.

[204] X. Wang, Z. Li, P. Xu, Y. Xu, X. Gao, and H.-H. Chen, "Spectrum sharing in cognitive radio networksan auction-based approach," *IEEE Transactions on Systems, Man, and Cybernetics, Part B (Cybernetics)*, vol. 40, no. 3, pp. 587–596, Dec. 2010.

[205] R. Mochaourab, B. Holfeld, and T. Wirth, "Distributed channel assignment in cognitive radio networks: Stable matching and Walrasian equilibrium," *IEEE Transactions on Wireless Communications*, vol. 14, no. 7, pp. 3924–3936, July 2015.

[206] F. Gul and E. Stacchetti, "Walrasian equilibrium with gross substitutes," *Journal of Economic Theory*, vol. 87, no. 1, pp. 95–124, July 1999.

References

[207] H. W. Kuhn, "The Hungarian method for the assignment problem," *Naval Research Logistics (NRL)*, vol. 2, no. 1–2, pp. 83–97, Mar. 1955.

[208] S. Reidt, M. Srivatsa, and S. Balfe, "The fable of the bees: Incentivizing robust revocation decision making in ad hoc networks," in *The 16th ACM Conference on Computer and Communications Security*, Chicago, IL, Nov. 2009, pp. 291–302.

[209] C. Lima and G. T. F. De Abreu, "Game-theoretical relay selection strategy for geographic routing in multi-hop WSNs," in *IEEE 5th Workshop on Positioning, Navigation and Communication*, Hannover, Germany, Mar. 2008, pp. 277–283.

[210] V. Rodriguez and F. Jondral, "Simple adaptively-prioritised spatially-reusable medium access control through the dutch auction: Qualitative analysis, issues, challenges," in *IEEE Symposium on Communications and Vehicular Technology*, Delft, Netherlands, Nov. 2007, pp. 1–5.

[211] R. Cassady, *Auctions and Auctioneering*. Oakland: University of California Press, 1967.

[212] J.-S. Lee and B. Hoh, "Sell your experiences: A market mechanism based incentive for participatory sensing," in *IEEE International Conference on Pervasive Computing and Communications (PerCom)*, Mannheim, Germany, Mar. 2010, pp. 60–68.

[213] N. Edalat, W. Xiao, C. K. Tham, E. Keikha, and L. L. Ong, "A price-based adaptive task allocation for wireless sensor network," in *IEEE 6th International Conference on Mobile Ad Hoc and Sensor Systems*, Macau, China, Oct. 2009, pp. 888–893.

[214] Q. Liu, X. Xian, and T. Wu, "Game theoretic approach in routing protocol for cooperative wireless sensor networks," in *International Conference in Swarm Intelligence*, Chongqing, China, June, 2011, pp. 207–217.

[215] W. Mobile, "Real-time maps and traffic information based on the wisdom of the crowd." Available: http://solsie.com/2009/09/real-time-maps-and-traffic-information-based-on-thewisdom- of-the-crowd/

[216] C. H. Liu, P. Hui, J. W. Branch, C. Bisdikian, and B. Yang, "Efficient network management for context-aware participatory sensing," in *SECON*, Salt Lake City, UT, June 2011, pp. 116–124.

[217] Y. Tian, Y. Gu, E. Ekici, and F. Özgüner, "Dynamic critical-path task mapping and scheduling for collaborative in-network processing in multi-hop wireless sensor networks," in *International Conference on Parallel Processing Workshops*, Columbus, OH, Aug. 2006, pp. 215–222.

[218] Y. Wang, G. Attebury, and B. Ramamurthy, "A survey of security issues in wireless sensor networks," *IEEE Communications Surveys and Tutorials*, vol. 8, no. 2, pp. 2–23, 2006.

[219] S. Buchegger and J. Y. Le Boudec, "Performance analysis of the confidant protocol," in *Proceedings of the 3rd ACM International Symposium on Mobile Ad Hoc Networking & Computing*, Lausanne, Switzerland, June 2002, pp. 226–236.

[220] G. E. Bolton and A. Ockenfels, "ERC: A theory of equity, reciprocity, and competition," *American Economic Review*, vol. 90, no. 1, pp. 166–193, Mar. 2000.

[221] W. Vickrey, "Counterspeculation, auctions, and competitive sealed tenders," *Journal of Finance*, vol. 16, no. 1, pp. 8–37, Mar. 1961.

[222] W.-Y. Lin, G.-Y. Lin, and H.-Y. Wei, "Dynamic auction mechanism for cloud resource allocation," in *IEEE/ACM International Conference on Cluster, Cloud and Grid Computing*. Melbourne, Australia: IEEE Computer Society, May 2010, pp. 591–592.

[223] N. Edalat, C.-K. Tham, and W. Xiao, "An auction-based strategy for distributed task allocation in wireless sensor networks," *Computer Communications*, vol. 35, no. 8, pp. 916–928, May 2012.

[224] S. Di, C.-L. Wang, L. Cheng, and L. Chen, "Social-optimized win–win resource allocation for self-organizing cloud," in *International Conference on Cloud and Service Computing (CSC)*, Hong Kong, China, Dec. 2011, pp. 251–258.

[225] S. Di, C.-L. Wang, W. Zhang, and L. Cheng, "Probabilistic best-fit multi-dimensional range query in self-organizing cloud," in *International Conference on Parallel Processing (ICPP)*, Taipei, Taiwan, Sept. 2011, pp. 763–772.

[226] B. Pinkas, "Cryptographic techniques for privacy-preserving data mining," *ACM SIGKDD Explorations Newsletter*, vol. 4, no. 2, pp. 12–19, Dec. 2002.

[227] I. Stanojev and A. Yener, "Improving secrecy rate via spectrum leasing for friendly jamming," *IEEE Transactions on Wireless Communications*, vol. 12, no. 1, pp. 134–145, Jan. 2013.

[228] R. M. Corless, G. H. Gonnet, D. E. Hare, D. J. Jeffrey, and D. E. Knuth, "On the Lambert W function," *Advances in Computational Mathematics*, vol. 5, no. 1, pp. 329–359, Dec. 1996.

[229] E. Tekin and A. Yener, "The general Gaussian multiple-access and two-way wiretap channels: Achievable rates and cooperative jamming," *IEEE Transactions on Information Theory*, vol. 54, no. 6, pp. 2735–2751, 2008.

[230] M. J. Osborne and A. Rubinstein, *A Course in Game Theory*. Cambridge, MA: MIT Press, 1994.

[231] X. Wang, X. Chen, and W. Wu, "Towards truthful auction mechanisms for task assignment in mobile device clouds," in *IEEE INFOCOM*, Atlanta, GA, May 2017, pp. 1–9.

[232] H. Ahmadi, I. Macaluso, I. Gomez, L. DaSilva, and L. Doyle, "Virtualization of spatial streams for enhanced spectrum sharing," in *IEEE Global Communications Conference (GLOBECOM)*, Washington, DC, Dec. 2016, pp. 1–6.

[233] J. Dai, F. Liu, B. Li, B. Li, and J. Liu, "Collaborative caching in wireless video streaming through resource auctions," *IEEE Journal on Selected Areas in Communications*, vol. 30, no. 2, pp. 458–466, Feb. 2012.

[234] A. Mas-Colell, M. D. Whinston, J. R. Green, et al., *Microeconomic Theory*. New York: Oxford University Press, 1995, vol. 1.

[235] H. R. Varian and C. Harris, "The VCG auction in theory and practice," *American Economic Review*, vol. 104, no. 5, pp. 442–445, May 2014.

[236] L. M. Ausubel, P. Milgrom, et al., "The lovely but lonely vickrey auction," *Combinatorial Auctions*, pp. 22–26, 2006.

[237] T. Roughgarden, "Algorithmic game theory," *Communications of the ACM*, vol. 53, no. 7, pp. 78–86, 2010.

[238] S. Dobzinski and N. Nisan, "Mechanisms for multi-unit auctions," *Journal of Artificial Intelligence Research*, vol. 37, no. 1, pp. 85–98, Feb. 2010.

[239] J. Jia, Q. Zhang, Q. Zhang, and M. Liu, "Revenue generation for truthful spectrum auction in dynamic spectrum access," in *Proceedings of the Tenth ACM International Symposium on Mobile Ad Hoc Networking and Computing*. New Orleans, LA: ACM, May 2009, pp. 3–12.

[240] Q. Wu, M. Zhou, Q. Zhu, and Y. Xia, "VCG auction-based dynamic pricing for multigranularity service composition," *IEEE Transactions on Automation Science and Engineering*, vol. 15, no. 2, pp. 796–805, May 2017.

[241] M. Dong, G. Sun, X. Wang, and Q. Zhang, "Combinatorial auction with time-frequency flexibility in cognitive radio networks," in *IEEE INFOCOM*, Orlando, FL, Mar. 2012, pp. 2282–2290.

[242] K. Zhu and E. Hossain, "Virtualization of 5G cellular networks as a hierarchical combinatorial auction," *IEEE Transactions on Mobile Computing*, vol. 15, no. 10, pp. 2640–2654, Oct. 2016.

[243] C. Xu, L. Song, Z. Han, Q. Zhao, X. Wang, X. Cheng, and B. Jiao, "Efficiency resource allocation for device-to-device underlay communication systems: A reverse iterative combinatorial auction based approach," *IEEE Journal on Selected Areas in Communications*, vol. 31, no. 9, pp. 348–358, Sept. 2013.

[244] N. Nisan, "Bidding languages," in *Combinatorial auctions*. Cambridge, MA: MIT Press, 2006, pp. 1–21.

[245] L. Ausubel, P. Crampton, and P. Milgrom, "The clock-proxy auction: A practical combinatorial auction design," in *Handbook of Spectrum Auction Design*. Cambridge: Cambridge University Press, 2017, pp. 119–140.

[246] J. A. Bondy, U. S. R. Murty, et al., *Graph Theory with Applications*. London: Springer, 1976, vol. 290.

[247] A. Mu'Alem and N. Nisan, "Truthful approximation mechanisms for restricted combinatorial auctions," *Games and Economic Behavior*, vol. 64, no. 2, pp. 612–631, Nov. 2008.

[248] A. Pikovsky, "Pricing and bidding strategies in iterative combinatorial auctions," Ph.D. dissertation, Technische Universität München, 2008.

[249] T. Mullen and M. P. Wellman, "Market-based negotiation for digital library services," in *Second USENIX Workshop on Electronic Commerce*, vol. 13, Oakland, CA, Nov. 1996, pp. 259–269.

[250] S. R. Williams and M. A. Satterthwaite, *The Bayesian Theory of the k-Double Auction*. London: Routledge, 2018, pp. 99–124.

[251] R. P. McAfee, "A dominant strategy double auction," *Journal of Economic Theory*, vol. 56, no. 2, pp. 434–450, Apr. 1992.

[252] C. Chen and Y. Wang, "Sparc: Strategy-proof double auction for mobile participatory sensing," in *International Conference on Cloud Computing and Big Data*, Fuzhou, China, Dec. 2013, pp. 133–140.

[253] D. Yang, X. Fang, and G. Xue, "Truthful incentive mechanisms for k-anonymity location privacy," in *IEEE INFOCOM*, Turin, Italy, Apr. 2013, pp. 2994–3002.

[254] L. Xiang, G. Sun, J. Liu, X. Wang, and L. Li, "A discriminatory pricing double auction for spectrum allocation," in *IEEE Wireless Communications and Networking Conference (WCNC)*, Shanghai, China, Apr. 2012, pp. 1473–1477.

[255] A.-L. Jin, W. Song, P. Wang, D. Niyato, and P. Ju, "Auction mechanisms toward efficient resource sharing for cloudlets in mobile cloud computing," *IEEE Transactions on Services Computing*, vol. 9, no. 6, pp. 895–909, Nov. 2016.

[256] M. O. Rabin and C. Thorpe, "Time-lapse cryptography," Technical Report, TR-22-06 [Online]. (2006). Available: www.eecs.harvard.edu/~cat/tlc.pdf

[257] M. Bellare, R. Canetti, and H. Krawczyk, "Keying hash functions for message authentication," in *Annual International Cryptology Conference*, vol. 1109. Santa Barbara, CA: Springer, July 1996, pp. 1–15.

[258] L. Sweeney, "k-Anonymity: A model for protecting privacy," *International Journal of Uncertainty, Fuzziness and Knowledge-Based Systems*, vol. 10, no. 05, pp. 557–570, Oct. 2002.

[259] G. Iachello, I. Smith, S. Consolvo, G. D. Abowd, J. Hughes, J. Howard, F. Potter, J. Scott, T. Sohn, J. Hightower, et al., "Control, deception, and communication: Evaluating the deployment of a location-enhanced messaging service," in *International Conference on Ubiquitous Computing*. Tokyo: Springer, 2005, pp. 213–231.

[260] H. Kido, Y. Yanagisawa, and T. Satoh, "An anonymous communication technique using dummies for location-based services," in *International Conference on Pervasive Services*, Santorini, Greece, July 2005, pp. 88–97.

[261] X. Zhou and H. Zheng, "Trust: A general framework for truthful double spectrum auctions," in *IEEE INFOCOM*, Rio de Janeiro, Brazil, Apr. 2009, pp. 999–1007.

[262] X. Zhou, S. Gandhi, S. Suri, and H. Zheng, "eBay in the sky: Strategy-proof wireless spectrum auctions," in *Proceedings of the 14th ACM International Conference on Mobile Computing and Networking*, San Francisco, CA, Sept. 2008, pp. 2–13.

[263] D. C. Parkes, J. R. Kalagnanam, and M. Eso, "Achieving budget-balance with Vickrey-based payment schemes in exchanges," in *International Joint Conferences on Artificial Intelligence, and American Association for Artificial Intelligence*, San Francisco, CA, Aug. 2001, pp. 1161–1168.

[264] C. Boutilier, M. Goldszmidt, and B. Sabata, "Sequential auctions for the allocation of resources with complementarities," in *IJCAI*, Stockholm, Sweden, Aug. 1999, pp. 527–523.

[265] P. R. Milgrom and R. J. Weber, "A theory of auctions and competitive bidding," *Econometrica: Journal of the Econometric Society*, vol. 50, no. 5, pp. 1089–1122, Sept. 1982.

[266] R. P. Leme, V. Syrgkanis, and É. Tardos, "Sequential auctions and externalities," in *Proceedings of the Twenty-Third Annual ACM-SIAM Symposium on Discrete Algorithms*, Kyoto, Japan, Jan. 2012, pp. 869–886.

[267] T. Wang, R. Zhang, L. Song, Z. Han, H. Li, and B. Jiao, "Power allocation for two-way relay system based on sequential second price auction," *Wireless Personal Communications*, vol. 67, no. 1, pp. 47–62, Nov. 2012.

[268] J. Bae, E. Beigman, R. Berry, M. L. Honig, and R. Vohra, "On the efficiency of sequential auctions for spectrum sharing," in *International Conference on Game Theory for Networks*, Istanbul, Turkey, May 2009, pp. 199–205.

[269] T. Wang, L. Song, Z. Han, J. Zhang, and X. Zhang, "Dynamic resource allocation in cognitive radio two-way relay networks using sequential auctions," in *IEEE International Wireless Symposium (IWS)*, Beijing, China, Apr. 2013, pp. 1–4.

[270] S. M. Errapotu, H. Li, R. Yu, S. Ren, Q. Pei, M. Pan, and Z. Han, "Clock auction inspired privacy preserving emergency demand response in colocation data centers," *IEEE Transactions on Dependable and Secure Computing*, in press.

[271] T. Wang, L. Song, Z. Han, X. Cheng, and B. Jiao, "Power allocation using Vickrey auction and sequential first-price auction games for physical layer security in cognitive relay networks," in *IEEE International Conference on Communications (ICC)*, Ottawa, ON, Canada, June 2012, pp. 1683–1687.

[272] A. Wang, Y. Cai, X. Guan, and S. Wang, "Physical layer security for multiuser two-way relay using distributed auction game," in *International Conference on Information Science and Technology (ICIST)*, Yangzhou, China, Mar. 2013, pp. 1202–1207.

[273] R. Zhang, L. Song, Z. Han, and B. Jiao, "Improve physical layer security in cooperative wireless network using distributed auction games," in *IEEE Conference on Computer Communications Workshops (INFOCOM WKSHPS)*, Shanghai, China, June 2011, pp. 18–23.

[274] L. M. Ausubel, "An efficient ascending-bid auction for multiple objects," *American Economic Review*, vol. 94, no. 5, pp. 1452–1475, Dec. 2004.

[275] J. Huang, Z. Han, M. Chiang, and H. V. Poor, "Auction-based resource allocation for cooperative communications," *IEEE Journal on Selected Areas in Communications*, vol. 26, no. 7, pp. 1226–1237, Sept. 2008.

[276] R. Bapna, P. Goes, and A. Gupta, "A theoretical and empirical investigation of multi-item on-line auctions," *Information Technology and Management*, vol. 1, no. 1–2, pp. 1–23, Jan. 2000.

[277] M. Karaliopoulos, O. Telelis, and I. Koutsopoulos, "User recruitment for mobile crowdsensing over opportunistic networks," in *IEEE Conference on Computer Communications (INFOCOM)*, Kowloon, Hong Kong, Apr. 2015, pp. 2254–2262.

[278] B. Wang and K. R. Liu, "Advances in cognitive radio networks: A survey," *IEEE Journal of Selected Topics in Signal Processing*, vol. 5, no. 1, pp. 5–23, Feb. 2011.

[279] I. F. Akyildiz, B. F. Lo, and R. Balakrishnan, "Cooperative spectrum sensing in cognitive radio networks: A survey," *Physical Communication*, vol. 4, no. 1, pp. 40–62, Mar. 2011.

[280] W. Shi, L. Zhang, C. Wu, Z. Li, and F. Lau, "An online auction framework for dynamic resource provisioning in cloud computing," *ACM SIGMETRICS Performance Evaluation Review*, vol. 42, no. 1, pp. 71–83, Aug. 2014.

[281] Y. Sun, F. Liu, B. Li, B. Li, and X. Zhang, "Fs2you: Peer-assisted semi-persistent online storage at a large scale," in *IEEE INFOCOM*, Rio de Janeiro, Brazil, Apr. 2009, pp. 873–881.

[282] A. Davoli and A. Mei, "Triton: A peer-assisted cloud storage system," in *Proceedings of the First Workshop on Principles and Practice of Eventual Consistency*. Amsterdam, Netherlands: ACM, Apr. 2014, pp. 4:1–4:7.

[283] J. Zhao, X. Chu, H. Liu, Y.-W. Leung, and Z. Li, "Online procurement auctions for resource pooling in client-assisted cloud storage systems," in *IEEE Conference on Computer Communications (INFOCOM)*, Kowloon, China, Apr. 2015, pp. 576–584.

[284] F. Mulder, A. van der Avoird, and P. E. Wormer, "Anisotropy of long range interactions between linear molecules: H2-H2 and H2-HE," *Molecular Physics*, vol. 37, no. 1, pp. 159–180, Aug. 1979.

[285] A. Gopinathan, Z. Li, and C. Wu, "Strategyproof auctions for balancing social welfare and fairness in secondary spectrum markets," in *IEEE INFOCOM*, Shanghai, China, Apr. 2011, pp. 3020–3028.

[286] J. de Hoog, T. Alpcan, M. Brazil, D. A. Thomas, and I. Mareels, "A market mechanism for electric vehicle charging under network constraints," *IEEE Transactions on Smart Grid*, vol. 7, no. 2, pp. 827–836, Mar 2016.

[287] C. A. Holt Jr. and R. Sherman, "Waiting-line auctions," *Journal of Political Economy*, vol. 90, no. 2, pp. 280–294, Apr. 1982.

[288] G. Wu, P. Ren, and C. Zhang, "A waiting-time auction based dynamic spectrum allocation algorithm in cognitive radio networks," in *IEEE Global Telecommunications Conference*, Kathmandu, Nepal, Dec. 2011, pp. 1–5.

[289] A. M. Manelli and D. R. Vincent, "Bundling as an optimal selling mechanism for a multiple-good monopolist," *Journal of Economic Theory*, vol. 127, no. 1, pp. 1–35, Mar. 2006.

[290] Y. Giannakopoulos and E. Koutsoupias, "Duality and optimality of auctions for uniform distributions," in *Proceedings of the Fifteenth ACM Conference on Economics and Computation*, Palo Alto, CA, June 2014, pp. 259–276.

[291] A. C.-C. Yao, "Dominant-strategy versus Bayesian multi-item auctions: Maximum revenue determination and comparison," in *Proceedings of the ACM Conference on Economics and Computation*, Cambridge, MA, June 2017, pp. 3–20.

[292] K. Hornik, "Approximation capabilities of multilayer feedforward networks," *Neural Networks*, vol. 4, no. 2, pp. 251–257, Jan. 1991.

[293] D. P. Bertsekas, *Constrained Optimization and Lagrange Multiplier Methods*. New York: Academic Press, 2014.

[294] M. Crosby, P. Pattanayak, S. Verma, V. Kalyanaraman, et al., "Blockchain technology: Beyond bitcoin," *Applied Innovation*, vol. 2, no. 6–10, p. 71, June 2016.

[295] Bitcoin: A peer-to-peer electronic cash system [Online]. (2008). Available: https://bitcoin.org/bitcoin.pdf

[296] Z. Xiong, Y. Zhang, D. Niyato, P. Wang, and Z. Han, "When mobile blockchain meets edge computing," *IEEE Communications Magazine*, vol. 56, no. 8, pp. 33–39, Aug. 2018.

[297] Y. Jiao, P. Wang, D. Niyato, and Z. Xiong, "Social welfare maximization auction in edge computing resource allocation for mobile blockchain," in *2018 IEEE International Conference on Communications (ICC)*, Kansas City, MO, May 2018, pp. 1–6.

[298] Z. Xiong, S. Feng, W. Wang, D. Niyato, P. Wang, and Z. Han, "Cloud/fog computing resource management and pricing for blockchain networks," *IEEE Internet of Things Journal*, vol. 6, no. 3, pp. 4585–4600, Jun. 2019.

[299] N. Houy, "The bitcoin mining game," 2014, SSRN 2407834.

[300] [Online]. Available: www.tensorflow.org/

[301] T. H. Cormen, C. E. Leiserson, R. L. Rivest, and C. Stein, *Introduction to Algorithms*. Cambridge, MA: MIT Press, 2009.

[302] L. M. Ausubel and P. Cramton, "Vickrey auctions with reserve pricing," *Economic Theory*, vol. 23, no. 3, pp. 493–505, Apr. 2004.

[303] D. Gale and H. Nikaido, "The Jacobian matrix and global univalence of mappings," *Mathematische Annalen*, vol. 159, no. 2, pp. 81–93, Apr. 1965.

Index

allocation rule, 53
anonymity, 50
ascending clock auction, 216
ascending proxy auction, 165
asking price, 104
atomic bid, 161
auction, 65
augmented Lagrangian method, 240

baseband processing unit pool, 22
Bayesian–Nash equilibrium, 57, 85, 86
Bayesian–Nash incentive compatibility, 58
best response, 74, 78, 86, 223
bid-rigging, 51
bid shading, 103
bidding language, 161
bidding ring, 51
bidding strategy, 75, 84, 86, 89, 103
black hole attack, 87, 114
blockchain etwork, 244
budget balance, 59, 192, 203, 208, 213

cache hit rate, 41
caching placement, 41
client-assisted cloud, 227
clock-proxy auction, 167
cloud, 14
cloud-based video-on-demand system, 24
cloud computing, 18
cloud data center networking, 18
cloud networking, 18
cloud tenant, 20
cognitive radio, 30, 79, 153
combinatorial auction, 161, 170
common value auction, 66
complementary item, 159
computational efficiency, 203, 213
conclusion, 51
confidentiality, 49
conflict graph, 207
continuous double auction, 194
co-tier interference, 38
critical payment, 173, 174
critical value, 62, 174, 180

cross-tier interference, 38
cutoff value, 235

D2D communication, 31
data accuracy, 34
data aggregation, 34, 105
deep learning, 238
discriminatory pricing, 193
dominant strategy, 57
dominant-strategy incentive compatibility, 58
double auction, 189
Dutch auction, 67, 83

eavesdropper, 44, 130
economic efficiency, 56, 59, 192
edge computing, 18, 209
English auction, 67, 72
English–Dutch auction, 97
equilibrium, 55
equilibrium strategy, 84, 86, 90, 102
ex-post equilibrium, 78

fairness, 29, 56
false-name bids, 51
false-name-proof, 51
fault tolerant, 35
first-price sealed-bid auction, 54, 67, 100
first-price sealed-bid reverse auction, 104
fog computing, 244
forward auction, 66, 104
friendly jammer, 44

hammer price, 72
HetNet, 27

incentive compatibility, 56, 141, 237
incentive compatibility violation, 240
incentive cost, 35
individual rationality, 56, 58, 141, 156, 192, 203,
   208, 210, 213, 227, 237
individual rationality violation, 239
information privacy, 50
Internet of Things (IoT), 11, 14
intrusion detection system, 88
iterative combinatorial auction, 165

jammer, 44

k-anonymity location privacy, 50

load balancing, 37
loss function, 238

M2M communication, 33
malicious node, 87
malicious user, 44
MANET, 88
massive MIMO, 146, 147
McAfee's mechanism, 193
mechanism, 53
mechanism design, 53
Meyerson auction, 250
mmWave, 29
mobile cloud networking, 18
mobile crowdsensing network, 15, 16
mobile device cloud, 142
monotone transform function, 250, 251
monotonicity, 174, 229

Nash equilibrium, 57, 121, 122, 223
network lifetime, 35
network throughput, 37

online auction, 215, 224
open-cry auction, 67
opportunity cost, 232
optimal auction, 237
OR bid, 161
outage probability, 28
outstanding ask, 194
outstanding bid, 194

participatory sensing, 16, 106, 197
pay-as-bid, 193
pay-what-you-bid, 100
payment rule, 53
payoff, 85, 90, 102, 121
physical layer security, 131, 218
privacy preserving, 50
private value, 232
private value auction, 66

remote radio head, 22
revelation principle, 56

revenue equivalence, 85
reverse auction, 66, 104
risk appetite, 89
RSSI, 29

sealed-bid auction, 67
second-price sealed-bid auction, 54, 120
second-price sealed-bid reverse auction, 124
secrecy capacity, 130
secrecy rate, 217
selfish user, 44
sensing data, 14
sensing information, 14
share auction, 215, 221
signal, 74
single-round double auction, 189
social choice function, 58
social surplus, 60
social welfare, 59, 60, 141
spectrum efficiency, 28
substitutable item, 159
supervised learning, 239
supply and demand model, 190
symmetric equilibrium strategy, 77
symmetric model, 75

time value, 234
transformed bid, 250
truthful bidding, 122, 141
truthfulness, 56, 156, 204, 208, 213, 227, 229

uniform pricing, 192
user association, 37
utility, 57, 198, 202, 211, 219, 237

Vickrey–Clarke–Groves (VCG) auction, 135
virtual bids, 153, 155
virtual value, 153, 154

wait-line auction, 215, 232
Walrasian equilibrium, 79, 81
weighted independent set problem, 155
winner's curse, 103
wireless caching, 40
wireless sensor network, 15

XOR bid, 162